THIRD EDITION

ALGEBRA REVIEW

THIRD EDITION

ALGEBRA REVIEW

Michael Sullivan
Chicago State University

with contributions by
Julia Ledet
Louisiana State University

Upper Saddle River, New Jersey 07458

Editor-in-Chief: *Sally Yagan*
Senior Acquisitions Editor: *Eric Frank*
Project Manager: *Dawn Murrin*
Vice President and Director of Production and Manufacturing, ESM: *David W. Riccardi*
Executive Managing Editor: *Vince O' Brien*
Production Editor: *Donna King*
Art Director: *Jayne Conte*
Cover Designer: *Bruce Kenselaar*
Manufacturing Manager: *Trudy Pisciotti*
Manufacturing Buyer: *Lynda Castillo*

Pearson Prentice Hall
Pearson Education, Inc.
Upper Saddle River, NJ 07458

Printed in the United States of America

10 9 8 7 6 5 4 3 2 1

ISBN: 0-13-148006-5
ISBN: 0-13-149046-X

Pearson Education Ltd., *London*
Pearson Education Australia Pty. Ltd., *Sydney*
Pearson Education Singapore, Pte. Ltd.
Pearson Education North Asia Ltd., *Hong Kong*
Pearson Education Canada, Inc., *Toronto*
Pearson Educación de Mexico, S.A. de C.V.
Pearson Education—Japan, *Tokyo*
Pearson Education Malaysia, Pte. Ltd.

To all my students,
past, present, and future:
Thank you for all your help
and insight

Contents

Preface to the Instructor

The purpose of *Algebra Review* is to help students who have typically struggled with mastering certain key concepts covered in beginning and intermediate algebra courses build a firm foundation for success in college algebra and other college mathematics courses. Although *Algebra Review* could be used by a student who is learning the material for the first time, it is best used, as the title suggests, as an *algebra review*.

Algebra Review explains and provides examples, practice problems, and exercises of key topics covered in a beginning or intermediate algebra course. The text emphasizes and encourages understanding of the material rather than just memorizing a series of steps to complete a problem.

Content

- Chapter 1 reviews numbers and their properties including the operations of addition, subtraction, multiplication, and division. Various types of calculators are also discussed in Chapter 1. While one type of calculator is not favored over another, keystrokes are given for the TI-83, TI-30X IIS, and in a few select cases, the TI-30Xa. This allows students to compare some of the calculators that are currently available.
- Chapter 2 begins with Laws of Exponents and continues with polynomials covering operations with polynomials and factoring of polynomials.
- Chapter 3 reviews rational expressions starting with a review of fractions. The remainder of the chapter involves working with rational expressions where numerators and denominators are polynomials and other algebraic expressions.
- Chapter 4 revisits Laws of Exponents as they relate to negative and rational exponents. This chapter introduces radicals and shows the connection between radicals and rational exponents. Complex numbers are also briefly discussed.

Using This Material with Your Text

The material in *Algebra Review* expands upon some of the topics in the chapter titled **Review** (Chapter R or Appendix) in your text.

- Chapter 1 may be used to replace sections titled Real Numbers or Algebra Review.
- Chapter 2 may be used as a replacement for sections titled Polynomials and Factoring Polynomials in your text.
- Chapter 3 may be used to replace the section titled Rational Expressions in your text.
- Chapter 4 may be used to replace sections titled Square Roots; Radicals and Rational Exponents in your text.

Acknowledgments

I am deeply indebted to Julia Ledet of Louisiana State University. Her excellence as a teacher is evident in her many substantial and thoughtful contributions to this edition of the Algebra Review. Julia passes along her gratitude to Phoebe Rouse, Allysa McMills, and Ian McMills. I would like to thank Reena Kothari and Susan Saale for working, checking, and providing the detailed solutions to all the odd problems.

Michael Sullivan

Preface to the Student

Learning mathematics, like most things in life, takes practice. All too often, a student tries to learn mathematics by watching the teacher present the material and then by working a few homework problems. What athlete has mastered a sport by watching someone else play it? An athlete needs long hours of practice. You cannot learn to ice-skate or hit a homerun by watching someone else do it. Mathematics is no different. If you have struggled with math and have become discouraged, don't give up. Decide right now that this course will be different. **Put in the time and practice needed for success.** A guideline for success in mathematics follows:

- Look over topics cited in the 'Are You Prepared?' introductions. Work and check the 'Are You Prepared?' problems.
- Read over the material in the textbook before the teacher presents it in class. Even if you do not understand it, you will be better prepared to follow the classroom presentation.
- Attend class and take lecture notes. Ask questions.
- After class, carefully read the material in the textbook again. This time, take notes writing topics, vocabulary (with definitions), and examples. Copy and work the Practice Exercises as you encounter them. Check your answers to the Practice Exercises with the answers at the end of each section.
- After reading and taking notes, check for understanding by answering the Concepts and Vocabulary questions at the beginning of each exercise set.
- Copy and work the exercises at the end of each section. Don't just work two or three problems! At a minimum, work all the odd problems in a section and check your answers with those in the back of the book. If you are learning and working problems with certain formulas, begin each problem by writing the formula. This will help you remember formulas accurately.
- Ask your teacher to help you with the problems you had difficulty with. Bring your work with you to identify troublesome steps.
- Review all the material before taking a test. Work the problems in the chapter review at the end of each chapter. If your teacher gives you a study guide or practice test, be sure to work it also.

Don't get discouraged. Learning mathematics takes practice and perseverance. It takes time to get your "math muscles" in shape! Keep trying!

Best Wishes!
Michael Sullivan

THIRD EDITION

ALGEBRA REVIEW

1 Numbers and Their Properties

OUTLINE

The word *algebra* is derived from the Arabic word *al-jabr*. This word is part of the title of a ninth century work, "Hisâb al-jabr w'al-muqâbalah," written by Mohammed ibn Mûsâ al-Khowârizmî. The word *al-jabr* means "a restoration," a reference to the fact that if a number is added to one side of an equation, then it must also be added to the other side in order to "restore" the equality. The title of the work, freely translated, means "The science of reduction and cancellation." Of course, today, algebra has come to mean a great deal more.

1.1 Notations; Symbols

OBJECTIVES

1 Use Algebraic Symbols
2 Know Properties of Equality
3 Work with Exponents
4 Evaluate Numerical Expressions

1 Algebra can be thought of as a generalization of arithmetic in which letters may be used to represent numbers. You may recall seeing $\square + 2 = 7$ in elementary school. In algebra, we write $x + 2 = 7$ to represent the same problem. So, in a sense, you have been doing algebra for a long time.

Operations

We will use letters such as x, y, a, b, and c to represent numbers. The symbols used in algebra for the operations of addition, subtraction, multiplication, and division are $+$, $-$, $\cdot$, and /. The words used to describe the results of these operations are **sum, difference, product,** and **quotient.** Table 1 summarizes these ideas.

Table 1

Operation	Symbol	Words
Addition	$a + b$	Sum: a plus b
Subtraction	$a - b$	Difference: a less b or a minus b
Multiplication	$a \cdot b$, $(a) \cdot b$, $a \cdot (b)$, $(a) \cdot (b)$, ab, $(a)b$, $a(b)$, $(a)(b)$	Product: a times b
Division	a/b or $\frac{a}{b}$	Quotient: a divided by b

In algebra, we generally avoid using the familiar multiplication sign $\times$ because it resembles the letter x. Notice that when two expressions are placed next to each other without an operation symbol, as in ab, or in parentheses, as in $(a)(b)$, it is understood that the expressions, called **factors,** are to be multiplied together.

In algebra, we also prefer not to use the familiar division sign $\div$ because it leads to awkward expressions. However, in this book we will use the division sign on a limited basis in situations where it supports a better understanding of a concept.

We also try not to use mixed numbers in algebra. When mixed numbers are used, addition is understood; for example, $2\frac{3}{4}$ means $2 + \frac{3}{4}$. In algebra, use of a mixed number may be confusing because the absence of an operation symbol between two terms is generally taken to mean multiplication. The expression $2\frac{3}{4}$ is therefore written instead as 2.75, or as $\frac{11}{4}$.

The symbol =, called an **equal sign** and read as "equals" or "is," is used to express the idea that the number or expression on the left of the equal sign is equivalent to the number or expression on the right.

EXAMPLE 1 **Translating Verbal Expressions into Mathematical Expressions**

(a) The sum of 2 and 7 equals 9. In symbols, this statement is written as $2 + 7 = 9$.

(b) The product of 3 and 5 is 15. In symbols, this statement is written as $3 \cdot 5 = 15$. ◀

EXAMPLE 2 **Translating Verbal Expressions into Mathematical Expressions**

Write each of the following statements using symbols:

(a) The product of 2 and 4 equals 8.

(b) The sum of 4 and 5 is 18 divided by 2.

(c) The product of 3 and x is 12.

Solution (a) "The product of 2 and 4 equals 8" can be symbolized in any of the following ways:

$$2 \cdot 4 = 8 \quad (2) \cdot (4) = 8 \quad 2 \cdot (4) = 8 \quad (2) \cdot 4 = 8 \quad 2(4) = 8 \quad (2)(4) = 8$$

(b) One way to understand this type of problem is to write out the phrase and pick it apart word for word. For example:

"The sum of $\underbrace{\text{4 and 5}}$ is $\underbrace{\text{18 divided by 2}}$."

↑ add ↑ 4 + 5 ↑ equals ↑ 18 ÷ 2

So,
$$4 + 5 = \frac{18}{2}$$

(c) "The product of 3 and x is 12" is symbolized by

$$3x = 12$$ ◀

EXAMPLE 3 **Translating Mathematical Expressions into Verbal Expressions**

Describe each of the following in words:

(a) $3 + x = 7$

(b) $2x = 8 + 2$

Solution (a) The sum of 3 and x is 7.

(b) The product of 2 and x is the sum of 8 and 2. ◀

Practice Exercise 1* **1.** *Write in symbols:*

(a) t less 5 is 4.

(b) 36 is the product of x and y.

(c) x divided by 18 equals the sum of x and 2.

*As you proceed through this book, work each set of Practice Exercises and check your answers. Answers to all Practice Exercises are found at the end of the section.

2. *Describe in words:*

(a) $5x = 30$ (b) $x + 5 = 2x - 7$ (c) $\frac{x}{2} = 2x + 5$

3. In your own words, explain the term *factor*. Give an example.

Properties of Equality

2 We have used the equal sign to convey the idea that one expression is equivalent to another. Listed below are seven important properties of equality. In this list, a, b, and c represent numbers.

Addition Property

1. The **Addition Property** states that if $a = b$, then $a + c = b + c$. (This is the "restoration" principle referred to in the introduction to this chapter.)

Subtraction Property

2. The **Subtraction Property** states that if $a = b$, then $a - c = b - c$.

Multiplication Property

3. The **Multiplication Property** states that if $a = b$, then $a \cdot c = b \cdot c$.

Reflexive Property

4. The **Reflexive Property** states that a number always equals itself; that is, $a = a$.

Symmetric Property

5. The **Symmetric Property** states that if $a = b$, then $b = a$.

Transitive Property

6. The **Transitive Property** states that if $a = b$ and $b = c$, then $a = c$.

Principle of Substitution

7. The **Principle of Substitution** states that if $a = b$, then we may substitute b for a in any expression containing a.

These properties form the basis for much of what we do in algebra, as we shall see a little later when we use them to solve equations. For now, let's look at an example that illustrates these properties.

EXAMPLE 4 Working with Properties of Equality

(a) If $x = 5$, then by the Addition Property, we can add 3 to each side and write $x + 3 = 5 + 3$.

(b) If $x = 10$, then by the Subtraction Property, we can subtract 6 from each side and write $x - 6 = 10 - 6$.

(c) If $x = 10$, then by the Multiplication Property, we can multiply each side by 3 and write $x \cdot 3 = 10 \cdot 3$.

(d) $5 = 5$ by the Reflexive Property.

(e) If $3 = x$, then $x = 3$ by the Symmetric Property.

(f) If $x = a + b$ and $a + b = 2$, then $x = 2$ by the Transitive Property.

(g) If $x = 3$ and $y = x + 2$, then $y = 3 + 2$ by the Principle of Substitution.

(h) Since $8 + 2 = 10$, then $(8 + 2) \cdot x = 10 \cdot x$ by the Multiplication Property.

(i) If $x = 2$ and $y = 5$, then $x + y + 6 = 2 + 5 + 6 = 13$ by the Principle of Substitution. ◀

Practice Exercise 2 *In Problems 1–4, identify the Property of Equality used.*

1. If $9 = x$, then $x = 9$.

2. If $x = 14$, then $x \cdot 8 = 14 \cdot 8$.

3. If $x = 3 + 2b$ and $3 + 2b = 7$, then $x = 7$.

4. If $x = 7$ and $y = 3$, then $x + y = 7 + 3 = 10$.

◀

Exponents

3 When a number is multiplied by itself repeatedly, such as $2 \cdot 2 \cdot 2 \cdot 2 \cdot 2$ (the number 2 is used as a factor 5 times), we may use **exponent notation** to write the expression in a simpler, more compact form, as 2^5. In this notation, the number 2 that is being repeatedly multiplied is called the **base,** and 5, the number of times that 2 appears as a factor, is called the **exponent,** or **power.** The **value** of 2^5 is

$$2 \cdot 2 \cdot 2 \cdot 2 \cdot 2 = 32$$

EXAMPLE 5 Using Exponents

Write each expression using exponents. Name the base and exponent in each case.

(a) $3 \cdot 3 \cdot 3 \cdot 3$ (b) $4 \cdot 4$ (c) $5 \cdot 5 \cdot 5$ (d) $(-2)(-2)(-2)(-2)$

Solution (a) For $3 \cdot 3 \cdot 3 \cdot 3$, the number 3 appears as a factor 4 times. So, $3 \cdot 3 \cdot 3 \cdot 3 = 3^4$, where 3 is the base and 4 is the exponent.

(b) For $4 \cdot 4$, the number 4 appears as a factor 2 times. So, $4 \cdot 4 = 4^2$, where 4 is the base and 2 is the exponent.

(c) For $5 \cdot 5 \cdot 5$, the number 5 appears as a factor 3 times. So, $5 \cdot 5 \cdot 5 = 5^3$, where 5 is the base and 3 is the exponent.

(d) For $(-2) \cdot (-2) \cdot (-2) \cdot (-2)$, the number -2 appears as a factor 4 times. So, $(-2) \cdot (-2) \cdot (-2) \cdot (-2) = (-2)^4$, where -2 is the base and 4 is the exponent. ◀

EXAMPLE 6 Identifying Exponents and Bases

Name the base and exponent, and give the value of each expression.

(a) 3^2 (b) 2^3 (c) -4^2 (d) -3^4

Solution (a) For 3^2, the base is 3 and the exponent is 2. The value is $3^2 = 3 \cdot 3 = 9$.

(b) For 2^3, the base is 2 and the exponent is 3. The value is $2^3 = 2 \cdot 2 \cdot 2 = 8$.

(c) For -4^2, the exponent 2 applies only to the number that immediately precedes it—in this case, 4. That is, $-4^2 = -(4^2)$, so the base is 4. The value is $-(4 \cdot 4) = -16$.

(d) For -3^4, the base is 3 and the exponent is 4. The value of $-3^4 = -(3 \cdot 3 \cdot 3 \cdot 3) = -81$. ◀

WARNING: Exponents are used for repeated multiplications, such as $5 \cdot 5 \cdot 5 = 5^3$. Repeated additions, such as $5 + 5 + 5$, may be written as $3 \cdot 5$. ■

Practice Exercise 3

1. *Write each expression using exponents; name the base and the exponent.*

(a) $7 \cdot 7 \cdot 7 \cdot 7$ (b) $18 \cdot 18 \cdot 18$ (c) $0 \cdot 0 \cdot 0$

(d) $x \cdot x$ (e) $(-5) \cdot (-5) \cdot (-5)$

2. *Name the base and exponent, and give the value of each expression.*

(a) 4^3 (b) -2^4 (c) -5^3

3. What does 5^2 mean? What does 2^5 mean? What does $5 \cdot (2)$ mean? What does $2 \cdot (5)$ mean? Explain how each expression differs from the others. ◀

Order of Operations

4 Consider the expression $2 + 3 \cdot 6$. It is not clear whether we should add 2 and 3 to get 5, and then multiply by 6 to get 30; or first multiply 3 and 6 to get 18, and then add 2 to get 20. To avoid this ambiguity, we have the following agreement:

> We agree that whenever the two operations of addition and multiplication separate three numbers, the multiplication operation always will be performed first, followed by the addition operation.

For $2 + 3 \cdot 6$, we have

$$2 + 3 \cdot 6 = 2 + 18 = 20$$

EXAMPLE 7 Finding the Value of an Expression

Evaluate each expression:

(a) $3 + 4 \cdot 5$ (b) $8 \cdot 2 + 1$ (c) $2 + 2 \cdot 2$

Solution (a) $3 + 4 \cdot 5 = 3 + 20 = 23$ (↑ Multiply first.)

(b) $8 \cdot 2 + 1 = 16 + 1 = 17$ (↑ Multiply first.)

(c) $2 + 2 \cdot 2 = 2 + 4 = 6$ ◀

Suppose it is desired to first add 3 and 4 and then multiply the result by 5. How is this symbolized? The answer cannot be $3 + 4 \cdot 5$, because we have just agreed to multiply first when we see such an expression. The solution is to use grouping symbols such as parentheses ( ), brackets [ ], or braces { }. Using parentheses, we write $(3 + 4) \cdot 5$. Whenever grouping symbols appear in an expression, it means "perform the operations in grouping symbols first!" Whether you use parentheses, brackets, or braces, they are all grouping symbols and are all treated the same.

EXAMPLE 8 **Finding the Value of an Expression**

(a) $(5 + 3) \cdot 4 = 8 \cdot 4 = 32$

(b) $(4 + 5) \cdot (8 - 2) = 9 \cdot 6 = 54$ ◀

Practice Exercise 4 *In Problems 1–8, evaluate each expression:*

1. $5 + 3 \cdot 2 + 6$ **2.** $7 \cdot 2 + 3 \cdot 3$ **3.** $2 \cdot 3 + 4$ **4.** $2 + 3 \cdot 4$

5. $5 + 3 \cdot (2 + 6)$ **6.** $7 \cdot (2 + 3) \cdot 3$ **7.** $(2 \cdot 3) + 4$ **8.** $(2 + 3) \cdot 4$ ◀

When grouping symbols are nested, the inside grouping symbols are evaluated first, followed in order by the outer ones.

EXAMPLE 9 **Finding the Value of an Expression**

$$2 \cdot [5 + (3 + 5) \cdot 2] = 2 \cdot [5 + 8 \cdot 2] = 2 \cdot [5 + 16] = 2 \cdot [21] = 42$$

↑ $3 + 5 = 8$ ↑ Multiply $8 \cdot 2$ ◀

EXAMPLE 10 **Finding the Value of an Expression**

$$\begin{aligned} 3 \cdot \{2 + [4 + 5 \cdot (8 + 2)]\} &= 3 \cdot \{2 + [4 + 5 \cdot 10]\} \\ &= 3 \cdot \{2 + [4 + 50]\} \\ &= 3 \cdot \{2 + 54\} \\ &= 3 \cdot \{56\} \\ &= 168 \end{aligned}$$

Inside first
Second
Third ◀

Notice in the above expression that we used braces and brackets as well as parentheses to assist us in seeing the relative order of operations.

EXAMPLE 11 **Finding the Value of an Expression**

Evaluate each expression:

(a) $[2 \cdot (3 + 2)]^2$ (b) $4 + 2^3$

(c) $(4 + 2)^3$ (d) $(2^3 + 3 \cdot 4) \cdot (8 + 2)$

Solution (a) $[2 \cdot (3 + 2)]^2 = [2 \cdot 5]^2 = 10^2 = 100$

↑ Add the numbers in parentheses. ↑ Multiply the numbers in brackets.

(b) Remember that exponents are symbols for repeated multiplication, and we have agreed to multiply before adding.

$$4 + 2^3 = 4 + 8 = 12$$

(c) Remember, we perform the operation inside the parentheses first:

$$(4 + 2)^3 = 6^3 = 216$$

(d) We begin inside each set of parentheses:

$(2^3 + 3 \cdot 4) \cdot (8 + 2) = (8 + 3 \cdot 4) \cdot (10)$ Evaluate the exponent in the first set of parentheses and simplify the sum in the second set of parentheses.

$= (8 + 12) \cdot (10)$ Multiply before adding.

$= 20 \cdot 10$ Add 8 and 12.

$= 200$ Multiply. ◀

The list below summarizes the rules for the order of operations.

Rules for the Order of Operations

1. Begin within the innermost grouping symbols and work outward.
2. Evaluate exponents first.
3. Perform multiplications and divisions, working from left to right.
4. Perform additions and subtractions, working from left to right.

EXAMPLE 12 **Finding the Value of an Expression**

$4 + 3^2 - \frac{10}{2} = 4 + 9 - \frac{10}{2}$ Exponents first

$= 4 + 9 - 5$ Multiplications and divisions next

$= 13 - 5 = 8$ Additions and subtractions next, working left to right ◀

WARNING: Students often get confused with Step 3 in Rules for the Order of Operations. Some students think that all the multiplications are to be performed first and then all the divisions, when we actually perform multiplication and division in the order they appear from left to right. This is also true for Step 4; additions and subtractions are performed in the order they appear from left to right. ■

EXAMPLE 13 **Finding the Value of an Expression**

Evaluate each expression:

(a) $6 \div 3 \cdot 5$ (b) $12 - 5 + 4$

Solution (a) $6 \div 3 \cdot 5 = 2 \cdot 5 = 10$ (b) $12 - 5 + 4 = 7 + 4 = 11$ ◀

EXAMPLE 14 Finding the Value of an Expression

$[(5 \cdot 2 - 2^3) + 4^2] \cdot 2 = [(5 \cdot 2 - 8) + 4^2] \cdot 2$	Begin with innermost parentheses, and evaluate exponents.
$= [(10 - 8) + 4^2] \cdot 2$	Multiply inside innermost parentheses.
$= [2 + 4^2] \cdot 2$	Perform the subtraction within parentheses.
$= [2 + 16] \cdot 2$	Evaluate exponents within brackets.
$= 18 \cdot 2$	Perform the addition within brackets.
$= 36$	Multiply.

Practice Exercise 5 *Evaluate each expression:*

1. $[(3 + 2) \cdot 5 + 1] \cdot 2$ **2.** $(3^2 + 1) \cdot 2 + 1$ **3.** $2 \cdot (3 + 3^2) + 1$

To divide 18 by 2, we may write either $\frac{18}{2}$ or 18/2. The choice of notation is generally a matter of convenience. It is important to observe that for a quotient such as

$$\frac{22 + 8}{8 + 2}$$

the expression on top of the horizontal bar, called the **numerator,** and the expression beneath the bar, called the **denominator,** are treated as if they were enclosed within parentheses. That is,

$$\frac{22 + 8}{8 + 2} = \frac{(22 + 8)}{(8 + 2)} = \frac{30}{10} = 3$$

To write this same expression using the slash symbol for division, we *must* use parentheses:

$$\frac{22 + 8}{8 + 2} = (22 + 8)/(8 + 2) = 30/10 = 3$$

Do you see why? Ignoring parentheses, we would obtain

$$22 + 8/8 + 2 = 22 + 1 + 2 = 25$$

↑ Division is performed first.

which is quite different.

EXAMPLE 15 Finding the Value of an Expression

Evaluate each expression:

(a) $\frac{2 + 30}{2 + 6}$ (b) $\frac{(3^2 + 4^2) \cdot 4}{15 + 10/2}$

Solution (a) $\frac{2 + 30}{2 + 6} = \frac{32}{8} = 4$

(b) $\frac{(3^2 + 4^2) \cdot 4}{15 + 10/2} = \frac{(9 + 16) \cdot 4}{15 + 5} = \frac{25 \cdot 4}{20} = \frac{100}{20} = 5$

Practice Exercise 6 *Evaluate each expression:*

1. $\frac{7+2}{3}$ 2. $\frac{4^2+5}{7}$ 3. $\frac{(6+8/2)^2}{2}$

Answers to Practice Exercises

Practice Exercise 1

1. (a) $t - 5 = 4$ (b) $36 = xy$ (c) $\frac{x}{18} = x + 2$
2. (a) The product of 5 and x is 30.
   (b) The sum of x and 5 is twice x less 7.
   (c) x divided by 2 is the sum of twice x and 5.
3. Expressions that are multiplied together; for example, in $3 \cdot 5$, the numbers 3 and 5 are factors.

Practice Exercise 2

1. Symmetric Property 2. Multiplication Property
3. Transitive Property 4. Substitution Property

Practice Exercise 3

1. (a) 7^4; the base is 7; the exponent is 4
   (b) 18^3; the base is 18; the exponent is 3
   (c) 0^3; the base is 0; the exponent is 3
   (d) x^2; the base is x; the exponent is 2
   (e) $(-5)^3$; the base is -5; the exponent is 3
2. (a) The base is 4; the exponent is 3; $4^3 = 64$.
   (b) The base is 2; the exponent is 4; $-2^4 = -(2^4) = -16$.
   (c) The base is 5; the exponent is 3; $-5^3 = -(5 \cdot 5 \cdot 5) = -125$
3. 5^2 means $5 \cdot 5 = 25$; 2^5 means $2 \cdot 2 \cdot 2 \cdot 2 \cdot 2 = 32$; $5 \cdot (2)$ means $2 + 2 + 2 + 2 + 2 = 10$; $2 \cdot (5)$ means $5 + 5 = 10$; $5^2 \neq 2^5$; $5 \cdot (2) = 2 \cdot (5)$

Practice Exercise 4 1. 17 2. 23 3. 10 4. 14 5. 29 6. 105 7. 10 8. 20

Practice Exercise 5 1. 52 2. 21 3. 25

Practice Exercise 6 1. 3 2. 3 3. 50

1.1 Assess Your Understanding

Concepts and Vocabulary

1. The sum of x and 3 is the product of 4 and 5 may be written as ________.
2. If $x = 5$, then $x - 2 = 5 - 2$ follows from the ________ Property.
3. In the expression 9^4, the number 9 is called the ________ and 4 is called the ________.
4. The value of $6 + 2 \cdot 5 - 7$ is ________.
5. *True or False:* When simplifying $12 \div 3 \cdot 2$, the multiplication is performed first.
6. *True or False:* $-3^2 = -9$.

Exercises

In Problems 7–20, write each statement using symbols.

7. The sum of 3 and 2 equals 5.
8. The product of 5 and 2 equals 10.
9. The sum of x and 2 is the product of 3 and 4.
10. The sum of 3 and y is the sum of 2 and 2.

11. The product of 3 and y is the sum of 1 and 2.

12. The product of 2 and x is the product of 4 and 6.

13. x less 2 equals 6.

14. 2 less y equals 6.

15. x divided by 2 is 6.

16. 2 divided by x is 6.

17. Three times a number x is twice x less 2.

18. The sum of 2 times a number x and 4 is the product of x and 1.

19. The product of x and x equals twice x.

20. 3 times a number x less 5 equals x divided by 5.

In Problems 21–28, describe each statement in words.

21. $2 + x = 6$ **22.** $3 - x = 4$ **23.** $2x = 3 + 5$ **24.** $\frac{x}{2} = 5 \cdot 2$

25. $3x + 4 = 7$ **26.** $\frac{x}{2} + 1 = 4$ **27.** $\frac{y}{3} - 6 = 4$ **28.** $4 + 2x = 6$

In Problems 29–36, use a Property of Equality to answer each question.

29. If $5 = y$, why does $y = 5$?

30. If $x = y + 2$, why does $x + 5 = (y + 2) + 5$?

31. If $x = y + 2$ and $y + 2 = 10$, why does $x = 10$?

32. If $x + 3 = 10$, why does $(x + 3) + 2 = 10 + 2$?

33. If $y = 8 + 2$, why does $y = 10$?

34. If $y = 8 + 2$, why does $y - 2 = (8 + 2) - 2$?

35. If $2x = 10$, why does $2x = 2 \cdot 5$?

36. If $x = 5$, why does $3 \cdot x = 3 \cdot 5$?

In Problems 37–46, write each expression using exponents.

37. $2 \cdot 2 \cdot 2$ **38.** $3 \cdot 3$ **39.** $3 \cdot 3 \cdot 3 \cdot 3 \cdot 3$ **40.** $2 \cdot 2$ **41.** $4 \cdot 4 \cdot 4$

42. $6 \cdot 6 \cdot 6 \cdot 6$ **43.** $8 \cdot 8$ **44.** $10 \cdot 10 \cdot 10$ **45.** $x \cdot x \cdot x$ **46.** $y \cdot y \cdot y \cdot y$

In Problems 47–54, name the base and the exponent, and give the value of each expression.

47. 4^2 **48.** 5^3 **49.** 1^4 **50.** 0^3

51. -5^2 **52.** -1^8 **53.** -3^2 **54.** -2^3

In Problems 55–92, evaluate each expression.

55. $4 + 2 \cdot 5$ **56.** $8 - 2 \cdot 3$ **57.** $6 \cdot 2 - 1$ **58.** $8 \cdot 5 - 4$

59. $2 \cdot (3 + 4)$ **60.** $(8 + 1) \cdot 2$ **61.** $5 + \frac{4}{2}$ **62.** $\frac{8}{2} - 1$

63. $3^2 + 4 \cdot 2$ **64.** $(3^2 + 4) \cdot 2$ **65.** $8 + 2 \cdot 3^2$ **66.** $(8 + 2) \cdot 3^2$

67. $8 + (2 \cdot 3)^2$ **68.** $(8 + 2 \cdot 3)^2$ **69.** $3 \cdot 4 + 2^3$ **70.** $(3)(4 + 2^3)$

71. $3 \cdot (4 + 2)^3$

72. $(3 \cdot 4) + 2^3$

73. $[8 + (4 \cdot 2 + 3)^2] - (10 + 4 \cdot 2)$

74. $2[(10 + 2) \cdot 3] - (6 \cdot 3 + 4^2)$

75. $2^3(4 + 1)^2$

76. $(3 + 1)^2 \cdot 3^2$

77. $[(6 \cdot 2 - 3^2) \cdot 2 + 1] \cdot 2$

78. $[(4^2 - 2 \cdot 3) \cdot 2 + 4]$

79. $1 \cdot 2^2 + 2 \cdot 3^2 + 3 \cdot 4^2$

80. $2 \cdot 1^3 + 3 \cdot 2^3 + 4 \cdot 3^3$

81. $\frac{8 + 4}{2 \cdot 3}$ **82.** $\frac{2 - 1}{1 \cdot 1}$ **83.** $\frac{4^2 + 2}{3^3}$ **84.** $\frac{2^3 + 3^2 - 1}{2^3}$

85. $4 + 10/2 - 1$ **86.** $(4 + 10)/2 - 1$ **87.** $(4 + 10)/(2 - 1)$ **88.** $4 + 10/(2 - 1)$

89. $\frac{2 \cdot 3}{1} + \frac{3 \cdot 4}{2}$ **90.** $\frac{2 \cdot 3 + 3 \cdot 4}{1 + 2}$ **91.** $\frac{3^2 + 14/2}{2^2 + 4}$ **92.** $\frac{(2^3 + 4)/3}{(8 + 2)/5}$

93. Look at Problems 85, 86, 87, and 88. How are these problems different from each other?

1.2 Number Systems; Integers, Rational Numbers, Real Numbers

OBJECTIVES

1. Work with Sets
2. Classify Numbers
3. Express Rational Numbers as Decimals

Sets

1 When we want to treat a collection of similar, but distinct, objects as a whole, we use the idea of a *set*. A **set** is a collection of distinct objects, all having one or more properties in common. The objects of a set are called **elements** of the set.

For example, the set of digits consists of the collection of numbers 0, 1, 2, 3, 4, 5, 6, 7, 8, and 9. If we use the symbol D to denote the set of digits, then we can write

$$D = \{0, 1, 2, 3, 4, 5, 6, 7, 8, 9\}$$

In this notation, the braces { } are used to enclose the elements of the set. This method of denoting a set is called the **roster method**. A second way to denote a set is to use **set-builder notation**, where the set D of digits is written as

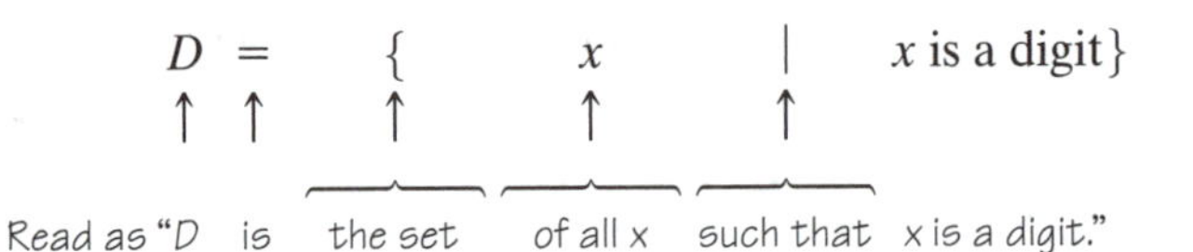

EXAMPLE 1 **Using Set-builder Notation and the Roster Method**

(a) $E = \{x|x \text{ is an even digit}\} = \{0, 2, 4, 6, 8\}$
(b) $O = \{x|x \text{ is an odd digit}\} = \{1, 3, 5, 7, 9\}$ ◀

The **symbol** $\in$ is used to indicate that an element belongs to a set, while the **symbol** $\notin$ is used to indicate that an element does not belong to the set. For example, if D is the set of digits, then

$1 \in D$ Read as "1 is an element of the set D."

$19 \notin D$ Read as "19 is not an element of the set D."

In listing the elements of a set, we generally do not list an element more than once, because the elements of a set are distinct. Multiple listings of the same element could cause confusion when we need to count the number of elements in the set. Also, the order in which the elements are listed does not matter. For example, $\{2, 3\}$ and $\{3, 2\}$ both represent the same set.

If two sets A and B have the same elements, then A is **equal to** B, written $A = B$.

EXAMPLE 2 **Identifying Equal Sets**

$$\{1, 3, 5, 7\} = \{3, 1, 7, 5\}$$

since each set contains the same elements. ◀

If every element of set A is also an element of set B, then we say A is a **subset** of B, written $A \subseteq B$.

EXAMPLE 3 **Identifying Subsets**

$$\{1, 3, 5\} \subseteq \{1, 3, 5, 7\}$$

because each element of the set $\{1, 3, 5\}$ is also an element of the set $\{1, 3, 5, 7\}$. ◀

Practice Exercise 1

1. Find another subset of $\{1, 3, 5, 7\}$.
2. Is it possible that your answer is different from someone else's? Explain.

◀

If a set has no elements, it is called the **empty set,** or **null set.** The empty set is denoted by the symbol $\emptyset$ or by $\{\ \}$. For example, the set of all people alive today who were born in the seventeenth century is the empty set.

The empty set is considered to be a subset of every set. Also, any set is a subset of itself.

EXAMPLE 4 **Finding Subsets**

List all the subsets of the set $\{1, 2, 3\}$.

Solution $\emptyset, \quad \{1\}, \quad \{2\}, \quad \{3\}, \quad \{1, 2\}, \quad \{2, 3\}, \quad \{1, 3\}, \quad \{1, 2, 3\}$ ◀

Practice Exercise 2

1. List all the subsets of $\{1\}$.
2. List all the subsets of $\{1, 2\}$.
3. Can you see a way to predict the number of subsets of a set in advance of listing them? **Hint:** Look for a pattern using your answers from Practice Exercises 1 and 2 (immediately above) and from the answer from Example 4.

◀

If the elements of a set A can be counted, the set A is called a **finite set.** The empty set $\emptyset$, which has no elements is also a finite set. Sets that are not finite are called **infinite sets.**

EXAMPLE 5 **Identifying a Finite Set**

The set D of digits $\{0, 1, 2, 3, 4, 5, 6, 7, 8, 9\}$ is a finite set. It contains 10 elements. ◀

Sets are often used in algebra to classify numbers.

Classification of Numbers

2 The **counting numbers,** or **natural numbers,** are the numbers 1, 2, 3, 4, ... (The first three dots, called **ellipses,** indicate that the pattern continues indefinitely. The last dot is the period.) As their name implies, these numbers are often used to count things. For example, there are 26 letters in the English alphabet; there are 100 cents in a dollar. Using set notation, the counting numbers are $\{1, 2, 3, 4, 5, \ldots\}$. The set of counting numbers is an infinite set.

The **whole numbers** are the numbers 0, 1, 2, 3, ..., that is, the counting numbers together with 0. In set notation, the whole numbers are $\{0, 1, 2, 3, 4, \ldots\}$. Is the set of whole numbers finite or infinite?

Practice Exercise 3 *Determine which sets are finite and which are infinite.*

1. $\{0\}$
2. $\emptyset$
3. $\{x \mid x \text{ is an even counting number}\}$

The counting numbers, although useful for counting, cannot be used in all situations. For example, suppose your checking account has $10 in it and you write a check for $15. How do you represent your balance? In accounting, you might write the balance as $(5) or as ⟨5⟩; in algebra, we represent the balance as −$5. When we expand the set of counting numbers to include the negative of each counting number and 0, we obtain the set of *integers*.

The **integers** are the numbers

$$\ldots, -3, -2, -1, 0, 1, 2, 3, \ldots$$

The set of integers $\{\ldots, -3, -2, -1, 0, 1, 2, 3, \ldots\}$ is an infinite set. Included among the integers are the counting numbers, sometimes referred to as **positive integers.** The set of counting numbers is a subset of the set of integers. When we expanded our number system from the counting numbers to the integers, we did so out of the need to describe numerical situations that could not be handled using just counting numbers. The integers allow us to solve problems requiring both positive and negative counting numbers and 0, such as problems involving profit/loss, height above/below sea level, temperature above/below 0°F, and so on.

But integers alone are not sufficient for *all* problems. For example, they do not answer the questions, "What part of a dollar is 38 cents?" or "What part of a pound is 5 ounces?" To answer such questions, we enlarge our number system to include *rational numbers*. For example, $\frac{38}{100}$ answers the question, "What part of a dollar is 38 cents?" and $\frac{5}{16}$ answers the question, "What part of a pound is 5 ounces?"

A **rational number** is a number that can be expressed as a quotient $\frac{a}{b}$ of two integers. The integer a is called the **numerator,** and the integer b, which cannot be 0, is called the **denominator.**

Using set-builder notation, the set of rational numbers is

$$\{x \mid x = \frac{a}{b}, \text{ where } a, b \text{ are integers and } b \neq 0\}$$

Examples of rational numbers are $\frac{3}{4}, \frac{5}{2}, \frac{0}{4}, -\frac{2}{3}, \frac{100}{3}$, and $\frac{5}{1}$. Notice that since $\frac{a}{1} = a$ for any integer a, it follows that the rational numbers contain all the integers making the integers a subset of the set of rational numbers.

3 Rational numbers may also be represented as **decimals.** For example, the rational numbers $\frac{3}{4}, \frac{5}{2}, -\frac{2}{3}$, and $\frac{7}{66}$ may be represented as decimals by merely carrying out the indicated division:

$$\frac{3}{4} = 0.75 \qquad \frac{5}{2} = 2.5 \qquad -\frac{2}{3} = -0.666\ldots = -0.\overline{6} \qquad \frac{7}{66} = 0.1060606\ldots = 0.1\overline{06}$$

Notice that the decimal representations of $\frac{3}{4}$ and $\frac{5}{2}$ terminate, or end. The decimal representations of $-\frac{2}{3}$ and $\frac{7}{66}$ do not terminate, but they do exhibit a pattern of repetition. For $-\frac{2}{3}$, the 6 repeats indefinitely, as indicated by the bar over the 6; for $\frac{7}{66}$, the block 06 repeats indefinitely, as indicated by the bar over the 06. It can be shown that every rational number may be represented by a decimal that either terminates or is nonterminating with a repeating block of digits, and vice versa.

EXAMPLE 6 Writing a Rational Number as a Decimal

Express each rational number as a decimal.

(a) $\frac{7}{8}$ (b) $\frac{10}{3}$ (c) $-\frac{3}{10}$ (d) $\frac{8}{7}$

Solution (a) Divide 7 by 8, obtaining

$$\begin{array}{r} .875 \\ 8\overline{)7.000} \\ \underline{64} \\ 60 \\ \underline{56} \\ 40 \\ \underline{40} \end{array} \qquad \frac{7}{8} = 0.875$$

(b) Divide 10 by 3, obtaining

$$\begin{array}{r} 3.33\ldots \\ 3\overline{)10.00} \\ \underline{9} \\ 10 \\ \underline{9} \\ 10 \\ \underline{9} \\ 1 \end{array} \qquad \frac{10}{3} = 3.333\ldots = 3.3\overline{3}$$

(c) $-\dfrac{3}{10} = -0.3$

(d) $\dfrac{8}{7} = 1.142857142857\ldots = 1.\overline{142857}$, where the block 142857 repeats ◀

Looking at the long division process in Example 6(b), it is easy to see we are caught in a loop that will keep repeating. So the 3's repeat in the quotient. Again, we indicate this with ellipses or a bar over the repeating block of digits.

Practice Exercise 4

1. *Express each rational number as a decimal.*

(a) $\dfrac{4}{3}$ (b) $\dfrac{3}{16}$ (c) $-\dfrac{17}{25}$ (d) $\dfrac{5}{6}$ ◀

Some decimals do not fit into either of the categories: terminating or nonterminating with a repeating block of digits. For example, the decimal $0.12345678910111213\ldots$, in which we write down the positive integers successively one after the other, will neither repeat (think about it) nor terminate. Such a decimal represents an **irrational number.** Every irrational number may be represented by a decimal that neither repeats nor terminates. In other words, irrational numbers cannot be written in the form $\frac{a}{b}$, where a, b are integers and $b \neq 0$.

Irrational numbers occur naturally. For example, consider the isosceles right triangle whose legs are each of length 1. See Figure 1. The number that equals the length of the hypotenuse is the positive number whose square is 2. It can be shown that this number, which we symbolize as $\sqrt{2}$, is not a rational number. The decimal representation for $\sqrt{2}$ begins with $1.414213\ldots$.

Figure 1

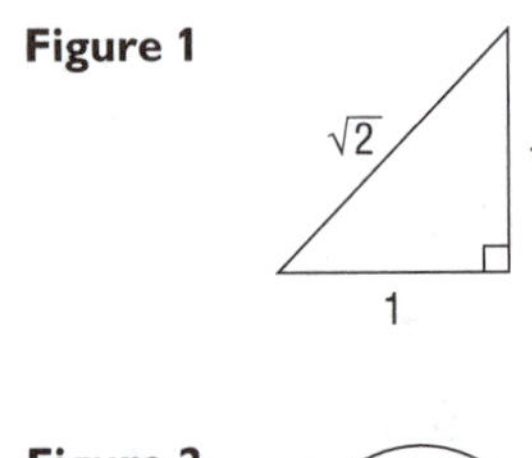

Also, the number that equals the ratio of the circumference C to the diameter d of any circle, denoted by the symbol π (the Greek letter pi), is an irrational number. See Figure 2. The decimal representation for π begins with $3.14159\ldots$.

Figure 2

$\pi = \dfrac{C}{d}$

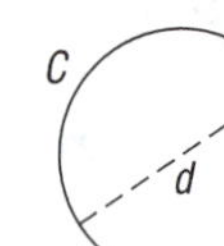

Other symbols for numbers that are irrational are $\sqrt{3}$, $\sqrt{5}$, $\sqrt{7}$, $\sqrt{6}$, $\sqrt[3]{2}$, $\sqrt[3]{3}$, and so on.

Together, the rational numbers and irrational numbers form the set of **real numbers.**

Figure 3 shows the relationship between various types of numbers.

To summarize, the decimal representation of a real number is always one of three types:

1. Terminating } Rational numbers
2. Nonterminating, repeating } Rational numbers
3. Nonterminating, nonrepeating } Irrational numbers

(All three types together: Real numbers)

Every decimal may be represented by a real number and, conversely, every real number may be represented by a decimal. It is this feature of the real numbers that gives them their practicality. In the physical world, many changing quantities, such as the length of a heated rod, the velocity of a falling object, and so on, are assumed to pass through every possible magnitude from the initial one to the final one as they change. Real numbers in

Figure 3

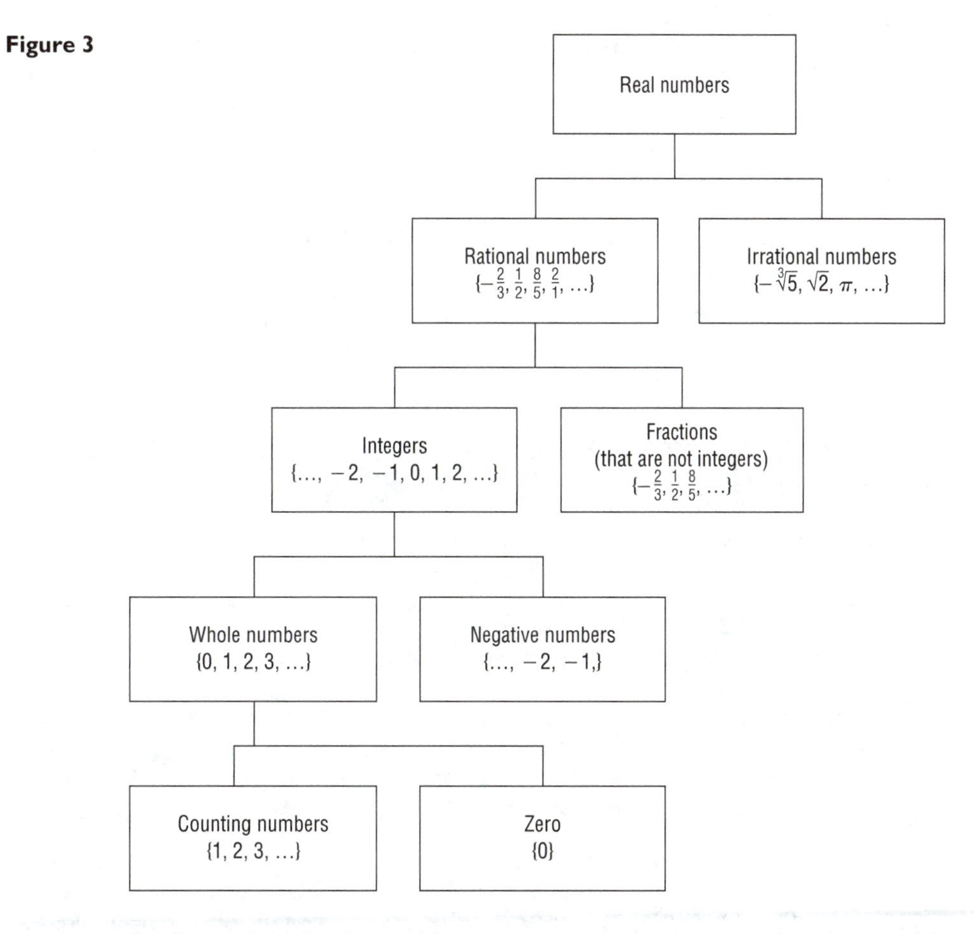

the form of decimals provide a convenient way to measure such quantities as they change.

In practice, irrational numbers are generally represented by approximations. After all, nonterminating decimals can never be completely written down. We must always stop and use a terminating decimal approximation. For example, using the symbol $\approx$ (read as "approximately equal to"), we can write

$$\sqrt{2} \approx 1.4142 \qquad \pi \approx 3.1416$$

We shall discuss the idea of approximation in more detail in the next section.

EXAMPLE 7 Classifying the Numbers in a Set

List the numbers in the set

$$\left\{-3, \frac{4}{3}, 0.12, \sqrt{2}, \pi, 2.151\overline{515}, 10\right\}$$

that are:

(a) Counting numbers
(b) Integers
(c) Rational numbers
(d) Irrational numbers
(e) Real numbers

Solution (a) 10 is the only counting number.

(b) -3 and 10 are integers.

(c) $-3, \frac{4}{3}, 0.12, 2.151\overline{515}$, and 10 are rational numbers.

(d) $\sqrt{2}$ and π are irrational numbers.

(e) All the numbers listed are real numbers. ◀

Finally, while it appears at this point that all numbers are real numbers, this is not the case. There are other numbers that we have not yet talked about that are not real, such as **complex numbers.** These numbers are discussed briefly in Section 4.5 and in more detail in your textbook.

HISTORICAL COMMENTS

The real number system has a history that stretches back at least to the ancient Babylonians (1800 BC). It is remarkable how much the ancient Babylonian attitudes resemble our own. As we stated in the text, the fundamental difficulty with irrational numbers is that they cannot be written as quotients of integers or, equivalently, as repeating or terminating decimals. The Babylonians wrote their numbers in a system based on 60 in the same way we write ours based on 10. They would carry as many places for π as the accuracy of the problem demanded, just as we now use approximations, such as

$$\pi \approx 3\frac{1}{7} \quad \text{or} \quad \pi \approx 3.1416 \quad \text{or} \quad \pi \approx 3.14159$$
$$\text{or} \quad \pi \approx 3.14159265359$$

depending on how accurate we need to be.

Things were very different for the Greeks, whose number system allowed only rational numbers. When it was discovered that $\sqrt{2}$ could not be expressed as a quotient of two integers, this was regarded as a fundamental flaw in the number concept. So serious was the matter that the Pythagorean Brotherhood (an early mathematical society) is said to have drowned one of its members for revealing this terrible secret. Greek mathematicians then turned away from the number concept, expressing facts about whole numbers in terms of line segments.

In astronomy, however, Babylonian methods, including the Babylonian number system, continued to be used. In 1585, Simon Stevin (1548–1620), probably using the Babylonian system as a model, invented the decimal system, complete with rules of calculation. [Others—for example, al-Kashi of Samarkand (d. 1424)—had made some progress in the same direction.] The decimal system so effectively conceals the difficulties of working with irrational numbers that the need for more logical precision began to be felt only in the early 1800's. Around 1880, Georg Cantor (1845–1918) and Richard Dedekind (1831–1916) gave precise definitions of real numbers. Cantor's definition, though more abstract and precise, has its roots in the decimal (and hence, Babylonian) numerical system.

Sets and set theory were a spin-off of the research that went into clarifying the foundations of the real number system. Set theory has developed into a large discipline of its own, and many mathematicians regard it as the foundation upon which modern mathematics is built. Cantor's discoveries that infinite sets also can be counted and that there are different sizes of infinite sets are among the most astounding results of modern mathematics.

Historical Problems

The Babylonian number system was based on 60 so that, 2,30 means $2 + \frac{30}{60} = 2.5$ and 4,25,14 means

$$4 + \frac{25}{60} + \frac{14}{60^2} = 4 + \frac{1514}{3600} = 4.42055555\ldots$$

1. What are the following numbers in Babylonian notation?

(a) $1\frac{1}{3}$ (b) $2\frac{5}{6}$

2. What are the following Babylonian numbers when written as fractions and as decimals?

(a) 2,20 (b) 4,52,30 (c) 3,8,29,44

Answers to Practice Exercises

Practice Exercise 1

1. Some possible subsets are $\{1, 5, 7\}$, $\{1, 3, 7\}$, $\{3, 5, 7\}$, $\{1, 3\}$, $\{1, 5\}$, $\{1, 3, 5, 7\}$, $\{1\}$

2. Yes, because there are several subsets.

Practice Exercise 2 **1.** $\emptyset, \{1\}$ **2.** $\emptyset, \{1\}, \{2\}, \{1, 2\}$
3. If a set has n elements, it will have 2^n subsets.

Practice Exercise 3 **1.** Finite **2.** Finite **3.** Infinite

Practice Exercise 4 **1.** (a) $1.333\ldots = 1.\overline{3}$ (b) 0.1875 (c) -0.68 (d) $0.8333\ldots = 0.83\overline{3}$

1.2 Assess Your Understanding

Concepts and Vocabulary

1. Two ways to denote a set are the __________ __________ and __________ __________.

2. *True or False:* $\{3\} \subseteq \{1, 2, 3, 4\}$

3. *True or False:* $\{3\} \in \{1, 2, 3, 4\}$

4. Is the set of whole numbers finite or infinite?

5. The numbers in the set $\left\{x \middle| x = \frac{a}{b}, \text{ where } a, b \text{ are integers and } b \neq 0\right\}$ are called __________ numbers.

6. *True or False:* No real number is both rational and irrational.

7. *True or False:* Every real number may be represented as a decimal.

8. *True or False:* $\frac{1}{3} = 0.3$

Exercises

In Problems 9–18, use the sets $A = \{1, 2, 3, 4, 5\}$ and $B = \{1, 2, 3\}$. Determine whether the given statement is true or false.

9. $1 \in A$ **10.** $4 \in A$ **11.** $4 \in B$ **12.** $5 \in B$ **13.** $A = B$
14. $A \subseteq B$ **15.** $B \subseteq A$ **16.** $\{2\} \subseteq B$ **17.** $\{1\} \subseteq A$ **18.** $\{4\} \subseteq B$

19. List all the subsets of the set $\{a, b\}$.

20. List all the subsets of the set $\{a, b, c\}$.

In Problems 21–28, determine whether the given set is finite or infinite.

21. $\{1, 3, 5, 7\}$ **22.** $\{2, 4, 6, 8\}$ **23.** $\{1, 3, 5, 7, \ldots\}$ **24.** $\{2, 4, 6, 8, \ldots\}$
25. $\{x | x \text{ is an even integer}\}$ **26.** $\{x | x \text{ is an odd integer}\}$ **27.** $\{x | x \text{ is an even digit}\}$ **28.** $\{x | x \text{ is an odd digit}\}$

In Problems 29–38, express each rational number as a decimal.

29. $\frac{1}{5}$ **30.** $\frac{1}{20}$ **31.** $\frac{7}{3}$ **32.** $\frac{8}{3}$ **33.** $-\frac{5}{8}$
34. $-\frac{1}{8}$ **35.** $\frac{4}{25}$ **36.** $-\frac{5}{6}$ **37.** $-\frac{3}{7}$ **38.** $\frac{8}{11}$

In Problems 39–42, list the numbers in each set that are
(a) Natural numbers
(b) Integers
(c) Rational numbers
(d) Irrational numbers
(e) Real numbers

39. $A = \left\{-6, \frac{1}{2}, -1.333\ldots \text{(the 3's repeat)}, \pi + 1, 2, 5\right\}$

40. $B = \left\{-\frac{5}{3}, 2.060606\ldots \text{(the block 06 repeats)}, 1.25, 0, 1, \sqrt{5}\right\}$

41. $C = \left\{\frac{x}{2} \middle| x \text{ is a digit}\right\}$

42. $D = \{2x | x \text{ is a digit}\}$

43. National Vital Statistics reported that an average 25-year-old male in 2000 can expect to live 50.6 more years, and an average 25-year-old female in 2000 can expect to live 55.4 more years.
(a) To what age can an average 25-year-old male expect to live?
(b) To what age can an average 25-year-old female expect to live?
(c) Who can expect to live longer, a male or a female? By how many years?

Source: National Center for Health Statistics, 2002.

44. A Gallup poll on "body image" found that two-thirds of college-educated women want a more muscular, "hard-bodied" look. If 300 college-educated women were involved in this poll, how many would have answered "yes" to the question, "Do you want a more muscular hard-bodied look?"

45. Women have only four-fifths as many red blood cells in each drop of their blood as men do. If a typical man has 1000 red cells in a drop of blood, how many red cells does a typical woman have in a drop of blood?

46. In 1900, the level of carbon dioxide in the atmosphere was 290 parts per million parts of atmosphere. What fraction of the atmosphere was carbon dioxide in 1900? Express this fraction as a decimal.

Source: Testimony of Syukuro Manabe, an atmospheric scientist at Princeton's Geophysical Fluid Dynamics Laboratory, before the Senate Energy and Natural Resources Committee, 1988.

47. In 2000, the level of carbon dioxide in the atmosphere was about 370 parts per million parts of atmosphere. What fraction of the atmosphere was carbon dioxide in 2000? Express this fraction as a decimal.

Source: **www.informationsphere.com,** 1998–2003.

48. Are there any real numbers that are both rational and irrational? Are there any real numbers that are neither? Explain your reasoning.

49. Explain why the sum of a rational number and an irrational number must be irrational.

50. What rational number does the repeating decimal 0.999 . . . equal?

51. Is there a positive real number "closest" to 0?

52. I'm thinking of a number! It lies between 1 and 10; its square is rational and lies between 1 and 10. The number is larger than π. Correct to one decimal place, name the number. Now think of your own number, describe it, and challenge a fellow student to name it.

53. In your own words, explain the difference between a digit, a counting number, and an integer.

54. A rational number is defined as the quotient of two integers. When written as a decimal, the decimal will either repeat or terminate. When given a rational number in lowest terms, we can tell from the denominator if its decimal representation will repeat or terminate. Make a list of rational numbers and their decimals. See if you can discover the pattern. Confirm your conclusion by consulting books on number theory at the library. Write a brief essay on your findings.

1.3 Approximations; Calculators

OBJECTIVES

1 Approximate Numbers
2 Use a Calculator

Approximations

1 In approximating decimals, we either *round off* or *truncate* to a given number of decimal places.* This number of places establishes the location of the *final digit* in the decimal approximation.

Truncation: Drop all the digits that follow the specified final digit in the decimal.

Rounding: Identify the final digit in the decimal. If the next digit is 5 or more, add 1 to the final digit; if the next digit is 4 or less, leave the final digit as it is. Now truncate following the final digit.

*Sometimes we say "correct to a given number of decimal places" instead of "truncate."

EXAMPLE 1 Approximating a Decimal to Two Places

Approximate 20.98752 to two decimal places by:

(a) Truncating (b) Rounding

Solution For 20.98752, the final digit is the 8, since it is two decimal places from the decimal point.

(a) To truncate, we remove all digits following the final digit 8. The truncation of 20.98752 to two decimal places is 20.98.

(b) The digit following the final digit 8 is the digit 7. Since 7 is 5 or more, we add 1 to the final digit 8. The rounded form of 20.98752 to two decimal places is 20.99. ◀

EXAMPLE 2 Approximating a Decimal to Two and Four Places

	Number	Rounded to Two Decimal Places	Rounded to Four Decimal Places	Truncated to Two Decimal Places	Truncated to Four Decimal Places
(a)	3.14159	3.14	3.1416	3.14	3.1415
(b)	0.056128	0.06	0.0561	0.05	0.0561
(c)	893.46125	893.46	893.4613	893.46	893.4612
(d)	29.54398	29.54	29.5440	29.54	29.5439

◀

Practice Exercise 1 *In Problems 1–4, write each number as a decimal to three decimal places by:*

(a) Rounding *(b) Truncating*

1. 9.3625 **2.** 0.0123 **3.** 58.8689 **4.** 0.0009

◀

WARNING: If you are unsure whether to round a number or truncate it, then round it. Also, never round off or truncate a decimal in the middle of a problem. Always wait until you have the final answer to round off or truncate to the desired number of decimal places. ■

Calculators

2 Calculators are finite machines. As a result, they are incapable of displaying decimals that contain a large number of digits. For example, some calculators are capable of displaying only eight digits. When a number requires more than eight digits, the calculator either truncates or rounds. To see how your calculator handles decimals, divide 2 by 3. How many digits do you see? Is the last digit a 6 or a 7? If it is a 6, your calculator truncates; if it is a 7, your calculator rounds.

There are different kinds of calculators. An **arithmetic** calculator can only add, subtract, multiply, and divide numbers; therefore, this type is not adequate for this course. **Scientific** calculators have all the capabilities of arithmetic calculators and also contain **function** keys such as LN, LOG, SIN, COS, TAN, INV, and so on. On some calculators these function keys are given in all capital letters and on others they are given in lowercase letters. **Graphing** calculators have all the capabilities of scientific calculators and contain a screen on which graphs can be displayed.

Another difference among calculators is the order in which various operations are performed. Some employ an **algebraic system,** whereas others use **reverse Polish notation (RPN).** Either system is acceptable, and the choice of which system to get is a matter of individual preference. Of course, no matter what calculator you purchase, be sure to study the instruction manual so that you use the calculator efficiently and correctly. In this book, our examples are worked using an algebraic system.

When the arithmetic involved in a given problem in this book is messy, we shall mark the problem with a , indicating that you should use a calculator. Calculators need not be used except on problems labeled with a . The examples marked with a and the answers provided in the back of the book for calculator problems have been found using a TI-83 and a TI-30X IIS. The TI-83 is a graphing calculator, while the TI-30X IIS is a scientific calculator. The keypads of both of these calculators are very similar and, as a result, for many problems you use the exact same keystrokes on each. Due to differences among calculators in the way they do arithmetic, your answers may vary slightly from those given in the text.

Calculator Operations

Arithmetic operations on your calculator are generally shown as keys labeled as follows:

Your calculator also has the following keys:

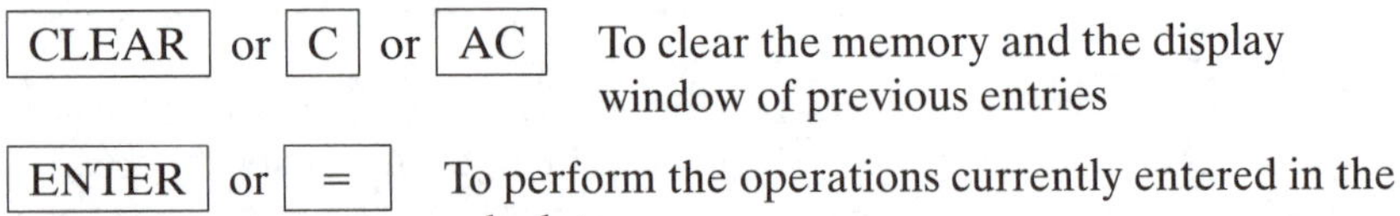

Now let's look at a few examples. Again, the keystrokes given are for the TI-83 and the TI-30X IIS. These keystrokes will also be valid for a variety of other calculators on the market.

EXAMPLE 3 Using a Calculator to Find the Value of an Expression

Evaluate: $8 + 9 \cdot 6$

Solution Keystrokes: 8 + 9 × 6 ENTER

Display: 8 + 9*6
62 ◀

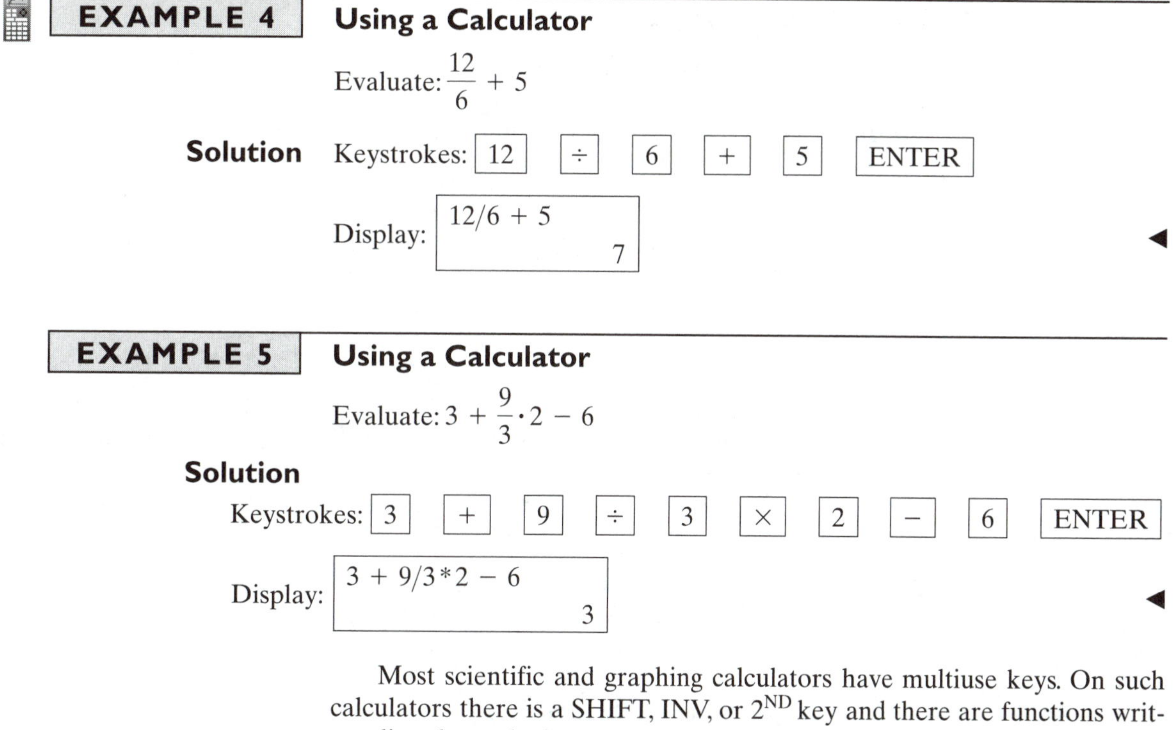

EXAMPLE 4 **Using a Calculator**

Evaluate: $\frac{12}{6} + 5$

Solution Keystrokes: [12] [÷] [6] [+] [5] [ENTER]

Display: 12/6 + 5 → 7

◀

EXAMPLE 5 **Using a Calculator**

Evaluate: $3 + \frac{9}{3} \cdot 2 - 6$

Solution

Keystrokes: [3] [+] [9] [÷] [3] [×] [2] [−] [6] [ENTER]

Display: 3 + 9/3*2 − 6 → 3

◀

Most scientific and graphing calculators have multiuse keys. On such calculators there is a SHIFT, INV, or 2ND key and there are functions written directly on the keys as well as functions written on the calculator above the keys. To access a function written on the calculator above a key, you must first press the SHIFT, INV, or 2ND key.

The next two examples are used to illustrate some of the differences among various calculators on the market.

EXAMPLE 6 **Using a Calculator**

Approximate π^2. Round the answer to two decimal places.

Solution A On most scientific calculators, there is a key labeled [x^2], used to square a number. Also, there may be a key labeled [π]; otherwise, some combination of keys gives the number π. On a TI-30X IIS the solution is

Keystrokes: [π] [x^2] [ENTER]

Display: π^2 → 9.868604401

Rounded to two decimal places, $\pi^2 = 9.87$.

Solution B For the TI-83, the solution is

Keystrokes: [2nd] [∧] [x^2] [ENTER]

Display: π^2 → 9.868604401

Again, rounded to two decimal places, $\pi^2 = 9.87$. ◀

EXAMPLE 7 **Using a Calculator**

Evaluate: $\sqrt{25} + 3 \cdot 4$

Solution A For the TI-30X IIS and the TI-83

Keystrokes: [2nd] [x^2] [25] [)] [+] [3] [×] [4] [ENTER]

Display: √(25) + 3*4 17

Notice with the TI-83 and the TI-30X IIS that once you press [2nd] [x^2], the calculator automatically opens a set of parentheses for you. You must close the parentheses after entering 25. If you do not close the parentheses, the calculator will automatically close parentheses at the end of the problem. In this instance, the calculator would compute $\sqrt{25 + 3 \cdot 4}$ instead of the problem given.

Solution B For a calculator such as the TI-30Xa, the solution is

Keystrokes: [25] [$\sqrt{x}$] [+] [3] [×] [4] [=]

Display: [25] [5] [3] [4] [17] ◀

Practice Exercise 2 *In Problems 1–5, evaluate each expression.*

1. $(2.3)^2$ **2.** $(0.21)^2$ **3.** $2 + 3 \cdot 4$ **4.** $2 \cdot 3 + 4$ **5.** $8 + 3 \cdot \frac{8}{2} - 4$ ◀

These examples should convince you that a calculator that uses the algebraic system does arithmetic according to the agreed-upon conventions regarding order of operations. However, there are times when you need to be careful.

The next example illustrates a use of the keys for parentheses,

[(] and [)]

EXAMPLE 8 **Using a Calculator**

Evaluate: $\frac{22 + 8}{8 + 2}$

Solution We mentioned earlier (in Section 1.1) that we treat this expression as if parentheses enclose the numerator and the denominator.

Keystrokes: [(] [22] [+] [8] [)] [÷] [(] [8] [+] [2] [)] [ENTER]

Display: (22 + 8)/(8 + 2) 3 ◀

Be careful! If you work the problem in Example 8 without using parentheses, you obtain $22 + 8/8 + 2 = 22 + 1 + 2 = 25$.

Your calculator should have the key [∧] or [x^y] or [y^x], which is used for computations involving exponents. The next two examples show how this key is used.

EXAMPLE 9 Evaluating Exponents with a Calculator

Evaluate 2^4.

Solution Keystrokes: [2] [∧] [4] [ENTER]

Display:
```
2 ^ 4
       16
```
◀

EXAMPLE 10 Evaluating Exponents with a Calculator

Evaluate $(2.3)^5$.

Solution Keystrokes: [2.3] [∧] [5] [ENTER]

Display:
```
2.3 ^ 5
        64.36343
```
◀

Next we look at how to use your calculator for a problem such as -4^2.

EXAMPLE 11 Evaluating Exponents with a Calculator

Evaluate -4^2.

Solution A Recall from Section 1.1 that $-4^2 = -(4 \cdot 4) = -16$. But on some calculators, like the TI-30Xa, if you use the following keystrokes you will get 16.

Keystrokes:	[4] ↘	[+/−] ↘	[x^2] ↘
Display:	[4]	[−4]	[16]

The calculator is actually computing $(-4)^2$. This is an example of where the calculator user must "outsmart" the calculator and realize the answer is -16.

Alternatively, you could use the following keystrokes on a calculator such as the TI-30Xa:

Keystrokes:	[4] ↘	[x^2] ↘	[+/−] ↘
Display:	[4]	[16]	[−16]

Solution B On the TI-83 and TI-30X IIS to compute -4^2 do the following:

Keystrokes: [(−)] [4] [x^2] [ENTER]

Display:

-4^2

-16

◀

The final word on calculators is to select the one you like, purchase it, and learn how to use it.

Practice Exercise 3 *In Problems 1–4, evaluate each expression.*

1. $(8.5)^4$ **2.** $(-2.5)^5$ **3.** $(95)^3$ **4.** $(1.2)^{10}$ ◀

Answers to Practice Exercises

Practice Exercise 1 **1.** (a) 9.363 (b) 9.362 **2.** (a) 0.012 (b) 0.012 **3.** (a) 58.869 (b) 58.868 **4.** (a) 0.001 (b) 0.000

Practice Exercise 2 **1.** 5.29 **2.** 0.0441 **3.** 14 **4.** 10 **5.** 16

Practice Exercise 3 **1.** 5220.0625 **2.** −97.65625 **3.** 857,375 **4.** 6.191736422

1.3 Assess Your Understanding

Concepts and Vocabulary

1. When we ________ a number, we drop all the digits that follow the specified final digit in the decimal.

2. *True or False:* When rounding 48.926 to two decimal places we obtain 48.92.

3. *True or False:* All calculators use the same keystroke sequence for computing $\sqrt{49}$.

4. Rounded to three decimal places, the value of $\sqrt{7} + 3\sqrt{11}$ is ________.

Exercises

In Problems 5–24, approximate each number (a) rounded and (b) truncated to three decimal places.

5. 18.9526 **6.** 25.86134 **7.** 28.65349 **8.** 99.05229 **9.** 0.06291

10. 0.05388 **11.** 9.9985 **12.** 1.0006 **13.** $\frac{3}{7}$ **14.** $\frac{5}{9}$

15. $\frac{521}{15}$ **16.** $\frac{81}{5}$ **17.** $\left(\frac{4}{9}\right)^2$ **18.** $(9-4)^2$ **19.** π

20. $\frac{\pi}{2} + 3$ **21.** $\pi + \frac{3}{2}$ **22.** π^2 **23.** π^3 **24.** π^4

In Problems 25–52, use a calculator to approximate each expression. Round your answer to two decimal places.

25. $(8.51)^2$ **26.** $(9.62)^2$ **27.** $4.1 + (3.2)(8.3)$ **28.** $(8.1)(4.2) + 6.1$

29. $(8.6)^2 + (6.1)^2$ **30.** $(3.1)^2 + (9.6)^2$ **31.** $8.6 + \frac{10.2}{4.2}$ **32.** $9.1 - \frac{8.2}{10.2}$

33. $\pi + \frac{\sqrt{2}}{8}$ **34.** $\sqrt{5} - 8\pi$ **35.** $\frac{\pi + 8}{10.2 + 8.6}$ **36.** $\frac{21.3 - \pi}{6.1 + 8.8}$

37. $(22.6 + 8.5)/81.3 + 21.2$ **38.** $(16.5 - 7.2) \cdot 51.2 - 18.6$

39. $22.6 + 8.5/81.3 + 21.2$ **40.** $16.5 - 7.2/51.2 - 18.6$

41. $(22.6 + 8.5)/(81.3 + 21.2)$ **42.** $(16.5 - 7.2)/(51.2 - 18.6)$

43. $22.6 + 8.5/(81.3 + 21.2)$ **44.** $16.5 - 7.2/(51.2 - 18.6)$

45. $(8.2)^5$ **46.** $(3.7)^4$ **47.** $(93.21)^3$ **48.** $(47.63)^3$

49. $(9.8 + 14.6)^4$ **50.** $(19.4 + 8.2)^3$ **51.** $(9.8)^4 + (14.6)^4$ **52.** $(19.4)^3 + (8.2)^3$

53. On your calculator, divide 8 by 0. What happens?

54. Examine two or three different calculators. Write a few paragraphs highlighting their similarities and differences. Tell which of the calculators you prefer and why.

55. The current time is 12 noon CST. What time (CST) will it be 12,997 hours from now?

1.4 Properties of Real Numbers

OBJECTIVE

1 Work with Properties of Real Numbers

1 As we pursue our study of algebra in this book, much of what we do will be based on certain basic properties of real numbers. These properties are given as equations involving addition and multiplication.

We begin with an example.

EXAMPLE 1 **Illustrating the Commutative Properties**

(a) $3 + 5 = 8$
$5 + 3 = 8$
$3 + 5 = 5 + 3$

(b) $2 \cdot 3 = 6$
$3 \cdot 2 = 6$
$2 \cdot 3 = 3 \cdot 2$ ◄

This example illustrates the **Commutative Property** of real numbers, which states that the order in which addition and multiplication takes place will not affect the final result.

Here, and in the properties that follow, a, b, and c represent real numbers.

Commutative Properties

$$a + b = b + a \quad \textbf{(1a)}$$

$$a \cdot b = b \cdot a \quad \textbf{(1b)}$$

EXAMPLE 2 **Illustrating the Associative Properties**

(a) $2 + (3 + 4) = 2 + 7 = 9$
$(2 + 3) + 4 = 5 + 4 = 9$
$2 + (3 + 4) = (2 + 3) + 4$

(b) $2 \cdot (3 \cdot 4) = 2 \cdot 12 = 24$
$(2 \cdot 3) \cdot 4 = 6 \cdot 4 = 24$
$2 \cdot (3 \cdot 4) = (2 \cdot 3) \cdot 4$ ◄

The way we add or multiply three real numbers will not affect the final result. This property is called the **Associative Property** of real numbers.

Associative Properties

$$a + (b + c) = (a + b) + c = a + b + c \quad \textbf{(2a)}$$

$$a \cdot (b \cdot c) = (a \cdot b) \cdot c = a \cdot b \cdot c \quad \textbf{(2b)}$$

Because of the Associative Properties, expressions such as $2 + 3 + 4$ and $3 \cdot 4 \cdot 5$ present no ambiguity, even though addition and multiplication are performed on one pair of numbers at a time. However, in evaluating any expression such as $3 \cdot 2 \cdot 7$, be sure to *mentally* place parentheses around one of the pairs of numbers and then proceed to multiply. For example,

$$3 \cdot 2 \cdot 7 = (3 \cdot 2) \cdot 7 = 6 \cdot 7 = 42$$

↑ ↑
Mentally

WARNING: Be sure *not* to do the following:

$$3 \cdot 2 \cdot 7 = (3 \cdot 2) \cdot (3 \cdot 7) = 6 \cdot 21 = 126$$

↑
Error here

While most people do not make this mistake on a simple problem, this is a fairly common mistake on more complicated algebra problems. ■

EXAMPLE 3 Illustrating the Distributive Property

Simplify: $2 \cdot (4 + 7)$

Solution While we would like to simplify the addition in parentheses first, on some problems this is not always possible. So we use a different approach. Study the following two ways of simplifying this problem.

$$2 \cdot (4 + 7) = 2 \cdot (11) = 22 \qquad \text{or} \qquad 2 \cdot (4 + 7) = 2 \cdot 4 + 2 \cdot 7 = 8 + 14 = 22$$

◀

This example as it is worked out on the right side illustrates the *Distributive Property*.

Distributive Property

$$a \cdot (b + c) = a \cdot b + a \cdot c \qquad \textbf{(3a)}$$

$$a \cdot c + b \cdot c = (a + b) \cdot c \qquad \textbf{(3b)}$$

Look at equation (3a). The **Distributive Property** derives its name from the fact that the number a is distributed over each of the numbers b and c in forming the sum. Let's look at two of the more important uses of this property.

EXAMPLE 4 Using the Distributive Property

(a) $2 \cdot (x + 3) = 2 \cdot x + 2 \cdot 3 = 2x + 6$ Use: To remove parentheses

↑ Use equation (3a)

(b) $6x + 8x = (6 + 8)x = 14x$ Use: To add two expressions

↑ Use equation (3b)

(c) $x \cdot (x + 3) = x \cdot x + x \cdot 3 = x^2 + 3x$ Use: To remove parentheses

↑ Use equation (3a)

◀

Practice Exercise 1 *In Problems 1–4, remove the parentheses in each expression and simplify, if possible.*

1. $6(x + 7)$ **2.** $(3 + y) \cdot 2$ **3.** $(3 + x) + 4$ **4.** $5 \cdot (3 \cdot 2)$

In Problems 5–7, name the property that justifies each statement.

5. $8 + (2 + 1) = (8 + 2) + 1$ **6.** $9 \cdot (4 + 7) = 9 \cdot 4 + 9 \cdot 7$

7. $3 \cdot 5 = 5 \cdot 3$

The real numbers 0 and 1 have unique properties, called **Identity Properties.**

EXAMPLE 5 Illustrating the Identity Properties

(a) $4 + 0 = 0 + 4 = 4$ (b) $3 \cdot 1 = 1 \cdot 3 = 3$

In general, we have:

Identity Properties

$$0 + a = a + 0 = a \qquad \textbf{(4a)}$$

$$a \cdot 1 = 1 \cdot a = a \qquad \textbf{(4b)}$$

We call 0 the **additive identity** and 1 the **multiplicative identity.**

For each real number a, there is a real number $-a$, called the **additive inverse** of a, having the following property:

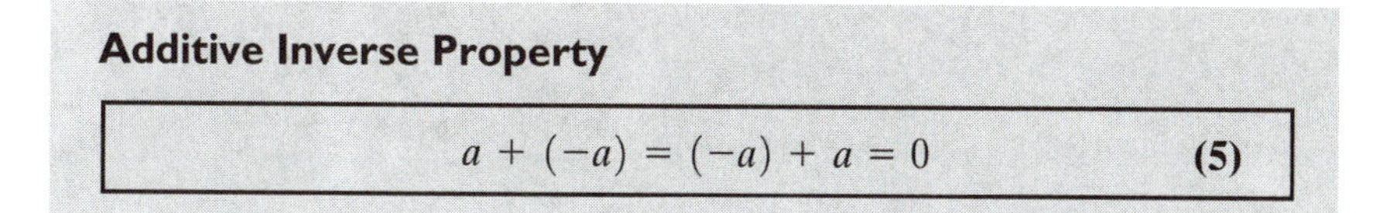

Additive Inverse Property

$$a + (-a) = (-a) + a = 0 \qquad \textbf{(5)}$$

The additive inverse of a is often called the *negative* of a, but the use of this term can be dangerous, because it suggests that the additive inverse is a negative number, which may not be the case. For example, the additive inverse of -3, or $-(-3)$, equals 3, a positive number.

EXAMPLE 6 Finding an Additive Inverse

(a) The additive inverse of 6 is -6, because $6 + (-6) = 0$.

(b) The additive inverse of -8 is $-(-8) = 8$, because $-8 + 8 = 0$.

Practice Exercise 2 *In Problems 1–4, find the additive inverse of each number.*

1. -2 **2.** 0 **3.** $\frac{2}{3}$ **4.** $-\frac{3}{4}$

For each *nonzero* real number a, there is a real number $\frac{1}{a}$, called the **multiplicative inverse** of a, having the following property:

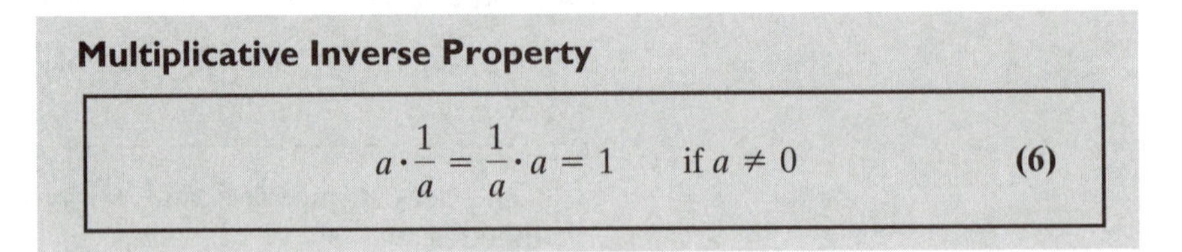

Multiplicative Inverse Property

$$a \cdot \frac{1}{a} = \frac{1}{a} \cdot a = 1 \qquad \text{if } a \neq 0 \qquad \textbf{(6)}$$

The multiplicative inverse $\frac{1}{a}$ of a nonzero real number a is also referred to as the **reciprocal** of a.

On a calculator, the key $\boxed{1/x}$ or $\boxed{x^{-1}}$ is used to find the reciprocal of a previously entered number.

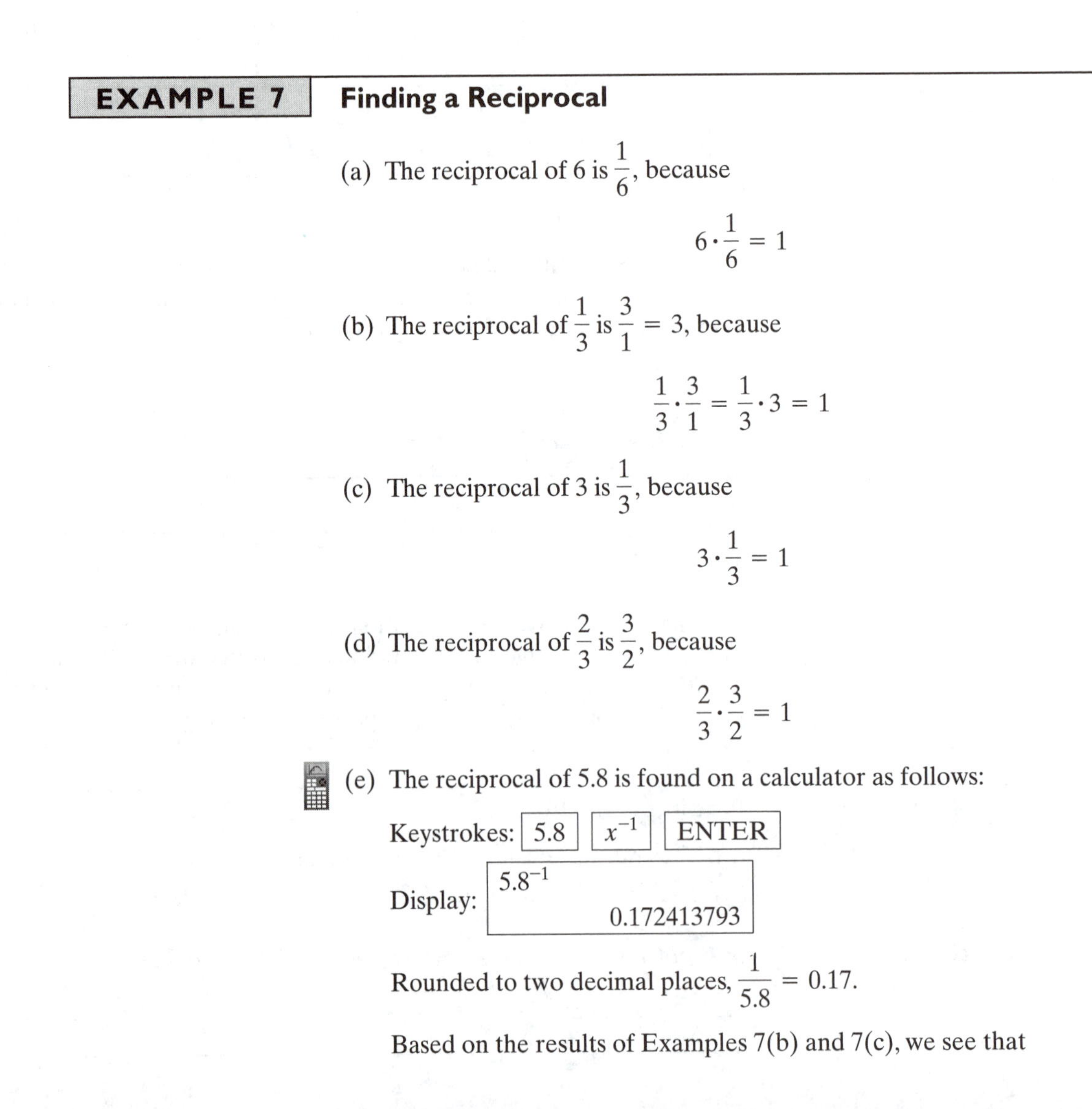

EXAMPLE 7 **Finding a Reciprocal**

(a) The reciprocal of 6 is $\frac{1}{6}$, because

$$6 \cdot \frac{1}{6} = 1$$

(b) The reciprocal of $\frac{1}{3}$ is $\frac{3}{1} = 3$, because

$$\frac{1}{3} \cdot \frac{3}{1} = \frac{1}{3} \cdot 3 = 1$$

(c) The reciprocal of 3 is $\frac{1}{3}$, because

$$3 \cdot \frac{1}{3} = 1$$

(d) The reciprocal of $\frac{2}{3}$ is $\frac{3}{2}$, because

$$\frac{2}{3} \cdot \frac{3}{2} = 1$$

(e) The reciprocal of 5.8 is found on a calculator as follows:

Keystrokes: $\boxed{5.8}$ $\boxed{x^{-1}}$ $\boxed{\text{ENTER}}$

Display:

5.8^{-1}
0.172413793

Rounded to two decimal places, $\frac{1}{5.8} = 0.17$. ◀

Based on the results of Examples 7(b) and 7(c), we see that

$$\frac{1}{1/a} = a \qquad a \neq 0 \qquad \textbf{(7)}$$

That is, the reciprocal of the reciprocal of a number is the number itself.

Practice Exercise 3

1. *Find the reciprocal of*

(a) 5 (b) $-\frac{1}{2}$ (c) $\frac{4}{7}$

The properties of real numbers listed above are often used to derive other properties. We list a few of them below. As before, a, b, and c denote real numbers in the properties listed.

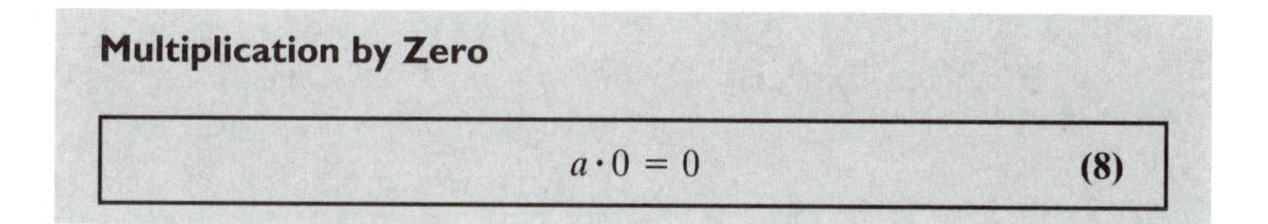

Multiplication by Zero

$$a \cdot 0 = 0 \qquad \textbf{(8)}$$

For any real number a, the product of a times 0 is always 0.

Division Properties

$$\frac{0}{a} = 0 \qquad \text{if } a \neq 0 \qquad \textbf{(9a)}$$

$$\frac{a}{a} = 1 \qquad \text{if } a \neq 0 \qquad \textbf{(9b)}$$

Note: Division by 0 is *not defined*. One reason is to avoid the following difficulty: We know that $\frac{10}{5} = 2$, because $5 \cdot 2 = 10$. So, $\frac{2}{0} = x$ means to find x such that $0 \cdot x = 2$. But $0 \cdot x$ equals 0 for all x, so there is *no* unique number x such that $\frac{2}{0} = x$. Division by 0 is *not defined*. On your calculator, divide 2 by 0. What do you get?

Zero-Product Property

$$\text{If } a \cdot b = 0, \text{ then } a = 0, \text{ or } b = 0, \text{ or both.} \qquad \textbf{(10)}$$

EXAMPLE 8 **Using the Zero-Product Property**

If $4x = 0$, then by the Zero Product Property, it follows that $4 = 0$ or $x = 0$. Since $4 \neq 0$, we conclude that $x = 0$. ◀

Cancellation Properties

If $a + c = b + c$, then $a = b$. **(11a)**

If $a \cdot c = b \cdot c$ and $c \neq 0$, then $a = b$. **(11b)**

If $b \neq 0$ and $c \neq 0$, then: $\dfrac{a \cdot c}{b \cdot c} = \dfrac{a}{b}$ **(11c)**

EXAMPLE 9 Using the Cancellation Properties

(a) If $x + 4 = 10$, then $x + 4 = 6 + 4$. By the Cancellation Property (11a), we may cancel the 4's to obtain $x = 6$.

(b) If $2x = 8$, then $2x = 2 \cdot 4$. By the Cancellation Property (11b), we may cancel the 2's to obtain $x = 4$.

(c) If $x \neq 0$, then: $\dfrac{3 \cdot x}{2 \cdot x} = \dfrac{3}{2}$ ◀

Practice Exercise 4 *In Problems 1–4, use the Zero-Product Property or a Cancellation Property to find x.*

1. $4x = 0$ **2.** $x + 2 = 5$ **3.** $3x = 6$ **4.** $x - 3 = -8$

In mathematics, it is important to understand not only *what* to do but also *why it is possible* to do it. The properties listed in this section explain why certain manipulations (many of which you have seen and done in earlier courses) can be done. The next example illustrates the use of several of the properties we have cited so far.

EXAMPLE 10 Simplifying an Expression Using Properties

STEP	JUSTIFICATION
$2x + 6 = 10$	Given
$2x + 6 = 4 + 6$	Principle of Substitution (substitute $4 + 6$ for 10)
$2x = 4$	Cancellation Property (cancel the 6's)
$2 \cdot x = 2 \cdot 2$	Principle of Substitution (substitute $2 \cdot 2$ for 4)
$x = 2$	Cancellation Property (cancel the 2's)

Answers to Practice Exercises

Practice Exercise 1 **1.** $6x + 42$ **2.** $6 + 2y$ **3.** $3 + x + 4 = x + 3 + 4 = x + 7$ **4.** $5 \cdot (3 \cdot 2) = 30$ **5.** Associative Property **6.** Distributive Property **7.** Commutative Property

Practice Exercise 2 **1.** 2 **2.** 0 **3.** $-\dfrac{2}{3}$ **4.** $\dfrac{3}{4}$

Practice Exercise 3 **1.** (a) $\dfrac{1}{5}$ (b) -2 (c) $\dfrac{7}{4}$

Practice Exercise 4 **1.** $x = 0$ **2.** $x = 3$ **3.** $x = 2$ **4.** $x = -5$

1.4 Assess Your Understanding

Concepts and Vocabulary

1. The fact that $4(3 + x) = 4 \cdot 3 + 4 \cdot x$ is a consequence of the ________ Property.
2. *True or False:* The Zero-Product Property states that the product of any number and zero equals zero.
3. *True or False:* We can use properties to simplify expressions.
4. *True or False:* In the expression $\frac{a}{b}$ it is acceptable for b to equal 0.

Exercises

In Problems 5–12, find the additive inverse and the reciprocal of each number.

5. 3 **6.** 8 **7.** -4 **8.** -6 **9.** $\frac{1}{4}$ **10.** $\frac{1}{5}$ **11.** 1 **12.** -1

In Problems 13–18, use the Distributive Property to remove the parentheses.

13. $3(x + 4)$
14. $5(2x + 1)$
15. $x(x + 3)$
16. $3x(x + 4)$
17. $4x \cdot (x + 4)$
18. $5x(x + 1)$

In Problems 19–34, use the property listed to fill in each blank.

19. Commutative Property: $x + \frac{1}{2} =$ ________
20. Commutative Property: $\frac{1}{4} \cdot x =$ ________
21. Associative Property: $x + (y + 3) =$ ________
22. Associative Property: $4 + (x + 4) =$ ________
23. Associative Property: $3 \cdot (2x) =$ ________
24. Associative Property: $\left(\frac{3}{7} \cdot 3\right) \cdot x =$ ________
25. Distributive Property: $(2 + a)x =$ ________
26. Distributive Property: $(a + 5)y =$ ________
27. Distributive Property: $8x + 5x =$ ________
28. Distributive Property: $ax + 4x =$ ________
29. Cancellation Property: If $x + \frac{3}{4} = x + y$, then ________
30. Cancellation Property: If $x + a = x + 5$, then ________
31. Cancellation Property: If $3 \cdot y = 3 \cdot 20$, then ________
32. Cancellation Property: If $a \cdot x = a \cdot \frac{9}{8}$, then ________
33. Zero-Product Property: If $ax = 0$, then ________
34. Zero-Product Property: If $x(x + 2) = 0$, then ________

In Problems 35–58, name the property that justifies each statement.

35. $2 + y = y + 2$
36. $a + 5 = 5 + a$
37. $2 \cdot (x + 3) = 2x + 2 \cdot 3$
38. $ax + bx = (a + b)x$
39. $2 \cdot \frac{1}{2} = 1$
40. $-1.8 + 1.8 = 0$
41. $\pi + 0 = \pi$
42. $1 \cdot \frac{1}{2} = \frac{1}{2}$
43. $\frac{7}{2} + \left(-\frac{7}{2}\right) = 0$
44. $\frac{1}{\sqrt{2}} \cdot \sqrt{2} = 1$
45. $2 \cdot (ax) = (2a) \cdot x$
46. $3 + (a + x) = (3 + a) + x$
47. If $8x = 8y$, then $x = y$.
48. If $xy = 0$, then $x = 0$ or $y = 0$, or both.
49. $\frac{1}{2} \cdot (4x) = \left(\frac{1}{2} \cdot 4\right)x$
50. $\frac{3}{4} + \left(\frac{1}{4} + x\right) = \left(\frac{3}{4} + \frac{1}{4}\right) + x$

51. $\frac{1}{2}\cdot x + \frac{3}{4}\cdot x = \left(\frac{1}{2} + \frac{3}{4}\right)\cdot x$

52. $6\cdot\left(\frac{x}{2} + \frac{1}{3}\right) = 6\cdot\frac{x}{2} + 6\cdot\frac{1}{3}$

53. $1\cdot\frac{1}{x} = \frac{1}{x}$

54. $\pi\cdot 1 = \pi$

55. If $x(x + 4) = 0$, then $x = 0$ or $x + 4 = 0$.

56. If $x^2 + 1 = 4 + 1$, then $x^2 = 4$.

57. $2\cdot(x^2 + 4) = 2x^2 + 2\cdot 4$

58. $3x^2 + 3\cdot 1 = 3(x^2 + 1)$

In Problems 59–64, fill in each blank with the property that justifies the step.

59. $3x + 8 = 20$ Given
$3x + 8 = 12 + 8$ ________
$3x = 12$ ________
$3x = 3\cdot 4$ ________
$x = 4$ ________

60. $4x + 10 = 38$ Given
$4x + 10 = 28 + 10$ ________
$4x = 28$ ________
$4x = 4\cdot 7$ ________
$x = 7$ ________

61. $x^2 + 14 = x(x + 2) + 4$ Given
$x^2 + 14 = (x\cdot x + x\cdot 2) + 4$ ________
$x^2 + 14 = (x^2 + x\cdot 2) + 4$ Change to exponent form.
$x^2 + 14 = x^2 + (x\cdot 2 + 4)$ ________
$14 = x\cdot 2 + 4$ ________
$10 + 4 = x\cdot 2 + 4$ ________
$10 = x\cdot 2$ ________
$5\cdot 2 = x\cdot 2$ ________
$5 = x$ ________
$x = 5$ ________

62. $x^2 + 4 = x(x + 2)$ Given
$x^2 + 4 = x\cdot x + x\cdot 2$ ________
$x^2 + 4 = x^2 + x\cdot 2$ ________
$4 = x\cdot 2$ ________
$2\cdot 2 = x\cdot 2$ ________
$2 = x$ ________
$x = 2$ ________

63. $(x + 2)(x + 3) = (x + 2)\cdot x + (x + 2)\cdot 3$ ________
$= x\cdot x + 2\cdot x + x\cdot 3 + 2\cdot 3$ ________
$= x^2 + 2\cdot x + x\cdot 3 + 2\cdot 3$ ________
$= x^2 + 2\cdot x + 3\cdot x + 2\cdot 3$ ________
$= x^2 + (2 + 3)x + 2\cdot 3$ ________
$= x^2 + 5x + 6$ ________

64. $(x + 1)^2 = (x + 1)(x + 1)$ ________
$= (x + 1)\cdot x + (x + 1)\cdot 1$ ________
$= x\cdot x + 1\cdot x + x\cdot 1 + 1\cdot 1$ ________
$= x^2 + 1\cdot x + x\cdot 1 + 1\cdot 1$ ________
$= x^2 + 1\cdot x + 1\cdot x + 1\cdot 1$ ________
$= x^2 + (1 + 1)\cdot x + 1\cdot 1$ ________
$= x^2 + 2x + 1$ ________

In Problems 65–68, two expressions are shown that, in general, are not equal. Give an example to show this. That is, pick numbers for a, b, and c and show that you get two different results for each expression.

65. $\dfrac{a+b}{a+c}$ $\quad \dfrac{b}{c}$ **66.** $a \cdot (b \cdot b)$ $\quad (a \cdot b) \cdot (a \cdot b)$ **67.** $a \cdot a$ $\quad 2a$ **68.** $(a+b)^2$ $\quad a^2 + b^2$

69. Both $\dfrac{a}{0}$ $\{a \neq 0\}$ and $\dfrac{0}{0}$ are undefined, but for different reasons. Write a paragraph or two explaining the different reasons.

1.5 The Real Number Line; Absolute Value; Addition and Subtraction with Real Numbers

OBJECTIVES

1. Work with the Real Number Line
2. Find Absolute Values
3. Add Real Numbers
4. Subtract Real Numbers

The Real Number Line

1 Real numbers can be represented by points on a line called the *real number line*. There is a one-to-one correspondence between real numbers and points on a line. That is, every real number corresponds to a point on the line, and each point on the line has a unique real number associated with it.

Pick a point on the line somewhere in the center, and label it O. This point, called the **origin,** corresponds to the real number 0. See Figure 4. The point 1 unit to the right of O corresponds to the number 1. The distance between 0 and 1 determines the **scale** of the number line. For example, the point associated with the number 2 is twice as far from O as 1 is. Notice that an arrowhead on the right end of the line indicates the direction in which the numbers increase. Figure 4 also shows the points associated with the irrational numbers $\sqrt{2}$ and π. Points to the left of the origin correspond to the real numbers -1, -2, and so on.

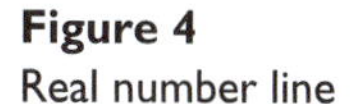

Figure 4
Real number line

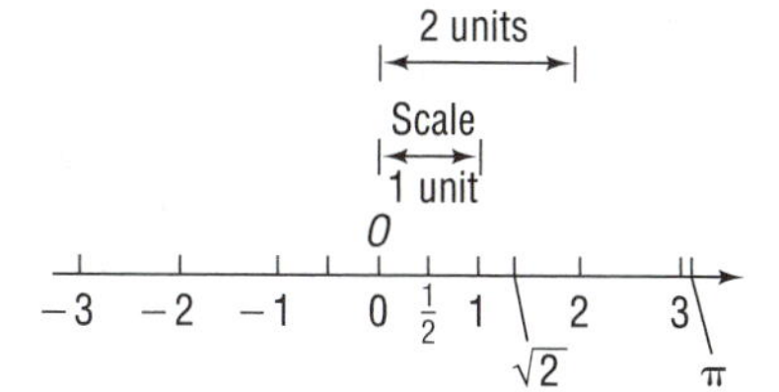

The real number associated with a point P is called the **coordinate** of P, and the line whose points have been assigned coordinates is called the **real number line.**

Often, we shall say "the point 5" on the real number line when we mean "the point whose coordinate is 5." This, however, should cause no problems.

Individual numbers can be graphed on the real number line by simply putting a dot at the location of the number.

EXAMPLE 1 **Graphing Points on the Real Number Line**

On the real number line, graph and label the points in the set $\{-4, 5, 0.5, -1.5, 2\}$.

Solution Figure 5 below shows the solution.

Figure 5

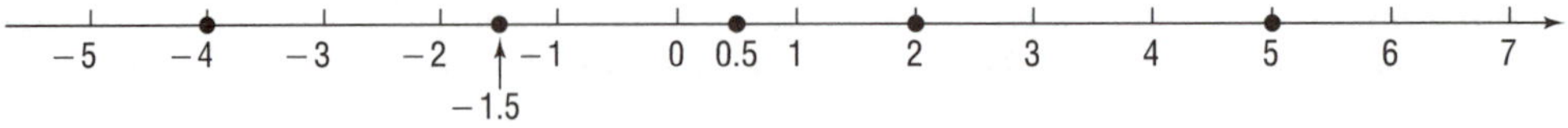

−5 −4 −3 −2 −1 0 0.5 1 2 3 4 5 6 7
−1.5

Practice Exercise 1 1. *On the real number line, graph and label the points with coordinates* $0, 1, -1, \frac{5}{2}, -2.5, \frac{7}{4}, 0.25$

The real number line consists of three classes of real numbers, as shown in Figure 6.

Figure 6

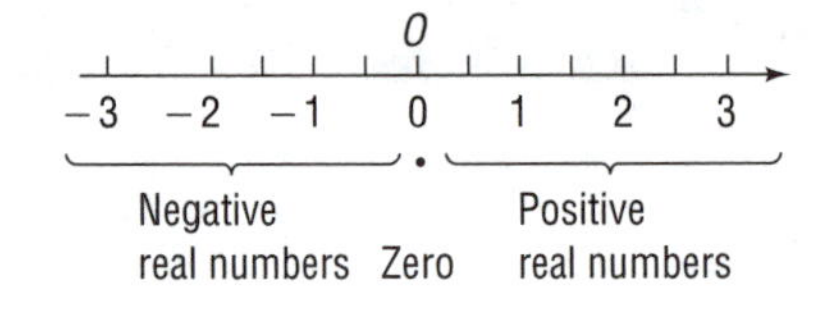

1. The **negative real numbers** are the coordinates of points to the left of the origin O.
2. The real number **zero** is the coordinate of the origin O.
3. The **positive real numbers** are the coordinates of points to the right of the origin O.

The real number line provides a convenient way to illustrate the geometric relationship between a number and its additive inverse. See Figure 7.

Figure 7

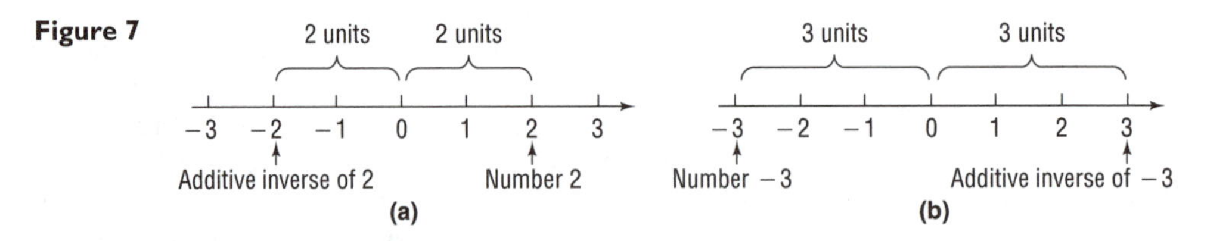

Notice that a number and its additive inverse are coordinates of points that are the same distance from the origin: one of them to the left of the origin and the other to the right of the origin. Because a number and its additive inverse are the same distance from the origin, but on opposite sides of the origin, they are sometimes referred to as **opposites** of each other.

EXAMPLE 2 **Finding Additive Inverses**

(a) Find the additive inverse of 5.
(b) Find the additive inverse of −5.
(c) Find the opposite of −7.
(d) Find the opposite of 7.

Solution (a) The additive inverse of 5 is −5.
(b) The additive inverse of −5 is 5.
(c) The opposite of −7 is 7.
(d) The opposite of 7 is −7.

Example 2 illustrates a more general property. The additive inverse of the additive inverse of a real number is the number itself. That is, if a is a real number, then we have the following theorem:

$$-(-a) = a \quad \textbf{(1)}$$

EXAMPLE 3 **Using Formula (1)**

(a) $-(-8) = 8$ (b) $-(-3) = 3$ ◀

Absolute Value

2 The **absolute value** of a number a is the distance from the origin O to the point whose coordinate is a. For example, the point whose coordinate is -4 is 4 units from O. The point whose coordinate is 3 is 3 units from O. See Figure 8. So, the absolute value of -4 is 4, and the absolute value of 3 is 3. In symbols, we write this as $|-4| = 4$ and $|3| = 3$.

Figure 8

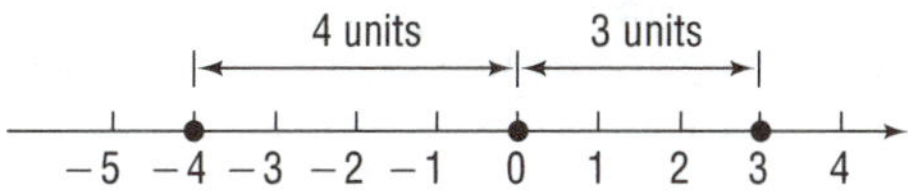

4 units 3 units

−5 −4 −3 −2 −1 0 1 2 3 4

−4 is a distance of 4 units from the origin
3 is a distance of 3 units from the origin

EXAMPLE 4 **Finding the Absolute Value of a Number**

(a) $|8| = 8$ (b) $|0| = 0$ (c) $|-15| = 15$ ◀

A more formal definition of absolute value is given below:

The **absolute value** of a real number a, denoted by $|a|$, is defined by the rules

$$|a| = a \quad \text{if } a \text{ is positive or if } a = 0 \quad \textbf{(2a)}$$

$$|a| = -a \quad \text{if } a \text{ is negative} \quad \textbf{(2b)}$$

For example, since -4 is negative, rule (2b) must be used to get $|-4| = -(-4) = 4$

Practice Exercise 2 *In Problems 1–6, simplify each expression.*

1. $-(-7)$ **2.** $-\left(-\frac{1}{4}\right)$ **3.** $|9|$ **4.** $|-6|$ **5.** $|0|$ **6.** $\left|-\frac{2}{3}\right|$ ◀

Adding Real Numbers

3 The real number line provides a mechanism for finding the sum of two real numbers. The idea is best illustrated through some examples.

EXAMPLE 5 Finding the Sum of Two Numbers

Find each sum:

(a) $5 + 2$ (b) $5 + (-2)$ (c) $-5 + 2$ (d) $-5 + (-2)$

Solution (a) Of course, we all know $5 + 2 = 7$. But let's see how the real number line can be used to get the answer. Look at Figure 9, where we begin at the point whose coordinate is 5. Now, since we want to add 2 to 5, we move 2 units *to the right* of 5. This places us at the point whose coordinate is 7, illustrating that $5 + 2 = 7$.

Figure 9

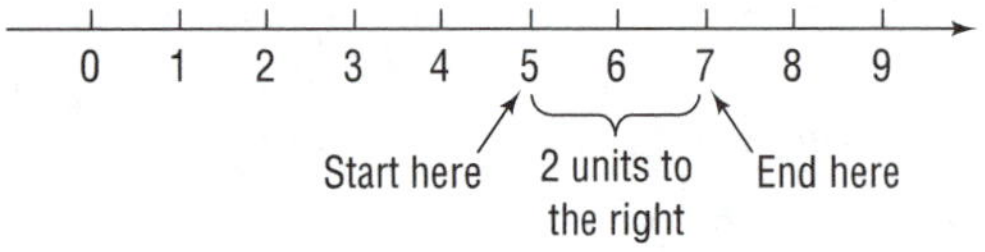

0 1 2 3 4 5 6 7 8 9
Start here
2 units to the right
End here

(b) To find $5 + (-2)$, we look at Figure 10. We again begin at 5. But, since we are adding -2 to 5, we move 2 units *to the left* of 5. This places us at 3, so $5 + (-2) = 3$.

Figure 10

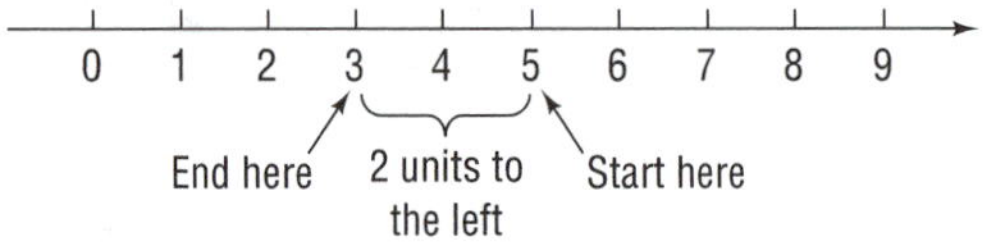

0 1 2 3 4 5 6 7 8 9
End here
2 units to the left
Start here

(c) To find $-5 + 2$, we begin at the point whose coordinate is -5, and, since we are adding 2 to -5, we move 2 units *to the right* of -5. See Figure 11. This brings us to -3. As a result, $-5 + 2 = -3$.

Figure 11

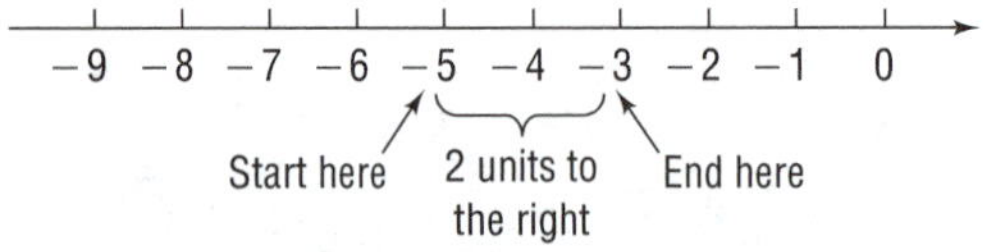

−9 −8 −7 −6 −5 −4 −3 −2 −1 0
Start here
2 units to the right
End here

(d) To find $-5 + (-2)$ we again start at -5. Since we are adding -2 to -5 we move 2 units *to the left* of -5. See Figure 12. This brings us to -7. As a result, $-5 + (-2) = -7$.

Figure 12

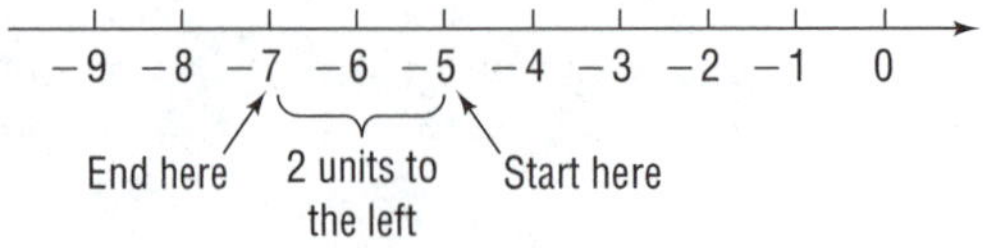

−9 −8 −7 −6 −5 −4 −3 −2 −1 0
End here
2 units to the left
Start here

◄

The real number line provides a visual aid to explain the procedure for adding positive and negative numbers. It is also helpful to put this process into words.

1. To add two positive numbers, add the two numbers; the answer will be positive. For example, $5 + 2 = 7$.
2. To add two negative numbers, first add the absolute values of each number. The answer to the original addition problem will be the additive inverse of this sum. For example, to find $-5 + (-2)$ add $|-5| = 5$ and $|-2| = 2$ to obtain $5 + 2 = 7$. Then $-5 + (-2) = -7$.
3. To add a positive number and a negative number, first find the absolute value of each number. Then subtract the smaller absolute value from the larger absolute value. Select for the sign of your answer the sign of the number in the original problem that had the largest absolute value. For example,

$$5 + (-2) = |5| - |2| = 3$$
$$2 + (-5) = -\left(|5| - |2|\right) = -(5 - 2) = -3.$$

With practice and persistence, adding real numbers will become automatic. Whether you use the real number line to visualize an addition problem or whether you use the process put into words, practice (without a calculator) is the key to success.

Practice Exercise 3 *In Problems 1–8, find each sum:*

1. $5 + (-8)$ **2.** $-4 + 9$ **3.** $(-7) + (-8)$ **4.** $-6 + 0$

5. $3 + 8$ **6.** $3 + (-8)$ **7.** $-3 + 8$ **8.** $(-3) + (-8)$

Subtracting Real Numbers

4 We begin with some examples.

EXAMPLE 6 **Comparing Addition and Subtraction Problems**

(a) $10 + (-3) = 7$ and $10 - 3 = 7$

(b) $10 + (-4) = 6$ and $10 - 4 = 6$

(c) $8 + (-8) = 0$ and $8 - 8 = 0$

(d) $9 + (-5) = 4$ and $9 - 5 = 4$

By comparing each addition problem with its corresponding subtraction problem in Example 6, we see that subtraction problems can be reformulated as addition problems. This is supported with the following rule:

If a and b are real numbers, the **difference** $a - b$, also read, "a less b" or "a minus b," is defined as

$$a - b = a + (-b) \quad \textbf{(3)}$$

In other words, to subtract b from a, add the opposite of b to a.

After reformulating the subtraction problem as an addition problem, we follow the rules for addition.

EXAMPLE 7 **Subtracting Real Numbers**

(a) $8 - 2 = 8 + (-2)$ — Add the additive inverse of 2.
$= 6$ — Add

(b) $5 - 9 = 5 + (-9)$ — Add the additive inverse of 9.
$= -4$ — Add

(c) $-10 - 5 = -10 + (-5)$ — Add the additive inverse of 5
$= -15$ — Add

(d) $6 - (-3) = 6 + [-(-3)]$ — Add the additive inverse of −3
$= 6 + 3$ — $-(-3) = 3$
$= 9$ — Add

(e) $-4 - (-2) = -4 + [-(-2)]$ — Add the additive inverse of −2
$= -4 + 2$ — $-(-2) = 2$
$= -2$ — Add ◀

EXAMPLE 8 **Subtracting Real Numbers**

$8 - (4 - 9) - 5 = 8 - [4 + (-9)] - 5$ — Start with parentheses
$= 8 - [-5] - 5$ — $4 + (-9) = -5$
$= 8 + [-(-5)] - 5$ — Perform leftmost subtraction
$= 8 + 5 - 5$
$= 13 - 5$ — Perform operations from left to right
$= 8$ ◀

If you have trouble with subtracting positive and negative numbers, then write in the step showing the addition of the additive inverse. After you master subtraction, you may omit this step.

Practice Exercise 4 *In Problems 1–11, perform the indicated operations*

1. $6 - 11$ **2.** $-3 - 8$ **3.** $7 - (-4)$
4. $-4 - (-7)$ **5.** $0 - (-3)$ **6.** $4 - (6 - 11) - 12$
7. $5 - 12$ **8.** $5 - (-12)$ **9.** $-5 - 12$
10. $-5 - (-12)$ **11.** $4.82 - (-6.19) - 8.35$ ◀

Answers to Practice Exercises

Practice Exercise 1 **1.**

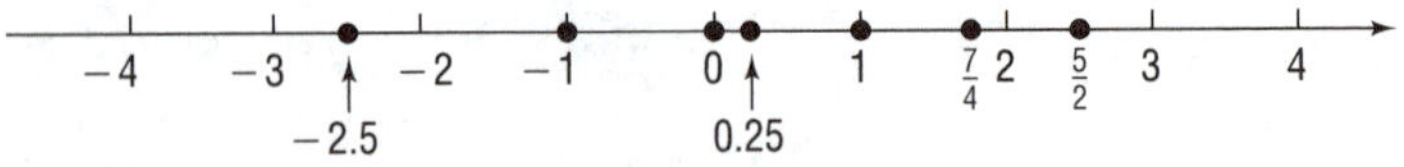

Practice Exercise 2 **1.** 7 **2.** $\frac{1}{4}$ **3.** 9 **4.** 6 **5.** 0 **6.** $\frac{2}{3}$

Practice Exercise 3 **1.** −3 **2.** 5 **3.** −15 **4.** −6 **5.** 11 **6.** −5 **7.** 5 **8.** −11

Practice Exercise 4 **1.** −5 **2.** −11 **3.** 11 **4.** 3 **5.** 3 **6.** −3 **7.** −7 **8.** 17 **9.** −17 **10.** 7 **11.** 2.66

1.5 Assess Your Understanding

Concepts and Vocabulary

1. *True or False:* Each point of the real number line has a unique real number associated with it.

2. The __________ numbers are to the left of 0 and the __________ numbers are to the right of 0 on the real number line.

3. *True or False:* $|10| = -10$

4. The value of $-11 + 4$ is __________.

5. To reformulate $8 - 13$ as an addition problem we would write __________.

6. The value of $-3 - 8$ is __________.

Exercises

In Problems 7–10, graph and label the following sets of numbers on a real number line.

7. $\left\{-4, 3, 0, \frac{1}{2}, -1.6\right\}$ **8.** $\{-3, -2, -1, 0, 1, 2\}$ **9.** $\left\{-7, 5, 0, 4.5, -\frac{5}{2}\right\}$ **10.** $\left\{3.5, \frac{1}{4}, -\frac{9}{2}, -2.3\right\}$

For Problems 11–16, find the additive inverse of each number.

11. -8 **12.** 12 **13.** 23 **14.** -1 **15.** -1.8 **16.** 4.9

In Problems 17–114, find the value of each expression.

17. $-(-3)$ **18.** $-(-10)$ **19.** $-(-\pi)$ **20.** $-(-29)$

21. $-\left(-\frac{1}{2}\right)$ **22.** $|9|$ **23.** $\left|\frac{3}{4}\right|$ **24.** $|-12|$

25. $|-25|$ **26.** $|-(-4)|$ **27.** $|-(-2)|$ **28.** $|44|$

29. $|0|$ **30.** $9 + 7$ **31.** $6 + 3$ **32.** $-9 + (-7)$

33. $-8 + (-3)$ **34.** $-7 + (-5)$ **35.** $-6 + (-3)$ **36.** $-4 + (-3)$

37. $-12 + (-3)$ **38.** $-4 + (-1)$ **39.** $-2 + (-6)$ **40.** $-10 + (-4)$

41. $-5 + (-2)$ **42.** $-6 + (-8)$ **43.** $9 + (-7)$ **44.** $8 + (-3)$

45. $-9 + 7$ **46.** $-8 + 3$ **47.** $6 + (-3)$ **48.** $4 + (-3)$

49. $-6 + 3$ **50.** $-4 + 3$ **51.** $12 + (-3)$ **52.** $7 + (-5)$

53. $-12 + 3$ **54.** $-7 + 5$ **55.** $4 + (-1)$ **56.** $2 + (-6)$

57. $-4 + 1$ **58.** $-2 + 6$ **59.** $(-6) + 4$ **60.** $8 + (-2)$

61. $1 + (-3)$ **62.** $(-3) + 2$ **63.** $(-8) + (-2)$ **64.** $(-4) + (-3)$

65. $-8 + (-2)$ **66.** $-4 + (-3)$ **67.** $(-4.5) + 1$ **68.** $2.5 + (-1)$

69. $0.5 + (-5)$ **70.** $-1 + (1.5)$ **71.** $-4.5 + (-1.5)$ **72.** $(-2.5) + (-0.5)$

73. $8.3 + (-1.6)$ **74.** $-1.2 + 8.6$ **75.** $9 - 7$ **76.** $2 - 6$

77. $-9 - 7$ **78.** $-2 - 6$ **79.** $-9 - (-7)$ **80.** $-2 - (-6)$

81. $9 - (-7)$ **82.** $2 - (-6)$ **83.** $10 - 17$ **84.** $8 - 12$

85. $-10 - 17$ **86.** $-8 - 12$ **87.** $-10 - (-17)$ **88.** $-8 - (-12)$

89. $10 - (-17)$ **90.** $8 - (-12)$ **91.** $4 - 7$ **92.** $5 - 4$

93. $-4 - 7$ **94.** $-5 - 4$ **95.** $-4 - (-7)$ **96.** $-5 - (-4)$

97. $4 - (-7)$ **98.** $5 - (-4)$ **99.** $14 - 8$ **100.** $-18 - 5$

101. $-14 - 8$
102. $18 - (-5)$
103. $14 - (-8)$
104. $-18 - (-5)$
105. $-14 - (-8)$
106. $18 - 5 + 4$
107. $14 - 8 + 4$
108. $18 - 5 - 4$
109. $14 - 8 - 4$
110. $-16 - (2 - 6) + 4$
111. $-8 - (4 - 8) - 1$
112. $14 - (3 - 6) - (4 - 8)$
113. $4 - (3 - 6) - (-2 - 6)$
114. $-(6 - 8) - 2$

1.6 Multiplication and Division with Real Numbers

PREPARING FOR THIS SECTION *Before getting started, review the following:*

- Order of Operations (Section 1.1, pp. 6–9)
- Addition of Real Numbers (Section 1.5, pp. 38–39)
- Subtraction of Real Numbers (Section 1.5, pp. 39–40)

Now work the 'Are You Prepared?' problems on page 46.

OBJECTIVES

1 Multiply Real Numbers
2 Divide Real Numbers
3 Evaluate Expressions with Real Numbers

Multiplication of Real Numbers

1 We have already observed that multiplication involving two positive integers can be thought of as repeated addition. For example,

$$3 \cdot 4 = 4 + 4 + 4 = 12 \quad \text{and} \quad 4 \cdot 5 = 5 + 5 + 5 + 5 = 20.$$

These examples illustrate that multiplying two positive real numbers yields a positive result.

Now consider $3 \cdot (-8) = (-8) + (-8) + (-8) = -24$. This indicates that the product of a positive number and a negative number is a negative number. Furthermore, since multiplication is commutative, $3 \cdot (-8) = (-8) \cdot 3 = -24$. Hence, it doesn't matter if the negative number is listed first or second; when multiplying a positive number and a negative number, you will get a negative number. This is further illustrated by observing the patterns in the following multiplication problems.

$$\left.\begin{array}{l} 3 \cdot 4 = 12 \\ 3 \cdot 3 = 9 \\ 3 \cdot 2 = 6 \\ 3 \cdot 1 = 3 \\ 3 \cdot 0 = 0 \end{array}\right\} \text{As the second factor decreases by 1, the product decreases by 3.}$$

$$\left.\begin{array}{l} 3 \cdot (-1) = -3 \\ 3 \cdot (-2) = -6 \\ 3 \cdot (-3) = -9 \end{array}\right\} \text{Continuing the pattern, we again see that the product of a positive number and a negative number must be negative.}$$

To determine the result of multiplying two negative numbers, we will use a pattern like the one above.

$$\left.\begin{aligned}(-4)\cdot(2) &= -8\\(-4)\cdot(1) &= -4\\(-4)\cdot(0) &= 0\end{aligned}\right\}$$ As the second factor decreases by 1, the product increases by 4.

$$\left.\begin{aligned}(-4)\cdot(-1) &= 4\\(-4)\cdot(-2) &= 8\end{aligned}\right\}$$ Continuing the pattern we see that the product of two negative numbers is a positive number.

These examples suggest the following general rules of signs for products of real numbers:

Rules of Signs for Products

If a and b are real numbers, then

$$a(-b) = -(ab) \qquad (-a)(b) = -(ab) \qquad (-a)(-b) = ab \qquad \textbf{(1)}$$

EXAMPLE 1 **Finding the Product of Two Numbers**

(a) $(-6)\cdot 4 = -(6\cdot 4) = -24$
(b) $8\cdot(-6) = -(8\cdot 6) = -48$
(c) $(-7)(-5) = 7\cdot 5 = 35$
(d) $(-3.1)(5.3) = -(3.1\cdot 5.3) = -16.43$ ◀

Practice Exercise 1 *In Problems 1–4, find each product.*

1. $(-9)(-6)$ **2.** $(1)(-8)$ **3.** $(-1)(-5)$ **4.** $(-3)(2)$ ◀

Multiplication of real numbers can be summarized as follows:

Theorem

The product of two positive real numbers is a positive real number.

The product of a positive real number and a negative real number is a negative real number.

The product of a negative real number and a positive real number is a negative real number.

The product of two negative real numbers is a positive real number.

Now look once more at equations (1). In the middle equation, let $a = 1$. Since $-a = -1$, the equation takes the form

$$(-1)\cdot(b) = -(1\cdot b) = -b$$

That is, the product of -1 and a real number equals the additive inverse of the real number.

2 Division of Real Numbers

If b is a nonzero real number, the quotient $\frac{a}{b}$, also read as "a divided by b" or "the ratio of a to b," is defined as

$$\frac{a}{b} = a \cdot \frac{1}{b} \qquad b \neq 0 \tag{2}$$

Since every division problem can be written as an equivalent multiplication problem by using equation (2), the rules for signs of quotients of real numbers are similar to those for products of real numbers.

Rules of Signs for Quotients

If a and b are real numbers and $b \neq 0$, then

$$\frac{a}{-b} = -\frac{a}{b} \qquad \frac{-a}{b} = -\frac{a}{b} \qquad \frac{-a}{-b} = \frac{a}{b} \tag{3}$$

EXAMPLE 2 **Applying the Rules of Signs**

(a) $\frac{2}{-3} = -\frac{2}{3}$ (b) $\frac{-8}{7} = -\frac{8}{7}$

(c) $\frac{-3}{-5} = \frac{3}{5}$ (d) $\frac{-\sqrt{5}}{3} = -\frac{\sqrt{5}}{3} \approx -0.7454$ ◀

Practice Exercise 2 *In Problems 1–4, simplify each quotient:*

1. $\frac{-7}{-9}$ **2.** $\frac{4}{-5}$ **3.** $\frac{-6}{-5}$ **4.** $\frac{-2}{8.51}$ ◀

Look closely at equations (3). By using the substitution principle, we have $\frac{a}{-b} = -\frac{a}{b} = \frac{-a}{b}$. For example, $\frac{-5}{9} = -\frac{5}{9} = \frac{5}{-9}$, showing that the placement of a negative sign in a fraction does not change the value of the fraction.

Division of real numbers can be summarized as follows:

Theorem

The quotient of two positive real numbers is a positive real number.

The quotient of a positive real number and a negative real number is a negative real number.

The quotient of a negative real number and a positive real number is a negative real number.

The quotient of two negative real numbers is a positive real number.

◀

3 The next three examples combine many of the ideas presented in this chapter. After working through each example, check your answer by reworking the problem using a calculator.

EXAMPLE 3 Evaluating an Expression

(a) $14 - 2\cdot(-8) = 14 - [-(2\cdot 8)]$ Multiply first.

$= 14 - (-16)$ $-(2\cdot 8) = -16$

$= 14 + [-(-16)]$ Change subtraction to addition.

$= 14 + 16$ $-(-16) = 16$

$= 30$ Add.

(b) $-8 - (-3)\cdot(4) = -8 - [-(3\cdot 4)]$ Multiply first.

$= -8 - (-12)$ $-(3\cdot 4) = -12$

$= -8 + [-(-12)]$ Change subtraction to addition.

$= -8 + 12$ $-(-12) = 12$

$= 4$ Add. ◀

EXAMPLE 4 Evaluating an Expression

$[(-4)^2 - 8] - 2^2\cdot(4 - 13) = [16 - 8] - 2^2\cdot(4 - 13)$ Work inside brackets; evaluate exponents first.

$= [16 + (-8)] - 2^2\cdot[4 + (-13)]$ Change subtraction in grouping symbols to addition.

$= [8] - 2^2\cdot[-9]$ Perform addition in grouping symbols.

$= 8 - 4\cdot[-9]$ Evaluate exponent.

$= 8 - [-(4\cdot 9)]$ Evaluate multiplication; use the fact that $4\cdot(-9) = -(4\cdot 9)$.

$= 8 - [-36]$ $-(4\cdot 9) = -36$

$= 8 + [-(-36)]$ Change subtraction to addition.

$= 8 + 36$ $-(-36) = 36$

$= 44$ Add. ◀

EXAMPLE 5 Evaluating Exponents

Evaluate:

(a) -4^2 (b) $(-4)^2$

Solution (a) Recall from Section 1.1 that the exponent 2 applies only to the number that immediately precedes it, in this case, 4. Then,

$$-4^2 = -(4\cdot 4) = -16$$

(b) Since the -4 is in parentheses with the exponent outside the parentheses, we have

$$(-4)^2 = (-4)(-4) = 16$$ ◀

Notice how your calculator handles a problem like Example 5(a). For details, see Example 11 in Section 1.3. To solve Example 5(b) on the

TI-83 or the TI-30X IIS use the following keystrokes:

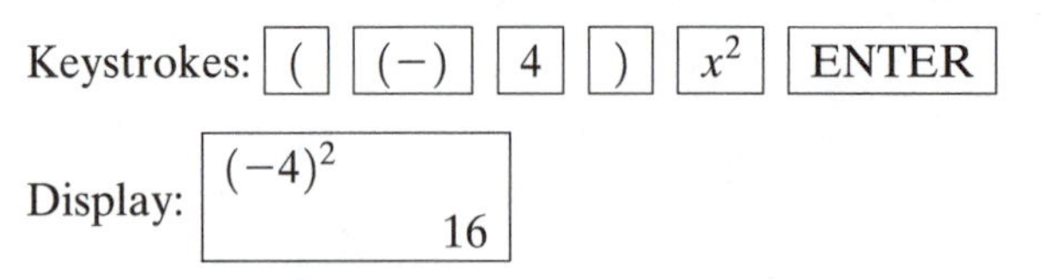

Practice Exercise 3 *In Problems 1–5, find the value of each expression:*

1. $18 - 4\cdot(-8)$ **2.** $-5 - (-3)\cdot(2)$

3. $[(-5)^2 - 2\cdot 6] - 2^3\cdot(3 - 5)$ **4.** $(-3)^4$ **5.** -3^4

In practice, many of the steps in the above examples are performed mentally and simultaneously, shortening the whole process considerably. For example, the solution to Example 3(b) may be simplified as

$$-8 - (-3)\cdot(4) = -8 - (-12) = -8 + 12 = 4$$

or as

$$-8 - (-3)\cdot(4) = -8 + 3\cdot 4 = -8 + 12 = 4$$

or as

$$-8 - (-3)\cdot(4) = -8 + 12 = 4$$

The more proficient you become, the shorter the process becomes.

Answers to Practice Exercises

Practice Exercise 1 **1.** 54 **2.** −8 **3.** 5 **4.** −6

Practice Exercise 2 **1.** $\frac{7}{9}$ **2.** $-\frac{4}{5}$ **3.** $\frac{6}{5}$ **4.** −0.235017626

Practice Exercise 3 **1.** 50 **2.** 1 **3.** 29 **4.** 81 **5.** −81

1.6 Assess Your Understanding

'Are You Prepared?' *Answers are given at the end of these exercises. If you get a wrong answer, read the pages listed in parentheses.*

1. Evaluate $12 + 8 \cdot 3 - 9$ (pp. 6–9)

2. Evaluate $5 + 3(9 - 4) - 2^3$ (pp. 6–9)

3. $-6 + 13 =$ ________ (pp. 38–39)

4. $5 - 17 =$ ________ (pp. 39–40)

Concepts and Vocabulary

5. *True or False:* When multiplying two negative numbers, the result is a negative number.

6. The quotient of a positive real number and a negative real number is ________ real number.

7. The expression $-6 - 3(8 + 4) + 10$ simplifies to ________.

8. *True or False:* $(-5)^2 = -5^2$

Exercises

In Problems 9–74, find the value of each expression.

9. 5(3) **10.** 8(9) **11.** (−5)(−3) **12.** (−8)(−9)

13. (5)(−3) **14.** (8)(−9) **15.** (−5)(3) **16.** (−8)(9)

17. $9(7)$ **18.** $7(6)$ **19.** $(-9)(-7)$ **20.** $(-7)(-6)$

21. $(9)(-7)$ **22.** $(7)(-6)$ **23.** $(-9)(7)$ **24.** $(-7)(6)$

25. $(-2)^6$ **26.** $(-3)^4$ **27.** -2^6 **28.** -3^4

29. $-2\cdot(3+4)+2^3$ **30.** $-4\cdot(3+8)-3^2$ **31.** $4\cdot(3-5)-6$ **32.** $6\cdot(12-14)-8$

33. $-4\cdot(-5)\cdot(-2)$ **34.** $(-3)\cdot(-4)\cdot(-2)$ **35.** $-5\cdot[-7+(-3)]$ **36.** $[-6+(-2)]\cdot 3$

37. $[9-(-12)]\cdot(-3)$ **38.** $-4\cdot(-12-3)$ **39.** $(-25-116)\cdot 0$ **40.** $0\cdot(-142+184)$

41. $-15-3\cdot(-4)$ **42.** $-3\cdot(-5)-(-10)$ **43.** $1-2\cdot(-4)-6$ **44.** $-4-3\cdot(4-6)$

45. $-4^2\cdot(4-8)-(6-8)$ **46.** $1-2^3\cdot(-3-5)-(-4+1)$

47. $(-4)^2\cdot[4-8-(6-8)]$ **58.** $1-(-2)^3\cdot[-3-5-(-4+1)]$

49. $1-[1-3^2+(-4)^2]$ **50.** $-8-2^3-[1-4^2\cdot(0)]$

51. $5\cdot(-2^3)-(-3^2)$ **52.** $5\cdot(-2)^3-(-3)^2$

53. $\frac{-56}{-8}$ **54.** $\frac{-28}{-7}$ **55.** $\frac{-12}{-4}$ **56.** $\frac{-25}{-5}$

57. $\frac{35}{-7}$ **58.** $\frac{48}{-8}$ **59.** $\frac{55}{-5}$ **60.** $\frac{36}{-4}$

61. $\frac{-81}{9}$ **62.** $\frac{-18}{9}$ **63.** $\frac{-6}{1}$ **64.** $\frac{-8}{2}$

65. $\frac{14}{-1}$ **66.** $\frac{-63}{7}$ **67.** $\frac{-4-8}{6-10}$ **68.** $\frac{4-(-6)}{-3-7}$

69. $\frac{2-3+(-4)}{-2-3-4}$ **70.** $\frac{1-2+3}{-1+2-3}$ **71.** $(-2-8)/(-2+3)$

72. $(8-10)/(-1-1)+4$ **73.** $-2-8/(-2)+3$ **74.** $8-10/(-2)+4$

75. Compare the answers to Problems 71 and 73. Are they the same? Why or why not?

76. Compare the answers to Problems 72 and 74. Are they the same? Why or why not?

The following discussion relates to Problems 77–84: In economic forecasting, statisticians sometimes employ the ***Sharpe ratio,*** *a simple measure of return versus volatility. The formula is calculated as follows:*

$$\text{Sharpe ratio} = \frac{ER - RFR}{SD}$$

where ER = Expected return, RFR = Risk-Free Rate, and SD = Standard Deviation. For example, in June 2003 the S&P 500 had a Sharpe ratio of −.021. In Problems 77–84, calculate the Sharpe ratio under the given conditions.

77. $ER = .05$, $RFR = .10$, $SD = .01$ **78.** $ER = .05$, $RFR = .12$, $SD = .01$

79. $ER = .20$, $RFR = .05$, $SD = .01$ **80.** $ER = .30$, $RFR = .05$, $SD = .01$

81. $ER = .10$, $RFR = .06$, $SD = .015$ **82.** $ER = .12$, $RFR = .06$, $SD = .015$

83. $ER = .05$, $RFR = .10$, $SD = .015$ **84.** $ER = .05$, $RFR = .12$, $SD = .015$

---

85. In the first quarter of its fiscal year, a company posted earnings of $1.20 per share. During the second and third quarters, it posted losses of $0.75 per share and $0.30 per share, respectively. In the fourth quarter, it earned a modest $0.20 per share. What were the annual earnings per share of this company?

86. At the beginning of the month, Mike had a balance of $210 in his checking account. During the next month, he deposited $80, wrote a check for $120, made another deposit of $25, wrote two checks for $60 and $32, and was assessed a monthly service charge of $5. What was his balance at the end of the month?

87. Find the value of $1 + 2 + 3 + 4 + \cdots + 99$. [*Hint:* Regroup the terms as $(1 + 99) + (2 + 98) + \cdots$ and proceed from there.]

88. Find the value of $1 + 3 + 5 + 7 + \cdots + 99$.

89. Find the value of $-2 - 4 - 6 - 8 - \cdots - 98$.

90. Find the value of $1 - 2 + 3 - 4 + 5 - 6 + \cdots + 97 - 98 + 99$.

91. Fill in reasons for each step:

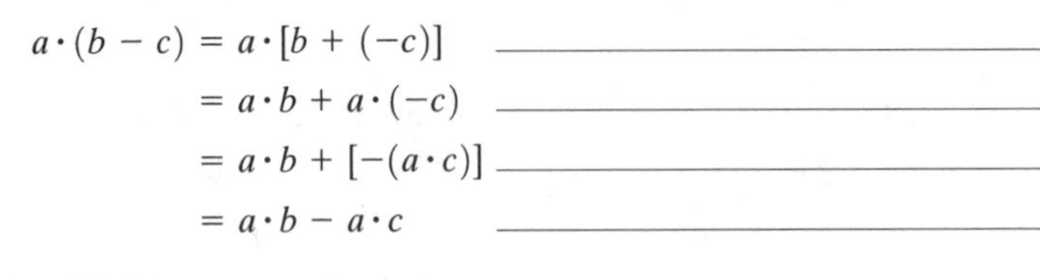

$$\begin{aligned} a \cdot (b - c) &= a \cdot [b + (-c)] \quad \underline{\qquad\qquad\qquad} \\ &= a \cdot b + a \cdot (-c) \quad \underline{\qquad\qquad\qquad} \\ &= a \cdot b + [-(a \cdot c)] \quad \underline{\qquad\qquad\qquad} \\ &= a \cdot b - a \cdot c \quad \underline{\qquad\qquad\qquad} \end{aligned}$$

This proves that multiplication distributes over subtraction.

'Are You Prepared?' Answers

1. 27 **2.** 12 **3.** 7 **4.** -12

1.7 Inequalities; Distance

OBJECTIVES

1 Use Inequality Symbols
2 Find Distance on the Real Number Line
3 Evaluate Expressions that Contain Absolute Value

Inequalities

1 An important property of the real number line follows from the fact that given two numbers (points) a and b, either a is to the left of b, a equals b, or a is to the right of b. See Figure 13.

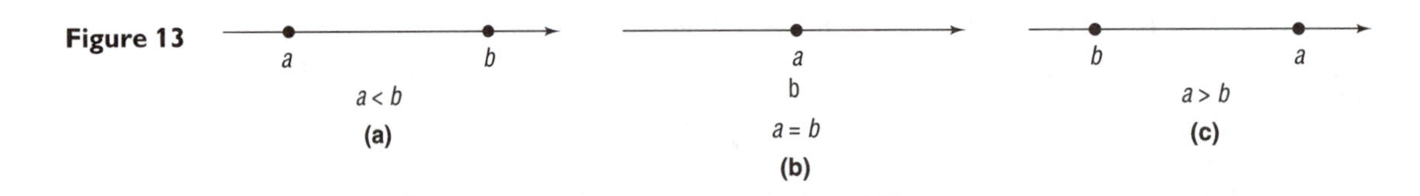

Figure 13

If a is to the left of b, we say "a is less than b" and write $a < b$. If a is to the right of b, we say "a is greater than b" and write $a > b$. If a is at the same location as b, we write $a = b$. If a is either less than or equal to b, we write $a \leq b$. Similarly, $a \geq b$ means a is either greater than or equal to b. Collectively, the symbols $<, >, \leq, \geq$ are called **inequality symbols.**

Note that $a < b$ and $b > a$ mean the same thing. It does not matter whether we write $2 < 3$ or $3 > 2$.

EXAMPLE 1 **Using Inequality Symbols**

(a) $3 < 7$ (b) $-8 > -16$ (c) $-6 < 0$
(d) $-8 < -4$ (e) $4 > -1$ (f) $8 > 0$ ◀

In Example 1(a), we conclude that $3 < 7$, either because 3 is to the left of 7 on the real number line or because the difference $7 - 3 = 4$, a positive real number.

Similarly, we conclude in Example 1(b) that $-8 > -16$ either because -8 lies to the right of -16 on the real number line or because the difference $-8 - (-16) = -8 + 16 = 8$, a positive real number.

Look again at Example 1. You may find it useful to observe that the inequality symbol always points in the direction of the smaller number.

Practice Exercise 1 *In Problems 1–4, fill in the blank with the correct inequality symbol, $<$ or $>$.*

1. 5 ______ -2 **2.** -2 ______ 3 **3.** -8 ______ 0

4. 0 ______ -3

An **inequality** is a statement in which two expressions are related by an inequality symbol. The expressions are referred to as the **sides** of the inequality. Statements of the form $a < b$ or $b > a$ are called **strict inequalities,** while statements of the form $a \leq b$ or $b \geq a$ are called **nonstrict inequalities.**

Based on the discussion so far, we conclude that:

$a > 0$ is equivalent to a is positive	**(1a)**
$a < 0$ is equivalent to a is negative	**(1b)**

We sometimes read $a > 0$ by saying that "a is positive." If $a \geq 0$, then either $a > 0$ or $a = 0$, and we may read this as "a is nonnegative." Similarly we sometimes read $a < 0$ as "a is negative," and we may read $a \leq 0$ as "a is not positive."

Distance

Recall from Section 1.5, that *absolute value* was defined as the distance a point is from the origin on a number line. For example, $|-4| = 4$ and $|3| = 3$. This is illustrated in Figure 14.

Figure 14

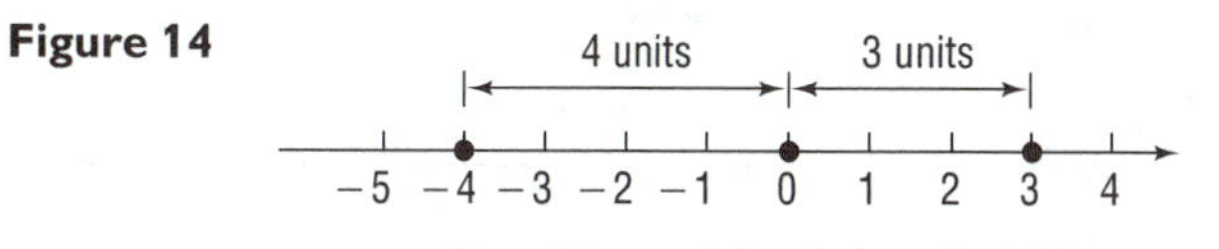

EXAMPLE 2 **Computing Absolute Value**

(a) $|8| = 8$ (b) $|0| = 0$ (c) $|-15| = 15$

(d) $|3 - 2| = |1| = 1$ (e) $|2 - 5| = |-3| = 3$

Look again at Figure 14. The distance from -4 to 3 is 7 units. This distance is the difference $3 - (-4)$, obtained by subtracting the smaller coordinate from the larger. However, since $|3 - (-4)| = |7| = 7$ and $|-4 - 3| = |-7| = 7$, we can use absolute value to calculate the distance between two points without being concerned about which is smaller.

If P and Q are two points on a real number line with coordinates a and b, respectively, the **distance between P and Q,** denoted by $d(P, Q)$, is

$$d(P, Q) = |b - a| \qquad \textbf{(2)}$$

and the distance from Q to P is

$$d(Q, P) = |a - b| \qquad \textbf{(3)}$$

Since $|b - a| = |a - b|$, it follows that $d(P, Q) = d(Q, P)$.

EXAMPLE 3 Finding Distance on a Number Line

Let P, Q, and R be points on a real number line with coordinates -5, 7, and -3, respectively. Find the distance:

(a) Between P and Q

(b) Between Q and R

Solution See Figure 15.

(a) $d(P, Q) = |7 - (-5)| = |12| = 12$

(b) $d(Q, R) = |-3 - 7| = |-10| = 10$

Figure 15

P R Q
−5 −4 −3 −2 −1 0 1 2 3 4 5 6 7
$d(P, Q) = |7 - (-5)| = 12$
$d(Q, R) = |-3 - 7| = 10$

◀

Practice Exercise 2 *In Problems 1–3, let P, Q, and R be points on a real number line with coordinates 8, −2, and 5, respectively. Find each of the following:*

1. $d(Q, R)$ **2.** $d(Q, P)$ **3.** $d(R, P)$

◀

3 ✓

EXAMPLE 4 Evaluating Expressions Containing Absolute Value

Find the value of each expression:

(a) $|3 - 12|$ (b) $|2 + 4(-3)|$ (c) $|-5| + 4 \cdot 3 - |4|$

Solution (a) $|3 - 12| = |-9| = 9$

(b) $|2 + 4(-3)| = |2 + (-12)| = |-10| = 10$

(c) $|-5| + 4 \cdot 3 - |4| = 5 + 4 \cdot 3 - 4 = 5 + 12 - 4 = 13$ ◀

Example 4 illustrates that absolute value is also a grouping symbol. In Examples 4(a) and 4(b) we perform the operations inside the absolute value first and then evaluate the absolute value. In Example 4(c) we evaluate the absolute values first and then follow the order of operations.

Practice Exercise 3 *In Problems 1–4, find the value of each expression:*

1. $|8.3|$ **2.** $|3 - 5|$ **3.** $|4 - 8 \cdot 2|$ **4.** $\left|7 - \frac{6}{3}\right|$

HISTORICAL COMMENT

The concept of absolute value has been known for a long time, and various symbols for it have been used, but it was never of great theoretical importance until the reformulation of the basis of calculus in terms of approximation by Karl Weierstrass (1815–1897) about 1840. In his lectures, Weierstrass introduced the symbol $|a|$ for the absolute value of a, and the utility of the symbol quickly became obvious. Within 40 years the symbol was in use throughout the world.

Answers to Practice Exercises

Practice Exercise 1	**1.** $>$	**2.** $<$	**3.** $<$	**4.** $>$
Practice Exercise 2	**2.** 7	**2.** 10	**3.** 3	
Practice Exercise 3	**3.** 8.3	**2.** 2	**3.** 12	**4.** 5

1.7 Assess Your Understanding

Concepts and Vocabulary

1. An inequality of the form $a < b$ is called a __________ inequality.

2. *True or False:* $-4 \leq 2$

3. *True or False:* The distance between two points on the real number line is always greater than zero.

4. *True or False:* $|12 - 3| = |12 + 3|$

Exercises

In Problems 5–14, replace the question mark by <, >, or =, whichever is correct.

5. $\frac{1}{2}$? 0 **6.** 5 ? 6 **7.** -1 ? -2 **8.** -3 ? $\frac{5}{2}$ **9.** π ? 3.14

10. $\sqrt{2}$? 2.14 **11.** $\frac{1}{2}$? 0.5 **12.** $\frac{1}{3}$? 0.33 **13.** $\frac{2}{3}$? 0.67 **14.** $\frac{1}{4}$? 0.25

In Problems 15–24, write each statement as an inequality.

15. x is greater than 2.

16. x is greater than -5.

17. x is positive.

18. z is negative.

19. x is less than 2.

20. y is greater than -5.

21. x is less than or equal to 1.

22. x is greater than or equal to 2.

23. x is less than or equal to -3.

24. x is greater than or equal to -4.

In Problems 25–34, find the value of each expression.

25. $|\pi|$ **26.** $\left|-\frac{2}{3}\right|$ **27.** $\left|4 - \frac{8}{2}\right|$ **28.** $|2 - 4 \cdot 2|$ **29.** $|6 + 3(-4)|$

30. $|4(-2) - 1|$ **31.** $\left|\frac{8}{2} + \frac{2}{-1}\right|$ **32.** $\left|\frac{6}{3} + \frac{3}{-3}\right|$ **33.** $\left|\frac{9 \cdot (-3)}{5}\right|$ **34.** $\left|\frac{2 \cdot (-3)}{4}\right|$

In Problems 35–44 use the illustration to locate the coordinate of the point that is:
(a) 2 units to the left of P
(b) 3 units to the right of P

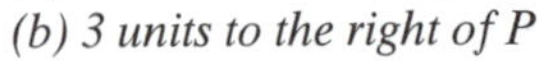

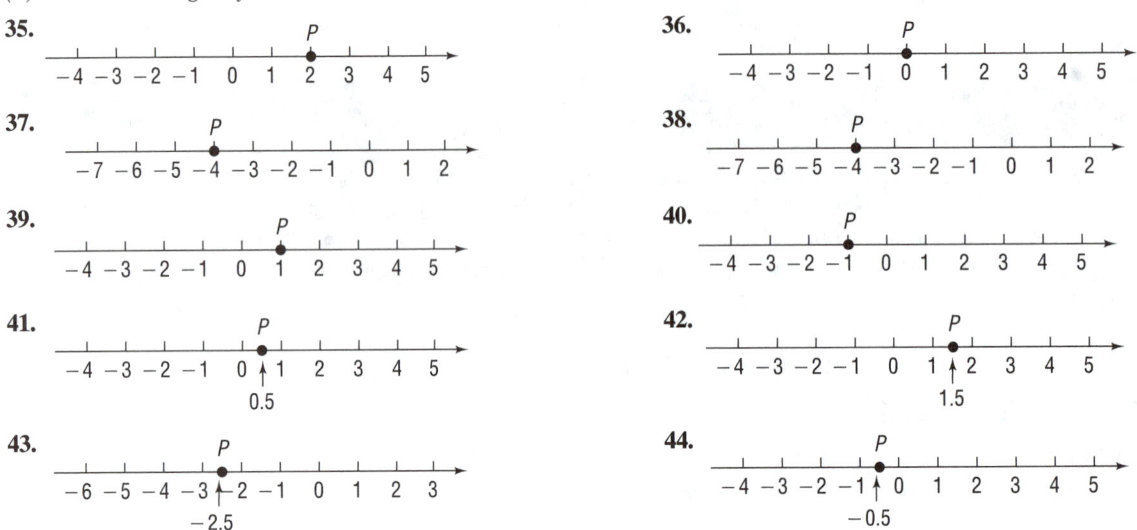

In Problems 45–54, use the real number line below to compute each distance.

A B C D E
−4 −3 −2 −1 0 1 2 3 4 5

45. $d(D, E)$ **46.** $d(C, E)$ **47.** $d(A, E)$ **48.** $d(D, B)$

49. $d(A, C)$ **50.** $d(D, A)$ **51.** $d(E, B)$ **52.** $d(B, A)$

53. $d(A, D) + d(B, E)$ **54.** $d(E, D) + d(D, B)$

55. Write a brief paragraph that illustrates the similarities and differences between "less than" (<) and "less than or equal" (≤).

1.8 Constants and Variables; Mathematical Models

OBJECTIVES

1 Evaluate Algebraic Expressions
2 Determine the Domain of a Variable
3 Use Formulas

Constants and Variables

As we said earlier, in algebra we use letters such as x, y, a, b, and c to represent numbers. If the letter used is to represent *any* number from a given set of numbers, it is called a **variable**. A **constant** is either a fixed number, such as 5 or $\sqrt{3}$, or a letter that represents a fixed (possibly unspecified) number.

Constants and variables are combined using the operations of addition, subtraction, multiplication, and division to form *algebraic expressions*. Examples of algebraic expressions include

$$x + 3 \qquad \frac{3}{1 - t} \qquad 7x - 2y$$

1 To evaluate an algebraic expression, substitute for each variable its numerical value.

EXAMPLE 1

Evaluating an Algebraic Expression

Evaluate each expression if $x = 3$ and $y = -1$.

(a) $x + 3y$ (b) $5xy$ (c) $\dfrac{3y}{2 - 2x}$ (d) $|-4x + y|$

Solution (a) Substitute 3 for x and -1 for y in the expression $x + 3y$.

$$x + 3y = 3 + 3(-1) = 3 + (-3) = 0$$

↑

$x = 3, y = -1$

(b) If $x = 3$ and $y = -1$, then

$$5xy = 5(3)(-1) = -15$$

(c) If $x = 3$ and $y = -1$, then

$$\frac{3y}{2 - 2x} = \frac{3(-1)}{2 - 2(3)} = \frac{-3}{2 - 6} = \frac{-3}{-4} = \frac{3}{4}$$

(d) If $x = 3$ and $y = -1$, then

$$|-4x + y| = |-4(3) + (-1)| = |-12 + (-1)| = |-13| = 13$$ ◀

Practice Exercise 1 *In Problems 1–4, evaluate each expression if $x = -5$ and $y = 2$.*

1. $7x + 4y$ **2.** $-3xy$ **3.** $\dfrac{2x + 5y}{x + y}$ **4.** $|2x - 3y|$ ◀

In algebra we use algebraic expressions as well as *formulas*. A **formula** is an algebraic expression that shows a specific relationship between quantities in the real world.

EXAMPLE 2

Area of a Circle

The formula for the area of a circle is

$$A = \pi r^2$$

In this formula, r is a variable representing the radius of the circle, A is a variable representing the area of the circle, and π is a constant (≈ 3.14). See Figure 16. ◀

Figure 16

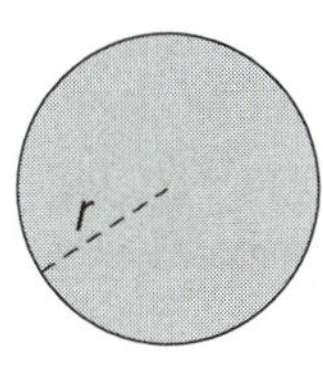

2 In working with expressions or formulas involving variables, the variables may be allowed to take on values from only a certain set of numbers. For example, in the formula for the area A of a circle of radius r, $A = \pi r^2$, the variable r is necessarily restricted to the positive real numbers since it is impossible to have a negative or zero radius for a circle. The variable A is also restricted to positive real numbers. Can you tell why?

Also, in the expression $\frac{1}{x}$, the variable x cannot take on the value of 0, since division by 0 is not defined.

The set of values that a variable may assume is called the **domain of the variable.**

EXAMPLE 3 **Finding the Domain of a Variable**

(a) The domain of the variable x in the expression

$$\frac{5}{x - 2}$$

is $\{x | x \neq 2\}$ since, if $x = 2$, the denominator becomes 0, which is not defined.

(b) In the formula for the circumference C of a circle of radius r,

$$C = 2\pi r$$

the domain of the variable r, representing the radius of the circle, is the set of positive real numbers. The domain of the variable C, representing the circumference of the circle, is also the set of positive real numbers. ◀

In describing the domain of a variable, we may use either set notation or words, whichever is more convenient.

Practice Exercise 2 *In each expression find the domain of the variable x.*

1. $\frac{8}{x}$ **2.** $2x + 5$ **3.** $\frac{6}{x + 4}$ ◀

Mathematical Models

3 Situations in the real world ordinarily do not present themselves in the form of a mathematical problem since the real world is normally far too complicated to be described precisely. To formulate real-world problems, we rely on a simplified version, called a **mathematical model.** Let's look at an example.

EXAMPLE 4 **Falling Objects**

If an object is dropped from a height of 64 feet above the ground, how far will it have fallen after 1 second? How long is it until the object strikes the ground?

Solution If the object is light and feathery, it will float to the ground; if the object is streamlined, the air will offer just a little resistance. To simplify the problem, we will take out these complications and assume that there is no air resistance at all—as if the object were falling in a vacuum. Laboratory experiments with objects falling in a vacuum have shown that the distance s an object will fall after t seconds is given by the formula

$$s = \frac{1}{2}gt^2$$

In this formula, g is a constant called the **acceleration due to gravity**, which is approximately 32 feet per second per second or $32\frac{\text{ft}}{\text{sec}^2}$ on Earth. This formula is based on a simplified version of the original problem and represents a mathematical model for the problem. In using this formula, the variable t may be any nonnegative real number.

To find out how far the object has fallen after 1 second, we substitute 1 for t in the formula $s = \frac{1}{2}gt^2$. Since $g = 32$, we obtain

$$s = \frac{1}{2}(32) \cdot 1^2 = 16 \text{ feet}$$

The object will have fallen 16 feet after 1 second.

The object strikes the ground when it has fallen 64 feet. To find out how long it takes for this to happen, we substitute 64 for s in the formula $s = \frac{1}{2}gt^2$:

$$\begin{aligned} s &= \frac{1}{2}gt^2 \\ 64 &= \frac{1}{2}(32)t^2 && s = 64;\ g = 32 \\ 64 &= 16t^2 && \text{Simplify.} \\ 16 \cdot 4 &= 16t^2 && 64 = 16 \cdot 4 \\ 4 &= t^2 && \text{Apply the Cancellation Property.} \end{aligned}$$

There are two numbers, -2 and 2, whose square is 4. However, the domain of the variable t is the set of nonnegative real numbers, so we discard -2.* It takes 2 seconds for the object to strike the ground. ◀

Answers to Practice Exercises

Practice Exercise 1 **1.** -27 **2.** 30 **3.** 0 **4.** 16

Practice Exercise 2 **1.** $\{x \mid x \neq 0\}$ **2.** All real numbers **3.** $\{x \mid x \neq -4\}$

*The solution -2 is sometimes called *extraneous*.

1.8 Assess Your Understanding

Concepts and Vocabulary

1. A __________ is a letter used in algebra to represent any number from a given set of numbers.

2. A __________ is an algebraic expression that shows a specific relationship between quantities in the real world.

3. *True or False:* The formula for the area of a circle is $A = \pi r^2$.

4. The set of values that a variable may assume is called the __________ of the variable.

5. In the expression $\frac{x}{x+2}$, the variable x cannot equal __________.

6. If $x = 4$ and $y = -2$, then the value of $x + 3y$ is __________.

Exercises

In Problems 7–14, evaluate each expression if $x = -2$ and $y = 3$.

7. $x + 2y$

8. $3x + y$

9. $5xy + 2$

10. $-2x + xy$

11. $\frac{2x}{x - y}$

12. $\frac{x + y}{x - y}$

13. $\frac{3x + 2y}{2 + y}$

14. $\frac{2x - 3}{y}$

In Problems 15–24, find the value of each expression if $x = 4$ and $y = -3$.

15. $|x + y|$ **16.** $|x - y|$ **17.** $|x| + |y|$ **18.** $|x| - |y|$ **19.** $\frac{|x|}{x}$

20. $\frac{|y|}{y}$ **21.** $|4x - 5y|$ **22.** $|3x + 2y|$ **23.** $||4x| - |5y||$ **24.** $3|x| + 2|y|$

In Problems 25–34, find the domain of the variable x in each expression.

25. $\frac{5}{x - 3}$ **26.** $\frac{-8}{x + 5}$ **27.** $\frac{x}{x + 4}$ **28.** $\frac{x - 2}{x - 6}$

29. $\frac{1}{x} + \frac{1}{x - 1}$ **30.** $\frac{3}{x(x - 4)}$ **31.** $\frac{x + 1}{x(x + 2)}$ **32.** $\frac{2}{x + 1} + \frac{3}{x + 2}$

33. The formula for the area A of a square with side of length x is $A = x^2$. What is the domain of the variable x? What is the domain of the variable A?

34. The formula for the perimeter P of a square with side of length x is $P = 4x$. What is the domain of the variable x? What is the domain of the variable P?

In Problems 35–44, express each statement as an equation involving the indicated variables and constants. What is the domain of each variable?

35. The area A of a rectangle is the product of its length l and its width w.

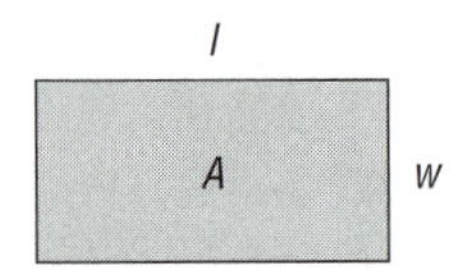

36. The perimeter P of a rectangle is twice the sum of its length l and its width w.

37. (a) The circumference C of a circle is π times its diameter d.

(b) The circumference C of a circle is 2 times π times the radius r of the circle.

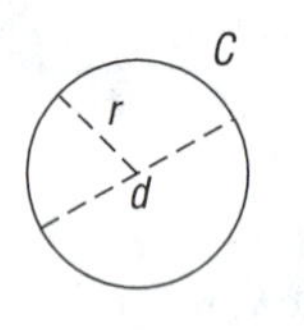

38. The area A of a triangle is one-half the product of its base b and its height h.

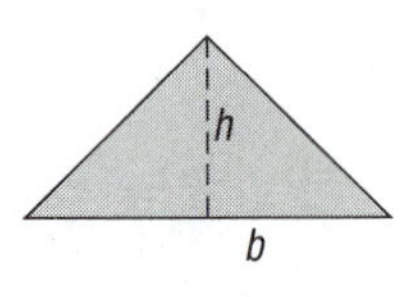

39. The area A of an equilateral triangle is $\frac{\sqrt{3}}{4}$ times the square of the length x of one side.

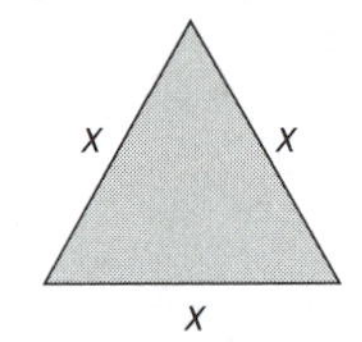

40. The perimeter P of an equilateral triangle is 3 times the length x of one side.

41. The volume V of a sphere $\frac{4}{3}$ times π times the cube of the radius r.

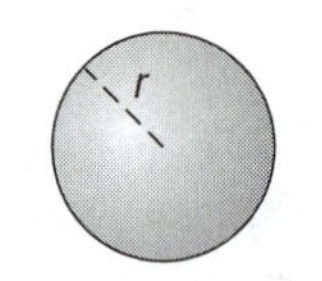

42. The surface area S of a sphere is 4 times π times the square of the radius r.

43. The volume V of a cube is the cube of the length x of a side.

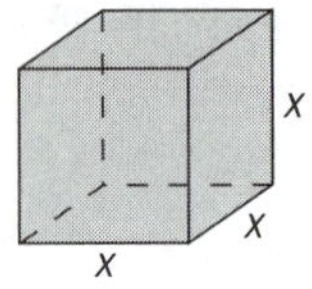

44. The surface area S of a cube is 6 times the square of the length x of a side.

In Problems 45–48, calculate the area and perimeter of a square with side of length x. Use the formula $A = x^2$ for the area A and $P = 4x$ for the perimeter P.

45. $x = 2$ inches **46.** $x = 4$ feet **47.** $x = 3$ meters **48.** $x = 5$ centimeters

In Problems 49–52, calculate the volume V and surface area S of a sphere whose radius is r. (See Problems 41 and 42 for the formulas.)

49. $r = 2$ meters
50. $r = 3$ centimeters
51. $r = 4$ inches
52. $r = 5$ feet

In Problems 53–56, use the formula $C = \frac{5}{9}(F - 32)$ *for converting degrees Fahrenheit into degrees Celsius to find the Celsius measure of each Fahrenheit temperature.*

53. $F = 32°$
54. $F = 212°$
55. $F = 68°$
56. $F = -4°$

57. An object is dropped from a height of 144 feet above the ground. How far has it fallen after 1 second? How far after 2 seconds? Assume no air resistance. [*Hint:* Refer to Example 4.]

58. An object is dropped from a height of 1600 feet above the ground. How far has it fallen after 1 second? How far after 2 seconds? How far after 5 seconds? How long does it take for the object to strike the ground? Assume no air resistance.

In Problems 59–64, use the following information: If an object is propelled vertically upward with an initial velocity of v feet per second, the height h of the object after t seconds, assuming no air resistance, is given by the formula

$$h = vt - \frac{1}{2}(32)t^2$$

Calculate the height h of the object under each of the following conditions:

59. Initial velocity = 50 feet per second; after 1 second
60. Initial velocity = 50 feet per second; after 2 seconds
61. Initial velocity = 50 feet per second; after 3 seconds
62. Initial velocity = 80 feet per second; after 1 second
63. Initial velocity = 80 feet per second; after 2 seconds
64. Initial velocity = 80 feet per second; after 3 seconds

65. The weekly production cost C of manufacturing x watches is given by the formula

$$C = 4000 + 2x,$$

where the variable C is in dollars.

(a) What is the domain of the variable x?
(b) What is the domain of the variable C?
(c) What is the cost of producing 1000 watches?
(d) What is the cost of producing 2000 watches?

Hint: To determine the domain of the variable C, take into account that there are minimum costs, called *fixed costs*, in this problem. **Fixed costs** are expenses that must be paid whether the company manufactures zero, one, two, or thousands of watches. To find the fixed costs, substitute zero for x.

66. The weekly production cost C of manufacturing x hand calculators is given by the formula

$$C = 3000 + 6x - \frac{x^2}{1000}$$

(a) What is the domain of the variable x?
(b) What is the domain of the variable C?
(c) What is the cost of producing 1000 hand calculators?
(d) What is the cost of producing 3000 hand calculators?

Chapter Review

Things to Know

Classification of numbers	Counting numbers	1, 2, 3, ...
	Whole numbers	0, 1, 2, 3, ...
	Integers	..., −2, −1, 0, 1, 2, ...
	Rational numbers	Quotients of two integers (the denominator cannot be 0); terminating or infinite, repeating decimals
	Irrational numbers	Infinite, nonrepeating decimals
	Real numbers	Rational or irrational numbers

Properties of real numbers	Commutative Properties	$a + b = b + a$ $\quad a \cdot b = b \cdot a$
	Associative Properties	$a + (b + c) = (a + b) + c$ $\quad a \cdot (b \cdot c) = (a \cdot b) \cdot c$
	Distributive Property	$a \cdot (b + c) = a \cdot b + a \cdot c$
	Identity Properties	$a + 0 = a$ $\quad a \cdot 1 = a$
	Inverse Properties	$a + (-a) = 0$ $\quad a \cdot \frac{1}{a} = 1$, where $a \neq 0$
	Zero-Product Property	If $ab = 0$, then $a = 0$ or $b = 0$ or both.
	Cancellation Properties	If $a + c = b + c$, then $a = b$. If $a \cdot c = b \cdot c$ and $c \neq 0$, then $a = b$. If $b \neq 0$ and $c \neq 0$, then: $\frac{a \cdot c}{b \cdot c} = \frac{a}{b}$
Absolute value	$\lvert a \rvert = a$ if $a \geq 0$ $\lvert a \rvert = -a$ if $a < 0$	

Objectives

Section		You should be able to . . .	Review Exercises
1.1	✓1	Use algebraic symbols (p. 2)	1–6
	✓2	Know properties of equality (p. 4)	29, 30, 33, 34
	✓3	Work with exponents (p. 5)	39, 40
	✓4	Evaluate numerical expressions (p. 6)	7–12
1.2	✓1	Work with sets (p. 12)	13–20
	✓2	Classify numbers (p. 14)	21, 22
	✓3	Express rational numbers as decimals (p. 15)	23–25
1.3	✓1	Approximate numbers (p. 20)	75–78
	✓2	Use a calculator (p. 21)	77, 78, 95, 96
1.4	✓1	Work with properties of real numbers (p. 27)	27, 28, 31, 32, 35, 36, 97
1.5	✓1	Work with the real number line (p. 35)	26
	✓2	Find absolute value (p. 37)	55–64
	✓3	Add real numbers (p. 38)	41, 43, 45, 54, 58
	✓4	Subtract real numbers (p. 39)	37, 38, 41, 43, 44, 49–52
1.6	✓1	Multiply real numbers (p. 42)	39, 40, 47
	✓2	Divide real numbers (p. 44)	65, 66
	✓3	Evaluate expressions with real numbers (p. 44)	37–66
1.7	✓1	Use inequality symbols (p. 48)	79–82
	✓2	Find distance on the real number line (p. 49)	83–86
	✓3	Evaluate expressions that contain absolute value (p. 50)	55–64
1.8	✓1	Evaluate algebraic expressions (p. 52)	67–74
	✓2	Determine the domain of a variable (p. 53)	87–92
	✓3	Use formulas (p. 54)	93, 94

Review Exercises

In Problems 1–6, write each statement using symbols.

1. The sum of x and 2 is 5.

2. The product of 3 and y is 9.

3. 3 less x is the sum of y and 2.

4. The product of 2 and x is x divided by 2.

5. Four times a number x is 2 less 3 times x.

6. The product of twice x and y is y divided by x.

In Problems 7–12, evaluate each expression.

7. $(2^3 + 5)\cdot 2$

8. $9 + 3\cdot 2^4$

9. $7 + (3\cdot 4 + 1)^2 + 3^2\cdot 4$

10. $(6 + 2\cdot 3)^2 + 2\cdot 3$

11. $\dfrac{3 + 5}{3 + 1}$

12. $\dfrac{8 + 4}{4 + 2}$

In Problems 13–20, determine whether the statement is true or false. Use the sets given below.

$$A = \{1, 3, 7, 9\} \qquad B = \{1, 2, 3, 4\}$$

13. $3 \in A$

14. $3 \notin B$

15. $4 \notin A$

16. $7 \in B$

17. $A \subseteq B$

18. $B \subseteq A$

19. A is a finite set.

20. $\{1, 3, 4\} \subseteq B$

In Problems 21 and 22, list the numbers in each set that are
(a) Natural numbers
(b) Integers
(c) Rational numbers
(d) Irrational numbers
(e) Real numbers

21. $A = \left\{-\sqrt{2}, 0, \frac{1}{2}, 1.25, 8.333\ldots, \pi/2, 8\right\}$

22. $B = \left\{-2.67, 1, \frac{5}{4}, \sqrt{42}, 9.313131\ldots, \frac{22}{7}\right\}$

In Problems 23–25, express each fraction as a decimal.

23. $\dfrac{3}{8}$

24. $\dfrac{5}{6}$

25. $\dfrac{7}{11}$

26. Graph on a real number line the set of numbers: $\left\{-4, 0, 3.2, \frac{-3}{2}, 5\right\}$

In Problems 27–36, name the property that justifies each statement.

27. $3x + ax = (3 + a)x$

28. $a + 4 = 4 + a$

29. If $x = 2$, then $x^3 + 1 = 2^3 + 1$.

30. If $3 = y$, then $y = 3$.

31. $4\cdot\dfrac{1}{4} = 1$

32. $4 + (-4) = 0$

33. If $2\cdot x = 2\cdot 3$, then $x = 3$.

34. If $x + 3 = 5 + 3$, then $x = 5$.

35. $x^2 + 0 = x^2$

36. $1\cdot\dfrac{1}{y} = \dfrac{1}{y}$

In Problems 37–66, find the value of each expression.

37. $3 - (8 - 10) - (-3 - 4)$

38. $(-4 - 3) - (2 - 3) - 8$

39. $-3^2\cdot(-2)^3$

40. $-4^2\cdot(-3)^2$

41. $0^2 + (-5 - 2)$

42. $4^2\cdot(8 - 2\cdot 0)$

43. $-4 - [2 - (-2)^3 + 3^2]$

44. $-6 - [-(2 - 3)^2]$

45. $1 - (-6 + 2 - 1) - (1 - 2^3)$

46. $-2^2\cdot 3 - 4^2$

47. $(-2)^2\cdot(3 - 4^2)$

48. $-(3 - 4^2)^2$

49. $-2 - (-3) - (-4)$

50. $-4 - (-3) - (-2)$

51. $1 - 2 - 3 - 4$

52. $-4 - 3 - 2 - 1$

53. $(1^2 - 2^2 + 3^2) - [(-1)^2 - (-2)^2 - (-3)^2]$

54. $(-1)^2 + (-2)^2 + (-3)^2 - [(-1)^2 - (-2^2) - (-3)^2]$

55. $|-2 - 3^2|$

56. $|4^2 - 3^2|$

57. $3^2 - |-4|$

58. $-3^2 + |-4|$

59. $|3 - 4| - |4 - 3|$

60. $|1 - 2| - |2 - 3| + |3 - 4|$

61. $|-2| - |-3|$

62. $|-2 - (-3)|$

63. $1 - |2^2 - (-2)^3|$

64. $-3 \cdot |2^3 - (-3)^2|$

65. $\dfrac{-5 - 13}{1 - 5}$

66. $\dfrac{-12}{4} - \dfrac{-18}{2}$

In Problems 67–74, evaluate each expression if $x = 3$ and $y = -5$.

67. $2x + 3y$

68. $3x - 2y$

69. $|x - y|$

70. $|x + y|$

71. $xy + x$

72. $xy - y$

73. $y^2 - x^2$

74. $|y^2|$

In Problems 75–78, write each number as a decimal rounded to two decimal places.

75. 183.4355

76. 21.4347

77. $\pi + 3.213$

78. $\sqrt{5} + \sqrt{3}$

In Problems 79–82, replace the question mark by <, >, or =, whichever is correct.

79. $-5 \ ? -8$

80. $\dfrac{3}{4} \ ? \ 0.75$

81. $-6 \ ? \ \dfrac{11}{2}$

82. $1.73 \ ? \ \sqrt{3}$

In Problems 83–86, use the real number line below to compute each distance.

A (−4), B (−1), C (0), D (2), E (5)

−4 −3 −2 −1 0 1 2 3 4 5 6

83. $d(C, B)$

84. $d(E, A)$

85. $d(D, E)$

86. $d(D, A)$

In Problems 87–92, determine the domain of the variable x in each expression.

87. $\dfrac{3 + x}{3 - x}$

88. $\dfrac{x^2}{x + 2}$

89. $\dfrac{1}{x^2}$

90. $\dfrac{x^2}{(x - 3)^2}$

91. $\dfrac{x + 7}{x - 5}$

92. $\dfrac{5}{2 - x}$

---

93. If an object is propelled vertically upward with an initial velocity of 144 feet per second, its height h after t seconds, assuming no air resistance, is

$$h = 144t - \frac{1}{2} \cdot 32t^2$$

(a) Find the height h of the object after 1 second.
(b) Find the height h of the object after 4 seconds.
(c) Find the height h of the object after 9 seconds.
(d) Find the height h of the object after 4.5 seconds.

94. The weekly production cost C of manufacturing x electric can openers is given by the formula $C = 6000 + 4x$, where the variable C is in dollars. What is the cost of manufacturing 200 can openers? What is the cost of manufacturing 1000 can openers?

95. Dan earns \$6.25 an hour. Last week he worked 30 hours. After deductions of \$18 for federal tax, \$6.50 for state tax, \$11.63 for FICA (Social Security), \$2.73 for Medicare, and \$10 for a charitable contribution, how much did he receive?

96. A certain cell phone account has a monthly usage allowance of 250 anytime minutes and 1250 night & weekend minutes. The base monthly fee for this account is \$45. After the monthly usage allowance is depleted, minutes are billed at \$0.35 each. A recent bill showed a usage of 299 anytime minutes and 713 night & weekend minutes. The bill also listed a special credit of \$13.80 as well as showing taxes and miscellaneous fees totaling \$8.05. What was the total amount due for the month?

---

97. Fill in each blank with the property that justifies that step.

$2x - 8 = 10$	Given
$2x + (-8) = 10$	______________
$2x + (-8) = 18 + (-8)$	______________
$2x = 18$	______________
$2x = 2 \cdot 9$	______________
$x = 9$	______________

2 Polynomials

OUTLINE

Polynomials are among the simplest of the algebraic expressions, and they assume an important role in algebra and in advanced mathematics. In this chapter, we define a polynomial and learn one way to evaluate polynomials. We shall also learn how to add, subtract, multiply, and divide polynomials. Techniques for factoring polynomials, which are use extensively in algebra and in advanced mathematics, are also developed.

2.1 Laws of Exponents

PREPARING FOR THIS SECTION *Before getting started, review the following:*

- Work with Exponents (Section 1.1, pp. 5–6)

Now work the 'Are You Prepared?' problems on page 69.

OBJECTIVE

1 Use the Laws of Exponents

We begin by reviewing the definition of an *exponent*.

If a is a real number and n is a positive integer, then the symbol a^n represents the product of n factors of a. That is,

$$a^n = \underbrace{a \cdot a \cdot \cdots \cdot a}_{n \text{ factors}} \qquad (1)$$

Here it is understood that $a^1 = a$.

Then $a^2 = a \cdot a$, $a^3 = a \cdot a \cdot a$, and so on. In the expression a^n, a is called the *base* and n is called the *exponent,* or *power*. We read a^n as "a raised to the power of n" or as "a raised to the nth power." We usually read a^2 as "a squared" and a^3 as "a cubed."

Laws of Exponents

1 Several general laws can be used when dealing with exponents. The first one we consider is used when multiplying two expressions that have the same base. For example:

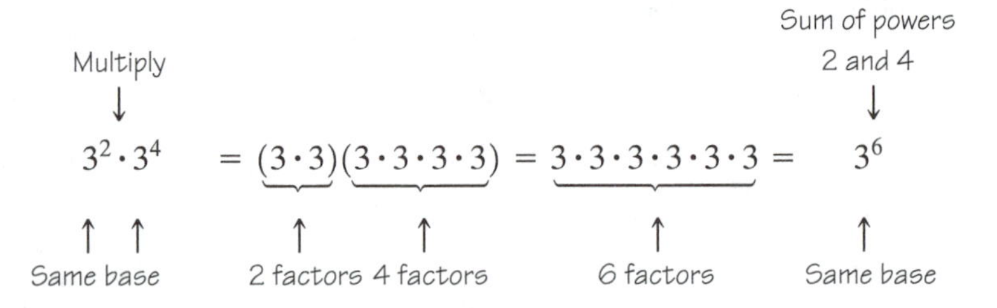

In general, if a is any real number and if m and n are positive integers, we have

$$a^m \cdot a^n = \underbrace{(a \cdot a \cdot \cdots \cdot a)}_{m \text{ factors}}\underbrace{(a \cdot a \cdot \cdots \cdot a)}_{n \text{ factors}} = \underbrace{a \cdot a \cdot \cdots \cdot a}_{m + n \text{ factors}} = a^{m+n}$$

To multiply two expressions having the same base, retain the base and add the exponents.

If a is a real number and m, n are positive integers, then

$$a^m \cdot a^n = a^{m+n} \qquad (2)$$

EXAMPLE 1 **Using Equation (2)**

(a) $2^2 \cdot 2^4 = 2^{2+4} = 2^6$ (b) $(-3)^2(-3)^3 = (-3)^{2+3} = (-3)^5$
(c) $4 \cdot 4^2 = 4^{1+2} = 4^3$ (d) $x^5 \cdot x^3 = x^{5+3} = x^8$ ◀

Notice that for the final answer in Example 1(a) we could write 2^6 or 64. Both are correct and the form of the final answer is just a matter of personal preference. Generally, if it is an easy (mental) calculation, the answer is multiplied out; however, since we are covering laws of exponents in this section, many of the answers are left in exponent form.

WARNING: When using equation (2), the bases *must* be the same. The expression $2^3 \cdot 3^2$ cannot be simplified using equation (2). Be sure *not* to write $2^3 \cdot 3^2$ as 6^5, which is incorrect. Also when using equation (2), the base does not change. For example $2^4 \cdot 2^3 = 2^{4+3} = 2^7$. A careless mistake would be to write $2^4 \cdot 2^3$ as 4^7. ■

Practice Exercise 1 *In Problems 1–4, use equation (2) to simplify each expression. Express your answer as an exponent.*

1. $3^2 \cdot 3^4$ **2.** $(-5)^1(-5)^3$ **3.** $3^2 \cdot 3^3$ **4.** $x^3 \cdot x^5 \cdot x$ ◀

Another Law of Exponents applies when an expression containing a power is itself raised to a power. For example,

$$(2^3)^4 = \underbrace{2^3 \cdot 2^3 \cdot 2^3 \cdot 2^3}_{\text{4 factors}} = \underbrace{\underbrace{(2 \cdot 2 \cdot 2)}_{\text{3 factors}}\underbrace{(2 \cdot 2 \cdot 2)}_{\text{3 factors}}\underbrace{(2 \cdot 2 \cdot 2)}_{\text{3 factors}}\underbrace{(2 \cdot 2 \cdot 2)}_{\text{3 factors}}}_{3 \cdot 4 = 12 \text{ factors}} = 2^{12}$$

In general, if a is any real number and if m and n are positive integers, then

$$\begin{aligned}(a^m)^n &= \underbrace{a^m \cdot a^m \cdot \cdots \cdot a^m}_{n \text{ factors}} \\ &= \underbrace{\underbrace{(a \cdot a \cdot \cdots \cdot a)}_{m \text{ factors}}\underbrace{(a \cdot a \cdot \cdots \cdot a)}_{m \text{ factors}} \cdots \underbrace{(a \cdot a \cdot \cdots \cdot a)}_{m \text{ factors}}}_{n \text{ groups of } m \text{ factors yields } m \cdot n \text{ factors}} \\ &= a^{mn}\end{aligned}$$

If an expression containing a power is itself raised to a power, retain the base and multiply the powers.

If a is a real number and m, n are positive integers, then

$$(a^m)^n = a^{mn} \qquad \textbf{(3)}$$

EXAMPLE 2 **Using Equation (3)**

(a) $[(-2)^3]^2 = (-2)^6 = 64$
(b) $(x^2)^5 = x^{(2)(5)} = x^{10}$
(c) $(7^2)^4 \cdot 7^3 = 7^{(2)(4)} \cdot 7^3 = 7^8 \cdot 7^3 = 7^{8+3} = 7^{11}$ ◀

Practice Exercise 2 *In Problems 1–3, simplify each expression. Express your answer as an exponent.*

1. $(2^3)^2$ **2.** $(x^3)^5$ **3.** $(y^4)^2 \cdot y$

The next law of exponents involves raising a product to a power. For example,

$$(2 \cdot 5)^3 = (2 \cdot 5)(2 \cdot 5)(2 \cdot 5) = (2 \cdot 2 \cdot 2)(5 \cdot 5 \cdot 5) = 2^3 \cdot 5^3$$

In general, if a and b are real numbers and if n is a positive integer, we have

$$(a \cdot b)^n = \underbrace{(ab) \cdot (ab) \cdot \cdots \cdot (ab)}_{n \text{ factors}} = \underbrace{(a \cdot a \cdot \cdots \cdot a)}_{n \text{ factors}}\underbrace{(b \cdot b \cdot \cdots \cdot b)}_{n \text{ factors}} = a^n \cdot b^n$$

When a product is raised to a power, the result equals the product of each factor raised to that power.

If a and b are real numbers and n is a positive integer, then

$$(a \cdot b)^n = a^n \cdot b^n \qquad \textbf{(4)}$$

EXAMPLE 3 Using Equation (4)

(a) $(2x)^3 = 2^3 \cdot x^3 = 8x^3$

(b) $(2x^3y)^4 = 2^4(x^3)^4y^4 = 16x^{12}y^4$

(c) $(-4y)^3 = (-4)^3 \cdot y^3 = -64y^3$

Practice Exercise 3 *In Problems 1–4, simplify each expression.*

1. $(3x)^2$ **2.** $(2x^2)^3$ **3.** $(2xy^3)^4$ **4.** $(2xy^3)^4(x^2y)$

The next two Laws of Exponents involve quotients and are analogous to equation (2), which involves products.

Consider the following problem:

$$\frac{2^6}{2^4} = \frac{2 \cdot 2 \cdot \cancel{2} \cdot \cancel{2} \cdot \cancel{2} \cdot \cancel{2}}{\cancel{2} \cdot \cancel{2} \cdot \cancel{2} \cdot \cancel{2}} = \frac{2^2}{1} = 2^2 \qquad \text{Cancel the 2's}$$

Notice that $\dfrac{2^6}{2^4} = 2^{6-4} = 2^2$.

In general, if a is any nonzero real number and m, n are positive integers with $m > n$, we have

$$\frac{a^m}{a^n} = \frac{\overbrace{a \cdot a \cdot \cdots \cdot a}^{m \text{ factors}}}{\underbrace{a \cdot a \cdot \cdots \cdot a}_{n \text{ factors}}} = \frac{\overbrace{\overbrace{\cancel{a \cdot a \cdot \cdots \cdot a}}^{n \text{ factors}} \cdot \overbrace{a \cdot a \cdot \cdots \cdot a}^{m - n \text{ factors}}}^{\text{Total of } m \text{ factors (since } m > n)}}{\underbrace{\cancel{a \cdot a \cdot \cdots \cdot a}}_{n \text{ factors}}}$$

$$= \overbrace{a \cdot a \cdot \cdots \cdot a}^{m - n \text{ factors}} = a^{m-n}$$

To divide two expressions having the same base, retain the base and subtract the exponent in the denominator from the exponent in the numerator.

If a is a nonzero real number and m, n are positive integers with $m > n$, then

$$\frac{a^m}{a^n} = a^{m-n} \qquad a \neq 0 \tag{5}$$

EXAMPLE 4 **Using Equation (5)**

(a) $\dfrac{3^7}{3^5} = 3^{7-5} = 3^2 = 9$ (b) $\dfrac{(-2)^8}{(-2)^5} = (-2)^{8-5} = (-2)^3 = -8$

(c) $\dfrac{4^3}{4^2} = 4^{3-2} = 4^1 = 4$ (d) $\dfrac{x^7}{x} = x^{7-1} = x^6$ ◀

Practice Exercise 4 *In Problems 1–5, simplify each expression.*

1. $\dfrac{5^8}{5^2}$ **2.** $\dfrac{3^2}{3}$ **3.** $\dfrac{(-6)^4}{(-6)}$ **4.** $\dfrac{7^5}{7^4}$ **5.** $\dfrac{x^{10}}{x^6}$ ◀

What happens if the exponent in the denominator is greater than the exponent in the numerator, as, for example, with $\dfrac{4^3}{4^5}$?

$$\frac{4^3}{4^5} = \frac{\not{4}\cdot\not{4}\cdot\not{4}\cdot 1}{\not{4}\cdot\not{4}\cdot\not{4}\cdot 4\cdot 4} = \frac{1}{4^2}$$

In general, we can say if a is any real number with $a \neq 0$ and m, m are positive integers with $n > m$, we have

$$\frac{a^m}{a^n} = \frac{\overbrace{a\cdot a\cdot a\cdot \cdots \cdot a}^{m \text{ factors}}}{\underbrace{a\cdot a\cdot a\cdot a\cdot \cdots \cdot a\cdot a}_{n \text{ factors}}} = \frac{\overbrace{a\cdot a\cdot a\cdot \cdots \cdot a}^{m \text{ factors}}\cdot 1}{\underbrace{(a\cdot a\cdot a\cdot \cdots \cdot a)}_{m \text{ factors}}\underbrace{(a\cdot a\cdot a\cdot \cdots \cdot a)}_{n-m \text{ factors}}} = \frac{1}{\underbrace{a\cdot a\cdot a\cdot \cdots \cdot a}_{n-m \text{ factors}}} = \frac{1}{a^{n-m}}$$

If a is any nonzero real number and m, n are positive integers with $n > m$, we have

$$\frac{a^m}{a^n} = \frac{1}{a^{n-m}} \qquad a \neq 0 \tag{6}$$

EXAMPLE 5 **Using Equation (6)**

(a) $\dfrac{(-3)^5}{(-3)^8} = \dfrac{1}{(-3)^{8-5}} = \dfrac{1}{(-3)^3} = \dfrac{1}{-27} = -\dfrac{1}{27}$ (b) $\dfrac{x^2}{x^4} = \dfrac{1}{x^{4-2}} = \dfrac{1}{x^2}$

(c) $\dfrac{2}{2^8} = \dfrac{1}{2^{8-1}} = \dfrac{1}{2^7}$ ◀

Practice Exercise 5 *In Problems 1–3, simplify each expression.*

1. $\dfrac{8^2}{8^5}$ **2.** $\dfrac{3}{3^6}$ **3.** $\dfrac{x^7}{x^{11}}$

The next Law of Exponents involves raising a quotient to a power and is analogous to equation (4) involving products. For example,

$$\left(\frac{3}{4}\right)^4 = \frac{3}{4}\cdot\frac{3}{4}\cdot\frac{3}{4}\cdot\frac{3}{4} = \frac{3^4}{4^4}$$

In general, if a and b are real numbers with $b \neq 0$ and n is a positive integer, we have

$$\left(\frac{a}{b}\right)^n = \underbrace{\frac{a}{b}\cdot\frac{a}{b}\cdot\frac{a}{b}\cdot\ \cdots\ \cdot\frac{a}{b}}_{n \text{ factors}} = \frac{a^n}{b^n}.$$

If a quotient is raised to a power, the result equals the numerator raised to that power divided by the denominator raised to that power.

If a and b are real numbers with $b \neq 0$ and n is a positive integer, then

$$\left(\frac{a}{b}\right)^n = \frac{a^n}{b^n} \qquad b \neq 0 \qquad \textbf{(7)}$$

EXAMPLE 6 Using Equation (7)

(a) $\left(\dfrac{3}{5}\right)^9 = \dfrac{3^9}{5^9}$ (b) $\left(\dfrac{x}{3}\right)^4 = \dfrac{x^4}{3^4}$ (c) $\left(\dfrac{3x}{2y}\right)^5 = \dfrac{(3x)^5}{(2y)^5} = \dfrac{3^5x^5}{2^5y^5}$

Practice Exercise 6 *In Problems 1–4, simplify each expression. Express your answer as a quotient of exponents.*

1. $\left(\dfrac{5}{7}\right)^3$ **2.** $\left(\dfrac{12}{17}\right)^2$ **3.** $\left(\dfrac{w}{z}\right)^6$ **4.** $\left(\dfrac{3u}{5v}\right)^4$

We can combine equations (5) and (6) as follows:

If a is a nonzero real number and m, n are positive integers, then

$$\frac{a^m}{a^n} = a^{m-n} \qquad a \neq 0 \qquad \textbf{(8)}$$

Do you see the difference between equation (5) and equation (8)? We took out the requirement that $m > n$.

The next example illustrates the reason behind defining $a^0 = 1$ for every nonzero real number a.

EXAMPLE 7 **Showing $a^0 = 1, a \neq 0$.**

Simplify $\frac{6^3}{6^3}$

Solution Using equation (8), $\frac{6^3}{6^3} = 6^{3-3} = 6^0$.

But, $\frac{6^3}{6^3} = \frac{\cancel{(6 \cdot 6 \cdot 6)} \cdot 1}{\cancel{(6 \cdot 6 \cdot 6)} \cdot 1} = \frac{1}{1} = 1$

Using the Transitive Property, we must have $6^0 = 1$. ◀

In general, if a is any nonzero real number and n is a positive integer, we have

$$\frac{a^n}{a^n} = a^{n-n} = a^0 \quad \text{and} \quad \frac{a^n}{a^n} = \frac{\overbrace{a \cdot a \cdot a \cdot \cdots \cdot a}^{n \text{ factors}}}{\underbrace{a \cdot a \cdot a \cdot \cdots \cdot a}_{n \text{ factors}}} = 1$$

This leads us to formulate the following definition:

If a is a nonzero real number, we define

$$a^0 = 1 \qquad a \neq 0 \qquad \textbf{(9)}$$

Notice in the definition of a^0 that the base a is not allowed to equal 0.

EXAMPLE 8 **Using Equation (9)**

(a) $2^0 = 1$ (b) $(-3)^0 = 1$ (c) $\pi^0 = 1$ (d) $-3^0 = -1$

(e) $x^0 = 1, x \neq 0$ ◀

In Example 8, can you explain the difference in the answers obtained in parts (b) and (d)? In part (b), -3 is raised to the 0 power, giving the answer 1; in part (d), 3 is raised to the 0 power to get 1, and then we find the opposite of 1 to get -1.

EXAMPLE 9 Simplify each of the following: (a) $(-2)^0(-2)^3$ (b) $\left(\frac{2x^0}{3y^5}\right)^4$

(a) **Solution A:** $(-2)^0(-2)^3 = (1)(-8) = -8$

Solution B: $(-2)^0(-2)^3 = (-2)^{0+3} = (-2)^3 = -8$

(b) **Solution A:** $\left(\frac{2x^0}{3y^5}\right)^4 = \left(\frac{2 \cdot 1}{3y^5}\right)^4 = \frac{2^4}{3^4(y^5)^4} = \frac{16}{81y^{20}}$

Solution B: $\left(\frac{2x^0}{3y^5}\right)^4 = \frac{2^4(x^0)^4}{3^4(y^5)^4} = \frac{16x^0}{81y^{20}} = \frac{16 \cdot 1}{81y^{20}} = \frac{16}{81y^{20}}$ ◀

Practice Exercise 7 *In Problems 1–7, simplify each expression.*

1. 5^0 **2.** $(-5)^0$ **3.** -5^0 **4.** $(3^4)^0$ **5.** $(-4y)^0$
6. $-4y^0$ **7.** $(2x^3y^0)^5$

EXAMPLE 10 Arriving at the Definition for Negative Exponents

Simplify $\dfrac{4^2}{4^5}$

Solution Using equation (8), we have

$$\frac{4^2}{4^5} = 4^{2-5} = 4^{-3}$$

Using equation (6), we have

$$\frac{4^2}{4^5} = \frac{1}{4^{5-2}} = \frac{1}{4^3}$$

Using the Transitive Property, we must have

$$4^{-3} = \frac{1}{4^3}$$

Based on Example 10, we define negative exponents as follows:

If a is a nonzero real number, and if n is a positive integer, then

$$a^{-n} = \frac{1}{a^n} \qquad a \neq 0 \qquad \textbf{(10)}$$

The Laws of Exponents involving negative exponents will be explored further in Section 4.1. Until then we will limit our discussion to expressions involving nonnegative integer exponents.

Answers to Practice Exercises

Practice Exercise 1 **1.** 3^6 **2.** $(-5)^4$ **3.** 3^5 **4.** x^9

Practice Exercise 2 **1.** 2^6 **2.** x^{15} **3.** y^9

Practice Exercise 3 **1.** $9x^2$ **2.** $8x^6$ **3.** $16x^4y^{12}$ **4.** $16x^6y^{13}$

Practice Exercise 4 **1.** 5^6 **2.** $3^1 = 3$ **3.** $(-6)^3 = -216$ **4.** $7^1 = 7$ **5.** x^4

Practice Exercise 5 **1.** $\dfrac{1}{8^3}$ **2.** $\dfrac{1}{3^5}$ **3.** $\dfrac{1}{x^4}$

Practice Exercise 6 **1.** $\dfrac{5^3}{7^3}$ **2.** $\dfrac{12^2}{17^2}$ **3.** $\dfrac{w^6}{z^6}$ **4.** $\dfrac{81u^4}{625v^4}$

Practice Exercise 7 **1.** 1 **2.** 1 **3.** -1 **4.** 1 **5.** 1 **6.** -4 **7.** $32x^{15}$

2.1 Assess Your Understanding

'Are You Prepared?' *Answers are given at the end of these exercises. If you get a wrong answer, read the pages listed in parentheses.*

1. In the expression 6^4, 6 is called the ________ and 4 is called the ________. (pp. 5–6)

2. Write in exponent form: $2 \cdot 2 \cdot 2 \cdot 2 \cdot 2 =$ ________. (pp. 5–6)

Concepts and Vocabulary

3. $5^0 =$ ________

4. $-5^0 =$ ________

5. *True or False:* $(a^m)^n = a^{mn}$ where $a \neq 0$ and m, n are positive integers.

6. *True or False:* $3^4 \cdot 3^6 = 9^{10}$

7. *True or False:* $\dfrac{7^9}{7^5} = 7^4$

8. *True or False:* $\left(\dfrac{x}{3}\right)^2 = \dfrac{x^2}{9}$

Exercises

In Problems 9–84, simplify each expression. Whenever an exponent is 0, we assume the base is not 0. We also assume no denominator equals 0.

9. $2^3 \cdot 2^2$ **10.** $3^2 \cdot 3$ **11.** $5^2 \cdot 5^4$ **12.** $8 \cdot 8^6$ **13.** $(-2)^3(-2)$

14 $(-6)^8(-6)^2$ **15.** $(-2)^2(-2)^3$ **16.** $(-4)^2(-4)$ **17.** $2^4 \cdot 2^0$ **18.** $3^0 \cdot 3^3$

19. $-2^4 \cdot 2$ **20.** $-3 \cdot 3^2$ **21.** $x^3 \cdot x^9$ **22.** $x^5 \cdot x$ **23.** $y \cdot y^3$

24. $z^7 \cdot z^3$ **25.** $(3^4)^5$ **26.** $(5^2)^4$ **27.** $(2^3)^2$ **28.** $(2^2)^3$

29. $(5^2)^1$ **30.** $(4^3)^1$ **31.** $[(-6)^8]^2$ **32.** $[(-3)^5]^4$ **33.** $-(6^8)^2$

34. $-(3^5)^4$ **35.** $(8^2)^0$ **36.** $(4^0)^5$ **37.** $(x^3)^9$ **38.** $(x^4)^2$

39. $(z^5)^0$ **40.** $(y^3)^4$ **41.** $(2x)^5$ **42.** $(3x)^4$ **43.** $(-3x)^4$

44. $(-5x)^3$ **45.** $(3xy)^3$ **46.** $(2xy)^5$ **47.** $(2x^3)^0$ **48.** $(3x^2)^0$

49. $(-3x^5)^4$ **50.** $(-5x^3)^3$ **51.** $(3x^4y)^3$ **52.** $(2x^2y^3)^5$ **53.** $\dfrac{2^{10}}{2^3}$

54. $\dfrac{(-3)^7}{(-3)^4}$ **55.** $\dfrac{(-5)^6}{(-5)^0}$ **56.** $\dfrac{8^6}{8}$ **57.** $\dfrac{4^3}{4}$ **58.** $\dfrac{x^7}{x^2}$

59. $\dfrac{y^5}{y^4}$ **60.** $\dfrac{z^3}{z^2}$ **61.** $\dfrac{5^4}{5^{10}}$ **62.** $\dfrac{3^4}{3^5}$ **63.** $\dfrac{6^8}{6^9}$

64. $\dfrac{(-7)^4}{(-7)^6}$ **65.** $\dfrac{(-2)^3}{(-2)^7}$ **66.** $\dfrac{x^0}{x^8}$ **67.** $\dfrac{x^2}{x^9}$ **68.** $\dfrac{y^3}{y^7}$

69. $\left(\dfrac{3}{5}\right)^2$ **70.** $\left(\dfrac{2}{9}\right)^3$ **71.** $\left(\dfrac{1}{2}\right)^4$ **72.** $\left(\dfrac{x}{2}\right)^5$ **73.** $\left(\dfrac{x}{-3}\right)^2$

74. $\left(\dfrac{2x}{3y}\right)^4$ **75.** $\left(\dfrac{x}{5y}\right)^2$ **76.** $\left(\dfrac{3}{2x}\right)^3$ **77.** $\left(\dfrac{x^3}{-3y}\right)^2$ **78.** $\left(\dfrac{3x^0}{5y^4}\right)^2$

79. $\left(\dfrac{3x^2}{5y^4}\right)^2$ **80.** $\left(\dfrac{3y}{2x^4}\right)^3$ **81.** 3^0 **82.** $(-4)^0$ **83.** -3^0

84. -4^0

85. In your own words, write a paragraph to justify the definition given in the text that $a^0 = 1$, if $a \neq 0$.

'Are You Prepared?' Answers

1. base; exponent **2.** 2^5

2.2 Monomials; Polynomials

PREPARING FOR THIS SECTION *Before getting started review the following:*

- Order of Operations (Section 1.1, pp. 6–10)

Now work the 'Are You Prepared?' problems on page 76.

OBJECTIVES

1. Recognize Monomials
2. Combine Like Terms
3. Recognize Binomials and Trinomials
4. Recognize Polynomials
5. Evaluate Polynomials

Monomials and polynomials are among the simplest of algebraic expressions and are used extensively in algebra. In this section, we define monomials and polynomials. In the remainder of the chapter, we learn how to add, subtract, multiply, divide, and factor them.

Monomials

1 We have described algebra as a generalization of arithmetic in which letters may be used to represent real numbers. Further, we said that a letter representing *any* real number is a **variable,** and a letter representing a fixed (possibly unspecified) real number is a **constant.** From now on, we shall use the letters at the end of the alphabet, such as x, y, and z, to represent variables and the letters at the beginning of the alphabet, such as a, b, and c, to represent constants. In the expressions $3x + 5$ and $ax + b$, it is understood that x is a variable and that a and b are constants, even though the constants a and b are unspecified. As you will find out, the context usually makes the intended meaning clear.

Now we introduce some basic terms.

A **monomial in one variable** is the product of a constant times a variable raised to a nonnegative integer power. That is, a monomial is of the form

$$ax^k \qquad \textbf{(1)}$$

where a is a constant, x is a variable, and $k \geq 0$ is an integer. The constant a is called the **coefficient** of the monomial. If $a \neq 0$, then k is called the **degree** of the monomial.

For simplicity, we usually say "monomial" instead of "monomial in one variable."

The domain of the variable x of a monomial ax^k is all real numbers if $k > 0$ and all nonzero real numbers if $k = 0$.

Let's look at some examples of monomials.

EXAMPLE 1 **Examples of Monomials**

Monomial	Coefficient	Degree	
(a) $6x^2$	6	2	
(b) $-\sqrt{2}x^3$	$-\sqrt{2}$	3	
(c) 3	3	0	Since $3 = 3 \cdot 1 = 3x^0, x \neq 0$
(d) $-5x$	-5	1	Since $-5x = -5x^1$
(e) x^4	1	4	Since $x^4 = 1 \cdot x^4$ ◀

Now let's look at some expressions that are not monomials.

EXAMPLE 2 **Expressions That Are Not Monomials**

(a) $3x^{1/2}$ is not a monomial, since the exponent of the variable x is $\frac{1}{2}$ and $\frac{1}{2}$ is not a nonnegative integer.

(b) $4x^{-3}$ is not a monomial, since the exponent of the variable x is -3 and -3 is not a nonnegative integer.

(c) $\frac{3}{x + 5}$ is not a monomial, since the variable x appears in the denominator, making it impossible to write the expression in the form $ax^k, k \geq 0$. ◀

Practice Exercise 1 *Determine which of the following expressions are monomials. For those that are, name the coefficient and degree. For those that are not monomials, state why.*

1. $5x^2$ **2.** $-x$ **3.** $\frac{1}{2}x^6$ **4.** $-5x^{3/4}$ **5.** $2x^{-3}$ **6.** $2x^3$

7. $\sqrt{3}x$ **8.** $\frac{1}{x^2}$

2 Two monomials ax^k and bx^k with the same degree and the same variable are called **like terms.** Such monomials when added or subtracted can be combined into a single monomial by using the Distributive Property.

EXAMPLE 3 **Combining Like Terms**

(a) $2x^2 + 5x^2 = (2 + 5)x^2 = 7x^2$
↑ Distributive property

(b) $8x^3 - 5x^3 = 8x^3 + (-5x^3) = 8x^3 + (-5)x^3 = [8 + (-5)]x^3 = 3x^3$
↑ Distributive Property

(c) $ax^3 + ax^3 = (a + a)x^3 = 2ax^3$

(d) $3x^2 + 2x^3$ cannot be combined to form a monomial, because $3x^2$ and $2x^3$ are not like terms (the degrees do not match).

(e) $5x^3 + 2y^3$ cannot be combined to form a monomial, because $5x^3$ and $2y^3$ are not like terms (the variables are not the same). ◀

The procedure we used to combine the monomials in Example 3(b) can be shortened by using the result of Problem 91, Exercise 1.6. That property states that if a, b, and c are real numbers, then

$$a \cdot b - a \cdot c = a \cdot (b - c) \quad (2)$$

We use this property in the next example.

EXAMPLE 4 **Combining Like Terms**

$$4x^5 - 10x^5 = (4 - 10)x^5 = -6x^5$$

↑ Use (2)

Examples 3 and 4 show that to add (or subtract) two monomials that are like terms, simply add (or subtract) their coefficients.

Practice Exercise 2 *In Problems 1–4, combine like terms, if possible. Give the coefficient and the degree of the answer if the terms can be combined.*

1. $7x^4 + 3x^4$ **2.** $5x^3 - x^3$ **3.** $2y^3 + 5y^2$ **4.** $z^7 - 5z^7$

3 If two monomials are not like terms, then they cannot be combined. The sum or difference of two such monomials is called a **binomial.** The sum or difference of three such monomials is called a **trinomial.**

EXAMPLE 5 **Examples of Binomials and Trinomials**

(a) $x^2 - 2$ is a binomial; it is the difference of two monomials, x^2 and 2.

(b) $x^3 - 3x + 5$ is a trinomial; it is the sum or difference of three monomials, x^3, $3x$, and 5.

(c) $2x^2 + 5x^2 + 2 = 7x^2 + 2$ is a binomial.

(d) $x^2 + y^2$ is a binomial; it is the sum of two monomials, x^2 and y^2.

Practice Exercise 3 *In Problems 1–3, identify each expression as a monomial, binomial, or trinomial.*

1. $3x^2 - 5x + 2$ **2.** $3x^2 - 5x^2 + 8x^2$ **3.** $3x^5 + 5x^3$

4 Definition of a Polynomial

A **polynomial** in one variable is an algebraic expression of the form

$$a_n x^n + a_{n-1} x^{n-1} + \cdots + a_1 x + a_0 \quad (3)$$

where $a_n, a_{n-1}, \ldots, a_1, a_0$ are constants,* called the **coefficients** of the polynomial, $n \geq 0$ is an integer, and x is a variable. If $a_n \neq 0$, it is called the **leading coefficient** and n is called the **degree** of the polynomial.

*The notation a_n is read as "a sub n." The number n is called a **subscript** and should not be confused with an exponent. We use subscripts in order to distinguish one constant from another when a large or undetermined number of constants is required.

The monomials that make up a polynomial are called its **terms.** If all the coefficients are 0, the polynomial is called the **zero polynomial,** which has no degree.

Polynomials are usually written in **standard form,** beginning with the nonzero term of highest degree and continuing with terms in descending order according to degree. If a power of x is missing, it is because its coefficient is zero.

EXAMPLE 6 **Describing Polynomials**

Write each of the following polynomials in standard form. List all the coefficients, starting with the leading coefficient, and give the degree of the polynomial.

(a) $3x^2 - 5$ (b) $8 - 2x + x^2$ (c) $5x + \sqrt{2}$ (d) 3 (e) 0

Solution

(a) $3x^2 - 5$ is in standard form. It can be rewritten as $3x^2 + 0x + (-5)$. The coefficients are $3, 0, -5$; the degree is 2.

(b) The polynomial $8 - 2x + x^2$ is not in standard form. We rearrange its terms to put it in standard form:

$$8 - 2x + x^2 = x^2 - 2x + 8$$

This can be rewritten as $1x^2 + (-2)x + 8$. The coefficients are $1, -2, 8$; the degree is 2.

(c) $5x + \sqrt{2}$ is in standard form. The coefficients are $5, \sqrt{2}$; the degree is 1.

(d) 3 is in standard form. It can be rewritten as $3x^0$. The only coefficient is 3; the degree is 0.

(e) This is the zero polynomial. The coefficient is 0; it has no degree. ◀

Although we have been using x to represent the variable in the above polynomials, letters such as y or z are also commonly used.

$3x^4 - x^2 + 2$ is a polynomial (in x) of degree 4.

$9y^3 - 2y^2 + y - 3$ is a polynomial (in y) of degree 3.

$z^5 + \pi$ is a polynomial (in z) of degree 5.

Practice Exercise 4 *In Problem 1–5, write each of the following polynomials in standard form. List all the coefficients, starting with the leading coefficients, and give the degree of the polynomial.*

1. $x^3 - 2x^5 + 5$ **2.** $3z^2 + z - 1$ **3.** -7 **4.** $\sqrt{5}x^2 - x$

5. $x - x^2 + x^3$

Monomials and Polynomials in Two or More Variables

A **monomial in two variables** x and y has the form ax^ny^m, where a is a constant, x and y are variables, and n and m are nonnegative integers. The **degree of a monomial** is the sum of the powers of the variables; that is, the degree of the monomial ax^ny^m, $a \neq 0$, is $n + m$.

EXAMPLE 7 **Examples of Monomials**

(a) x^2y^3 is a monomial in two variables with degree 5, since $2 + 3 = 5$.

(b) $2xy^3$ is a monomial in two variables with degree 4, since $1 + 3 = 4$. ◀

A **polynomial in two variables** x and y is the sum of one or more monomials in two variables. The **degree of a polynomial** in two variables is the highest degree of all the monomials with nonzero coefficients.

Polynomials in three or more variables are defined in a similar way.

EXAMPLE 8 **Examples of Polynomials**

(a) $3x^2 + 2x^3y + 5$ is a polynomial (a trinomial) in two variables with degree 4, since $2x^3y$ is the monomial with highest degree and its degree is 4.

(b) $\pi x^3 - y^2$ is a polynomial (a binomial) in two variables with degree 3, since πx^3 is the monomial with highest degree and its degree is 3.

(c) $x^4 + 4x^3y - xy^3 + y^4$ is a polynomial in two variables with degree 4.

(d) $x^2 + y^2 - z^2 + 4$ is a polynomial in three variables with degree 2.

(e) x^3y^2z is a polynomial (a monomial) in three variables with degree 6.

(f) $5x^2 - 4y^2 + x^3y + 2w^2x - 7z$ is a polynomial in four variables with degree 4. ◀

Practice Exercise 5 *In Problems 1–3, determine how many variables each polynomial contains, and give the degree of the polynomial.*

1. $2x^3yz^4 - 5x^3y^2$ **2.** $\sqrt{3}x^2y$ **3.** $x^3 + 3x^2y + 3xy^2 + y^3$ ◀

Evaluating Polynomials

5 Polynomials may be evaluated by substituting the given number for the variable(s) of the polynomial.

EXAMPLE 9 **Evaluating a Polynomial**

Evaluate the polynomial $3x^2 - 4x + 6$ for

(a) $x = 1$ (b) $x = 0$ (c) $x = -2$

Solution (a) Replace x by 1 in $3x^2 - 4x + 6$. The result is

$$\begin{aligned} 3x^2 - 4x + 6 &= 3\cdot(1)^2 - 4\cdot(1) + 6 && \text{Substitute } x = 1. \\ &= 3\cdot 1 - 4\cdot 1 + 6 && \text{Simplify exponents.} \\ &= 3 - 4 + 6 && \text{Multiply.} \\ &= 3 + (-4) + 6 && \text{Change subtraction to addition.} \\ &= -1 + 6 = 5 && \text{Add.} \end{aligned}$$

(b) Replace x by 0 in $3x^2 - 4x + 6$. The result is

$$\begin{aligned} 3x^2 - 4x + 6 &= 3\cdot(0)^2 - 4\cdot(0) + 6 \\ &= 3\cdot 0 - 4\cdot 0 + 6 \\ &= 0 - 0 + 6 = 6 \end{aligned}$$

(c) Replace x by -2 in $3x^2 - 4x + 6$. The result is

$$\begin{aligned} 3x^2 - 4x + 6 &= 3\cdot(-2)^2 - 4\cdot(-2) + 6 \\ &= 3\cdot 4 - 4\cdot(-2) + 6 \\ &= 12 + 8 + 6 = 26 \end{aligned}$$

◀

Practice Exercise 6 *In Problems 1–4, evaluate the polynomial* $2x^3 - 2x^2 + 3$ *at the number* x.

1. $x = -1$ **2.** $x = 0$ **3.** $x = 1$ **4.** $x = 2$

◀

The idea is the same for a polynomial in two (or more) variables.

EXAMPLE 10 Evaluating a Polynomial

Evaluate the polynomial $2x^2y - 5xy^2 + xy$ for

(a) $x = -1, y = 2$ (b) $x = 0, y = 5$

Solution (a) Substitute -1 for x and 2 for y in the polynomial.

$$\begin{aligned} 2x^2y - 5xy^2 + xy &= 2\cdot(-1)^2\cdot(2) - 5\cdot(-1)(2)^2 + (-1)\cdot(2) \\ &\uparrow \ x = -1, y = 2 \\ &= 2\cdot(1)\cdot(2) - 5\cdot(-1)\cdot(4) + (-1)\cdot(2) \\ &= 4 - (-20) + (-2) = 4 + 20 + (-2) = 22 \end{aligned}$$

(b) Substitute 0 for x and 5 for y in the polynomial.

$$2x^2y - 5xy^2 + xy \underset{\uparrow\, x = 0,\, y = 5}{=} 2\cdot(0)^2\cdot(5) - 5\cdot(0)\cdot(5)^2 + (0)\cdot(5) = 0 - 0 + 0 = 0$$

◀

EXAMPLE 11 Evaluating a Polynomial

Evaluate the polynomial $5x^2 + 4y^2 + 3z^2 - 10xyz$ for $x = 2, y = -3, z = 4$.

Solution Substitute 2 for x, -3 for y, and 4 for z in the polynomial:

$$\begin{aligned} 5x^2 + 4y^2 + 3z^2 - 10xyz &= 5\cdot(2)^2 + 4\cdot(-3)^2 + 3\cdot(4)^2 - 10\cdot(2)\cdot(-3)\cdot(4) \\ &\uparrow \ x = 2, y = -3, z = 4 \\ &= 5\cdot(4) + 4\cdot(9) + 3\cdot(16) - (-240) \\ &= 20 + 36 + 48 + 240 = 344 \end{aligned}$$

◀

Practice Exercise 7 *In Problems 1–4, evaluate each polynomial for the given values of the variables.*

1. $x^2 + xy - y^2$ for $x = 1, y = -1$

2. $x^3 - y^3$ for $x = 0, y = 2$

3. $x^2 + y^2 + z^2 - xyz$ for $x = 1, y = -1, z = 2$

4. $x^2y^2 - xz$ for $x = 1, y = -1, z = 0$

◀

Answers to Practice Exercises

Practice Exercise 1

1. monomial, coefficient 5, degree 2
2. monomial, coefficient −1, degree 1
3. monomial, coefficient $\frac{1}{2}$, degree 6
4. not a monomial because of the rational, non-integer, exponent
5. not a monomial because of the negative exponent
6. monomial, coefficient 2, degree 3
7. monomial, coefficient $\sqrt{3}$, degree 1
8. not a monomial because $\frac{1}{x^2} = x^{-2}$ so the exponent is negative

Practice Exercise 2

1. $10x^4$; coefficient 10; degree 4
2. $4x^3$; coefficient 4; degree 3
3. Not like terms
4. $-4z^7$; coefficient −4; degree 7

Practice Exercise 3 1. Trinomial 2. Monomial $(6x^2)$ 3. Binomial

Practice Exercise 4

1. $-2x^5 + x^3 + 5$; coefficients are −2, 0, 1, 0, 0, 5; degree 5
2. $3z^2 + z - 1$; coefficients are 3, 1, −1; degree 2
3. −7; coefficient −7; degree 0
4. $\sqrt{5}x^2 - x$; coefficients are $\sqrt{5}$, −1, 0; degree 2
5. $x^3 - x^2 + x$; coefficients are 1, −1, 1, 0; degree 3

Practice Exercise 5

1. Three variables; degree 8
2. Two variables; degree 3
3. Two variables; degree 3

Practice Exercise 6 1. −1 2. 3 3. 3 4. 11

Practice Exercise 7 1. −1 2. −8 3. 8 4. 1

2.2 Assess Your Understanding

'Are You Prepared?' *Answers are given at the end of these exercises. If you get a wrong answer, read the pages listed in parentheses.*

1. Simplify: $4 \cdot 3^2 - 5 \cdot 3 + 5$ (pp. 6–10)

2. Simplify: $-2 \cdot 5 - 3 \cdot 2^3$ (pp. 6–10)

Concepts and Vocabulary

3. *True or False:* $3x^{-5}$ is a monomial of degree −5.

4. The polynomial $5x^3 - 3x^2 - 2x - 4$ has degree ________ and the ________ coefficient is 5.

5. A ________ is a polynomial with three terms.

6. *True or False:* The polynomial $3x^5 - 5x^4 + 1$ has degree 3.

7. Combine like terms: $-4x^5 + x^5 =$ ________.

8. If $x = 2$, then $x^2 - 5x =$ ________.

Exercises

In Problems 9–28, tell whether the expression is a monomial. If it is, name the variable(s) and the coefficient and give the degree of the monomial. If it is not a monomial, tell why.

9. $5x^3$ **10.** $-4x^2$ **11.** $\frac{8}{x}$ **12.** $-8x$ **13.** $-2x^{-3}$

14. $-2xy^2$ **15.** $5x^4y^3$ **16.** $\frac{8x}{y}$ **17.** $-\frac{2x^2}{y^3}$ **18.** $\frac{1}{2}x^2$

19. $-\frac{1}{3}x^3$ **20.** $\frac{2}{x}$ **21.** $10x^{3/2}$ **22.** 1 **23.** −4

24. $x^2 + y^2$ **25.** $3x^2 + 4$ **26.** $5x^{-4}$ **27.** $2x^9y$ **28.** −7

In Problems 29–40, combine all like terms. Tell whether the final expression is a monomial, binomial, or trinomial.

29. $8x^3 + 4x^3$ **30.** $-2x^2 + 4x^2$ **31.** $5x - 2x + 4$

32. $14x^3 - 8x^3 + x$ **33.** $10x^2 - 21x^2 + x + 1$ **34.** $4x^5 - 3x^2 + 4x^2 + 1$

35. $4x^5 - 5x^2 + 5x - 3x^2$ **36.** $4 - 3x^2 + 5x - 10$ **37.** $4(x^2 - 5) + 5(x^2 + 1)$

38. $2(x^2 + x) - 3(x - x^2)$ **39.** $-3(2x - 1) - 5(3 - 2x)$ **40.** $7(x^2 - 2) + 3(x^2 + 5)$

In Problems 41–50, write each polynomial in standard form. List the coefficients, starting with the leading coefficient, and give the degree of the polynomial.

41. $3x^2 + 4x + 1$ **42.** $5x^3 - 3x^2 + 2x + 4$ **43.** $1 - x^2$ **44.** $x^3 - 1$

45. $9y^2 + 8y - 5$ **46.** $3z^4 + z^2 + 2$ **47.** $1 - x + x^2 - x^3$ **48.** $1 + x + x^2$

49. $\sqrt{2} - x$ **50.** $\pi x + 2$

In Problems 51–56, give the degree of each polynomial.

51. $xy^2 - 1 + x$ **52.** $x^3 - xy + y^3$ **53.** $x^2y + y^2z + z^2x - 3xyz$

54. $3 - (x^2 + y^2 + z^2)$ **55.** $2x^4y^3z + 5x^2y^2 - 7z^5 + 10$ **56.** $-6x^3y^2 + 17xy^4 - xyz$

In Problems 57–70, evaluate each polynomial for the given value(s) of the variable(s).

57. $x^2 - 1$ for $x = 1$ **58.** $x^2 - 1$ for $x = -1$ **59.** $x^2 + x + 1$ for $x = -1$

60. $x^2 - x + 1$ for $x = 1$ **61.** $3x - 6$ for $x = 2$ **62.** $4x - 12$ for $x = 3$

63. $5y^3 - 3y^2 + 4$ for $y = 2$ **64.** $z^5 - z^3 + 1$ for $z = 1$

65. $5x^2 - 3y^3 + z^2 - 4$ for $x = 2, y = -1, z = 3$ **66.** $15xyz - x^2 + 2y^2 + 4z^2$ for $x = 1, y = -1, z = 2$

67. $x^5 - x^4 + x^3 - x^2 + x - 1$ for $x = 1$ **68.** $x^{16} + 2x^8 + 1$ for $x = 0$

69. $x^3 + x^2 + x + 1$ for $x = -1.5$ **70.** $x^3 + x^2 + x + 1$ for $x = 1.5$

---

71. If an object is propelled vertically upward with an initial velocity of 160 feet per second, its height h after t seconds (neglecting air friction) is given by the formula

$$h = -16t^2 + 160t$$

(a) Find the height of the object after 3 seconds.
(b) Find the height of the object after 4 seconds.

72. The velocity v of an object propelled vertically upward with an initial velocity of 160 feet per second (neglecting air friction) is given by the formula

$$v = -32t + 160$$

(a) Find the velocity of the object after 1 second.
(b) Find the velocity of the object after 4 seconds.
(c) When is the velocity of the object equal to 0?
(d) What is the height of the object (see Problem 63) when its velocity is 0?

'Are You Prepared?' Answers

1. 26 **2.** −34

2.3 Addition and Subtraction of Polynomials

PREPARING FOR THIS SECTION *Before getting started, review the following:*

- Combine Like Terms (pp. 71–72)
- Multiplicative Identity (p. 29)
- Subtraction Property (pp. 39–40)
- Distributive Property (p. 28)

Now work the 'Are You Prepared?' problems on page 82.

OBJECTIVES

1 Add Polynomials
2 Subtract Polynomials

Polynomials are added and subtracted by combining like terms after applying Commutative and Associative Properties of addition.

1 Adding Polynomials

EXAMPLE 1 Adding Polynomials

Find the sum of the polynomials

$$3x^2 + 2x + 5 \quad \text{and} \quad 2x^2 - 3x + 2$$

Solution $(3x^2 + 2x + 5) + (2x^2 - 3x + 2)$

$= (3x^2 + 2x + 5) + [2x^2 + (-3)x + 2]$ — Rewrite subtraction as addition of the additive inverse.

$= 3x^2 + 2x + 5 + 2x^2 + (-3)x + 2$ — Remove parentheses.

$= (3x^2 + 2x^2) + [2x + (-3)x] + (5 + 2)$ — Group according to like terms using Commutative and Associative Properties.

$= 5x^2 + (-1)x + 7$ — Combine like terms.

$= 5x^2 - x + 7$ — Simplify. ◀

When you are comfortable with this process, you may condense or omit some of these steps. The method used above to find the sum of two polynomials is sometimes called **horizontal addition.** Notice that in adding two polynomials, we simply add the coefficients of like terms.

EXAMPLE 2 Adding Polynomials

Find the sum of the polynomials

$$8x^3 - 2x^2 + 6x - 2 \quad \text{and} \quad 3x^4 - 2x^3 + x^2 + x$$

Solution We shall find the sum in two ways.

Horizontal Addition: As before, we rewrite subtraction as addition of the additive inverse, group like terms, and combine like terms.

$$\begin{aligned}
&(8x^3 - 2x^2 + 6x - 2) + (3x^4 - 2x^3 + x^2 + x)\\
&= [8x^3 + (-2)x^2 + 6x + (-2)] + [3x^4 + (-2)x^3 + x^2 + x]\\
&= 3x^4 + [8x^3 + (-2)x^3] + [(-2)x^2 + x^2] + (6x + x) + (-2)\\
&= 3x^4 + 6x^3 + (-1)x^2 + 7x + (-2)\\
&= 3x^4 + 6x^3 - x^2 + 7x - 2
\end{aligned}$$

Vertical Addition: The idea here is to vertically line up the like terms in each polynomial and then add the coefficients.

$$\begin{array}{rrrrrr}
 & x^4 & x^3 & x^2 & x^1 & x^0\\
 & & 8x^3 & -\ 2x^2 & +\ 6x & -\ 2\\
(+) & 3x^4 & -\ 2x^3 & +\ x^2 & +\ x & \\
\hline
 & 3x^4 & +\ 6x^3 & -\ x^2 & +\ 7x & -\ 2
\end{array}$$
◀

The next example is shown in the horizontal format with condensed steps. Try to follow along.

EXAMPLE 3 **Adding Polynomials**

Find the sum: $2(x^2 - x + 1) + 3(2x^2 + x - 6)$

Solution $2(x^2 - x + 1) + 3(2x^2 + x - 6)$

$= (2x^2 - 2x + 2) + (6x^2 + 3x - 18)$ Distributive property

$= (2x^2 + 6x^2) + (-2x + 3x) + (2 - 18)$ Group like terms.

$= 8x^2 + x - 16$ Combine like terms. ◀

Practice Exercise 1 *In Problems 1–4, find each sum. Use either the horizontal or the vertical method.*

1. $(2x^3 + 8x^2 + 10x + 5) + (3x^3 - 8x^2 + 2x - 1)$
2. $(3x^2 - 5) + (8x^3 - x^2 - 3)$
3. $(5x^2 - 2x + x^3 + 5) + (3x^3 - x^4 + 7)$
4. $3(2x^2 + 4) + 2(-x^2 + x + 1)$

Subtracting Polynomials

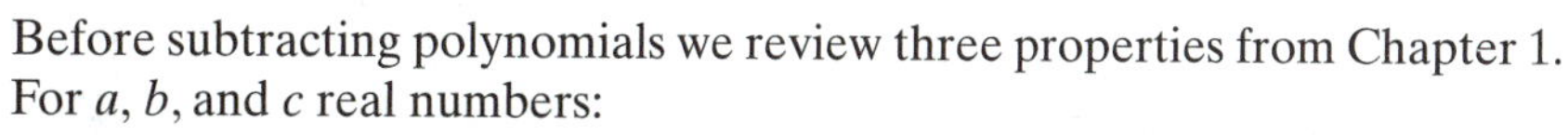

2 Before subtracting polynomials we review three properties from Chapter 1. For a, b, and c real numbers:

Multiplicative Identity $a \cdot 1 = 1 \cdot a = a$

Multiplying a real number (or any algebraic expression) by 1 yields the original expression.

Subtraction Property $a - b = a + (-b)$

Subtraction can always be rewritten as addition of the additive inverse.

Distributive Property $a(b + c) = ab + ac$

In addition to the distributive property as stated above we will use the **extended form of the distributive property**

$$a(b_1 + b_2 + \cdots + b_n) = a \cdot b_1 + a \cdot b_2 + \cdots + a \cdot b_n$$

where $a, b_1, b_2, \ldots, b_n$ are real numbers.

For simplicity, we will just say the *distributive property* when referring to either the distributive property or the extended form of the distributive property. No confusion should arise.

We use these properties to subtract polynomials.

EXAMPLE 4 **Subtracting Polynomials**

Find the difference: $(3x^2 + 2x + 5) - (2x^2 - 3x + 2)$

Solution $(3x^2 + 2x + 5) - (2x^2 - 3x + 2)$

$= (3x^2 + 2x + 5) - 1 \cdot (2x^2 - 3x + 2)$	Multiplicative Identity
$= (3x^2 + 2x + 5) + (-1)[2x^2 + (-3)x + 2]$	Change all subtraction to addition with the additive inverse.
$= (3x^2 + 2x + 5) + [(-1)(2x^2) + (-1)(-3)x + (-1)(2)]$	Distribute −1.
$= (3x^2 + 2x + 5) + [(-2)x^2 + 3x + (-2)]$	Perform multiplication.
$= 3x^2 + 2x + 5 + (-2)x^2 + 3x + (-2)$	Remove grouping symbols.
$= [3x^2 + (-2)x^2] + (2x + 3x) + [5 + (-2)]$	Group like terms.
$= x^2 + 5x + 3$	Combine like terms. ◀

The above example shows a lot of steps. Upon mastering the skill of subtraction, some of these can be omitted. A common question is "How did you know to multiply by 1?" The answer is "When I was learning algebra, someone showed me that step." You are not expected to know these things from birth. Learning algebra takes someone showing you how to do it and practice, lots of practice!

EXAMPLE 5 Subtracting Polynomials

Find the difference:

$$(7x^2 - 4x + 3) - (2x^3 - 5x^2 + 9x + 5)$$

Solution $(7x^2 - 4x + 3) - (2x^3 - 5x^2 + 9x + 5)$

$= (7x^2 - 4x + 3) - 1 \cdot (2x^3 - 5x^2 + 9x + 5)$	Multiplicative Identity
$= [7x^2 + (-4)x + 3] + (-1)[2x^3 + (-5)x^2 + 9x + 5]$	Change all subtraction to addition with the additive inverse.
$= 7x^2 + (-4)x + 3 + (-1)(2x^3) + (-1)(-5)x^2 + (-1)(9x) + (-1)(5)$	Distribute −1 and remove grouping symbols.
$= 7x^2 + (-4)x + 3 + (-2)x^3 + 5x^2 + (-9)x + (-5)$	Perform multiplication.
$= -2x^3 + (7x^2 + 5x^2) + [-4x + (-9)x] + [3 + (-5)]$	Group like terms.
$= -2x^3 + 12x^2 + (-13)x + (-2)$	Combine like terms.
$= -2x^3 + 12x^2 - 13x - 2$	Simplify. ◀

Notice the net result of distributing the −1 is that the sign of each term in the polynomial following the −1 is changed. We'll use this idea in the next example. Also, subtraction of polynomials, like addition, can be done horizontally or vertically. This too will be shown in the next example.

EXAMPLE 6 Subtracting Polynomials

Find the difference: $(3x^4 - 4x^3 + 6x^2 - 1) - (2x^4 - 8x^2 - 6x + 5)$

Solution *Horizontal Subtraction:*

$(3x^4 - 4x^3 + 6x^2 - 1) - (2x^4 - 8x^2 - 6x + 5)$

$= 3x^4 - 4x^3 + 6x^2 - 1 - 2x^4 + 8x^2 + 6x - 5$ Distribute the negative sign.

$= (3x^4 - 2x^4) + (-4)x^3 + (6x^2 + 8x^2) + 6x + (-1 - 5)$ Group like terms.

$= x^4 - 4x^3 + 14x^2 + 6x - 6$ Combine like terms.

Vertical Subtraction: Line up like terms in vertical columns.

$$\begin{array}{rrrrrr} & x^4 & x^3 & x^2 & x^1 & x^0 \\ & 3x^4 & -\ 4x^3 & +\ 6x^2 & & -\ 1 \\ (-) & 2x^4 & & -\ 8x^2 & -\ 6x & +\ 5 \\ \hline & x^4 & -\ 4x^3 & +\ 14x^2 & +\ 6x & -\ 6 \end{array}$$

◀

WARNING: You have to be extra careful when subtracting in vertical columns. For example, the third column in the above problem is $6x^2 - (-8x^2) = 6x^2 + 8x^2 = 14x^2$. It is easy to make a careless mistake by overlooking the subtraction sign on the left side of the problem. ■

The choice of which of these methods to use for adding and subtracting polynomials is left to you. To save space, we shall most often use the horizontal format.

EXAMPLE 7 **Subtracting Polynomials**

Find the difference: $3(-2x^2 + 6x + 1) - 2(4x^2 + x)$

Solution $3(-2x^2 + 6x + 1) - 2(4x^2 + x)$

$= 3(-2x^2 + 6x + 1) + (-2)(4x^2 + x)$ Rewrite subtraction as addition.

$= 3(-2x^2) + 3(6x) + 3(1) + (-2)(4x^2) + (-2)(x)$ Distribute 3 and −2.

$= -6x^2 + 18x + 3 + (-8)x^2 + (-2)x$ Multiply.

$= [-6x^2 + (-8)x^2] + [18x + (-2)x] + 3$ Group like terms.

$= -14x^2 + 16x + 3$ Combine like terms. ◀

Practice Exercise 2 *In Problems 1–3, find the difference:*

1. $(9x^3 - 8x^2 + x - 2) - (3x^3 + 5x^2 + x + 4)$
2. $(4x^2 + 5) - (3x^2 + x - 2)$
3. $-2(x^2 + 5x - 1) - 4(-2x^2 + 1)$

◀

Polynomials in Two or More Variables

Adding and subtracting polynomials in two or more variables is handled in the same way as polynomials in one variable.

EXAMPLE 8 **Subtracting Polynomials in Two Variables**

Find the difference: $(x^3 - 3x^2y + 3xy^2 - y^3) - (x^3 - x^2y + y^3)$

Solution $(x^3 - 3x^2y + 3xy^2 - y^3) - (x^3 - x^2y + y^3)$

$= x^3 - 3x^2y + 3xy^2 - y^3 - x^3 + x^2y - y^3$ Distribute the negative sign and remove parentheses.

$= x^3 + (-3x^2y) + 3xy^2 + (-y^3) + (-x^3) + x^2y + (-y^3)$ Change subtraction to addition.
$= [x^3 + (-x^3)] + (-3x^2y + x^2y) + 3xy^2 + [-y^3 + (-y^3)]$ Group like terms.
$= -2x^2y + 3xy^2 + (-2)y^3$ Combine like terms.
$= -2x^2y + 3xy^2 - 2y^3$ Simplify. ◀

Practice Exercise 3 *In Problems 1–2, perform the indicated operations.*

1. $(4x^3 + 2x^2y - 5xy^2 + y^3) + (x^3 - 3x^2y - 2xy^2 - y^3)$
2. $6(2x^2y - 3xy^2 + 1) - (5x^2y - 8xy^2)$ ◀

Application

In many businesses, the **cost** C of production can be expressed as a polynomial in which the variable x represents the number of items produced. Similarly, the **revenue** R obtained from sales can be expressed as a polynomial whose variable is the number of items sold. The **profit** P due to producing and selling x items is the difference $R - C$.

EXAMPLE 9 **Finding the Profit**

A company engaged in the manufacture of sunglasses has daily fixed costs from salaries and building operations of \$500. Each pair of sunglasses manufactured costs \$1 and is sold for \$2. If x pairs of sunglasses are manufactured and sold, the cost C and the revenue R are given by the formulas

$$C = \$500 + \$1 \cdot x = 500 + x \qquad R = \$2 \cdot x = 2x$$

Find the profit P.

Solution The profit P is the revenue R less the cost C. That is,

$$P = R - C = 2x - (500 + x) = 2x - 500 - x = x - 500$$ ◀

Answers to Practice Exercises

Practice Exercise 1 **1.** $5x^3 + 12x + 4$ **2.** $8x^3 + 2x^2 - 8$ **3.** $-x^4 + 4x^3 + 5x^2 - 2x + 12$ **4.** $4x^2 + 2x + 14$

Practice Exercise 2 **1.** $6x^3 - 13x^2 - 6$ **2.** $x^2 - x + 7$ **3.** $6x^2 - 10x - 2$

Practice Exercise 3 **1.** $5x^3 - x^2y - 7xy^2$ **2.** $7x^2y - 10xy^2 + 6$

2.3 Assess Your Understanding

'Are You Prepared?' *Answers are given at the end of these exercises. If you get a wrong answer, read the pages listed in parentheses.*

1. We can reformulate 4–17 as an addition problem by writing ______________. (pp. 39–40)

2. The multiplicative identity is ______________. (p. 29)

3. Use the Distributive Property to remove parentheses in the expression: $-2(3x - 9)$ (p. 28)

4. Combine like terms: $7x^2 - 2x^2 + 3x$ (pp. 71–72)

Concepts and Vocabulary

5. Polynomials are added and subtracted by combining ______________ ______________.

6. *True or False:* $(8x^2 - 4x + 1) - (5x^2 + 2x - 6) = 8x^2 - 4x + 1 - 5x^2 + 2x - 6$

Exercises

In Problems 7–54, perform the indicated operations. Express your answer as a polynomial in standard form.

7. $(3x + 2) + (5x - 3)$

8. $(5x - 3) + (2x + 6)$

9. $(5x^2 - x + 4) + (x^2 + x + 1)$

10. $(3x^2 + x - 10) + (2x^2 - 2x)$

11. $(8x^3 + x^2 - x) + (3x^2 + 2x + 1)$

12. $(-4x^4 - x^2 + 2) + (5x^3 - 2x^2 + x + 1)$

13. $(2x + 3y + 6) + (x - y + 4)$

14. $(4x - y + 1) + (3x + 2y + 5)$

15. $3(x^2 + 3x + 4) + 2(x - 3)$

16. $4(3x^2 - 4) + 2(x^2 + x - 2)$

17. $5(2x^3 - x^2 + 3x + 4) + 3(x^2 - 4x + 1)$

18. $4(3x^2 - 4x + 8) + 2(x^2 - 5)$

19. $(x^2 + x + 5) - (3x - 2)$

20. $(x^3 - 3x^2 + 2) - (x^2 - x + 1)$

21. $(x^3 - 3x^2 + 5x + 10) - (2x^2 - x + 1)$

22. $(x^2 - 3x - 1) - (x^3 - 2x^2 + x + 5)$

23. $(6x^5 + x^3 + x) - (5x^4 - x^3 + 3x^2)$

24. $(10x^5 - 8x^2) - (3x^3 - 2x^2 + 6)$

25. $(2x^3 + x^2 + 3x + 4) - (x^2 - 4x + 1)$

26. $(6x^2 + 7x + 4) - (8x^3 - 4x^2 + 12x + 5)$

27. $3(x^2 - 3x + 1) - 2(3x^2 + x - 4)$

28. $-2(x^2 + x + 1) - 6(-5x^2 - x + 2)$

29. $6(x^3 + x^2 - 3) - 4(2x^3 - 3x^2)$

30. $8(4x^3 - 3x^2 - 1) - 6(4x^3 + 8x - 2)$

31. $(x^2 - x + 2) + (2x^2 - 3x + 5) - (x^2 + 1)$

32. $(x^2 + 1) - (4x^2 + 5) + (x^2 + x - 2)$

33. $9(y^2 - 2y + 1) - 6(1 - y^2)$

34. $8(1 - y^3) - 2(1 + y + y^2 + y^3)$

35. $4(x^3 - x^2 + 1) - 3(x^2 + 1) + 2(x^3 + x + 2)$

36. $3(x^5 + x^3 - 1) + 4(1 - x^2) - 3(1 - x^3)$

37. $(2x^2 + 3y^2 - xy) + (x^2 + 2y^2 - 3xy)$

38. $(5x^2 + xy + 7y^2) - (1 - x^2 - y^2)$

39. $(4x^2 - 3xy + 2y^2) - (x^2 - y^2 + 4xy)$

40. $(3x^2 + 2xy - y^2) - (x^2 - xy - y^2)$

41. $(4x^2y + 5xy^2 + xy) - (3x^2y - 4xy^2 - xy)$

42. $(-3xy^2 + 4xy - 7y^2) - (3xy^2 + 8xy - 8y^2)$

43. $3(x^2 + 2xy - 4y^2) + 4(3x^2 - 2xy + 4y^2)$

44. $-2(x^2 - xy + y^2) + 5(2x^2 - xy + 2y^2)$

45. $-3(2x^2 - xy - 2y^2) - 5(-x^2 - xy + y^2)$

46. $2(-x^2 + xy + 2y^2) - 4(x^2 + y^2)$

47. $(5xy + 3yz - 4xz) + (4xy - 2yz - xz)$

48. $(-3xy + 2yz + 2xz) + (8xy + 2yz - xz)$

49. $(4xy - 2yz - xz) - (xy - yz - xz)$

50. $(xy + yz - xz) - (2xy + yz - 3xz)$

51. $3(2xy - yz + 3xz) + 4(-xy + 2yz - 3xz)$

52. $-2(xy - 3yz - 3xz) + 3(-2xy - yz + xz)$

53. $4(xy - 2yz + 3xz) - 2(-2xy + yz - 3xz)$

54. $-3(-xy - 2yz + 3xz) - 4(xy - yz + 2xz)$

In Problems 55–60, follow the instructions.

55. Find the sum of $x^3 + 3x^2 - x - 2$ and $-4x^3 - x^2 + x + 4$.

56. Find the sum of $2x^3 - x^2 + x - 4$ and $-3x^3 - x^2 - x - 2$.

57. Find $x^3 + 3x^2 - x - 2$ less $-4x^3 - x^2 + x + 4$.

58. Find $2x^3 - x^2 + x - 4$ less $-3x^3 - x^2 - x - 2$.

59. Find the sum of twice $4x^3 + x^2 - x - 2$ and three times $x^3 - 2x^2 + x - 2$.

60. Find twice $4x^3 + x^2 - x - 2$ less three times $x^3 - 2x^2 + x - 2$.

61. If $R = 10x$ and $C = 100 + 5x$, find $R - C$.

62. If $R = 6x$ and $C = 56 + 2x$, find $R - C$.

63. A producer sells items for \$3 each. If x items are produced and sold, the revenue R and cost C are given respectively by $R = 3x$ and $C = 2x + 100$. Find the profit P.

64. If the cost C to produce x items is $C = x^2 + 100x + 500$ and the revenue R due to selling x items is $R = 300x$, find the profit P.

65. If asked to add two polynomials, would you prefer to work the problem horizontally or vertically? Explain the reason for your choice.

'Are You Prepared?' Answers

1. $4 + (-17)$ **2.** 1 **3.** $-6x + 18$ **4.** $5x^2 + 3x$

2.4 Multiplication of Polynomials

PREPARING FOR THIS SECTION *Before getting started, review the following:*

- Laws of Exponents (pp. 62–64)
- Distributive Property (p. 28)
- Principle of Substitution (p. 4)

Now work the 'Are You Prepared?' problems on page 92.

OBJECTIVES

1 Multiply Monomials
2 Multiply a Monomial and a Polynomial
3 Multiply Polynomials
4 Multiply Two Binomials with the FOIL Method
5 Find Special Products

1 The product of two monomials ax^n and bx^m is obtained using the Laws of Exponents and the Commutative and Associative Properties. For example,

$$(2x^3)(5x^4) = (2 \cdot 5)(x^3 \cdot x^4) = 10x^{3+4} = 10x^7$$

If a, b, and x are real numbers and if m, n are positive integers, then

$$ax^n \cdot bx^m = a \cdot b \cdot x^n \cdot x^m = abx^{n+m}$$

EXAMPLE 1 **Multiplying Monomials**

Find the product of each of the following monomials.

(a) $2x^2$ and $6x^3$ (b) $-7x^4$ and $3x$

Solution (a) $(2x^2)(6x^3) = (2 \cdot 6) \cdot (x^2 \cdot x^3)$ Commutative and Associative Properties

$= 12x^{2+3}$ Multiply and use Laws of Exponents

$= 12x^5$

(b) $(-7x^4)(3x) = (-7 \cdot 3) \cdot (x^4 \cdot x)$ Commutative and Associative Properties

$= -21x^{4+1}$ Multiply and use Laws of Exponents

$= -21x^5$ ◀

2 For the product of a monomial and a polynomial, the Distributive Property is used.

EXAMPLE 2 **Multiplying a Monomial and a Polynomial**

Find the product of

(a) $2x^2$ and $6x^3 + 3$ (b) $4x^3$ and $2x^2 + x - 5$

Solution (a) $2x^2 \cdot (6x^3 + 3) = (2x^2 \cdot 6x^3) + (2x^2 \cdot 3)$ Distributive Property

$= 12x^5 + 6x^2$ Multiply and use Laws of Exponents

(b) $4x^3 \cdot (2x^2 + x - 5) = (4x^3 \cdot 2x^2) + (4x^3 \cdot x) - (4x^3 \cdot 5)$ Distributive Property

$= 8x^5 + 4x^4 - 20x^3$ Multiply and use Laws of Exponents ◀

Practice Exercise 1 *In Problems 1–3, find the product*

1. of $5x^3$ and $3x$ **2.** of $3x$ and $4x^3 + 2x^2 - 3$
3. of x^2 and $5x^2 - x$

3 Products of polynomials are found by repeated use of the Distributive Property and the Laws of Exponents. Again, you have a choice of horizontal or vertical format.

EXAMPLE 3 Multiply Polynomials

Find the product: $(2x + 5)(x^2 - x + 2)$

Solution *Horizontal Multiplication:* Here we think of one of the polynomials as if it were a single term, for example, let's say that $x^2 - x + 2 = P$. Then

$(2x + 5)(x^2 - x + 2) = (2x + 5) \cdot P$	Substitution Property
$= 2x \cdot P + 5 \cdot P$	Distributive Property
$= 2x \cdot (x^2 - x + 2) + 5 \cdot (x^2 - x + 2)$	Substitution Property, resubstitute $x^2 - x + 2$ for P.
$= (2x \cdot x^2 - 2x \cdot x + 2x \cdot 2) + (5 \cdot x^2 - 5 \cdot x + 5 \cdot 2)$	Distributive Property
$= 2x^3 - 2x^2 + 4x + 5x^2 - 5x + 10$	Multiply and use Laws of Exponents
$= 2x^3 + 3x^2 - x + 10$	Combine like terms.

The substitution step of representing one of the polynomials with a single letter is used in this example for clarity and can be omitted once you have practiced doing this type of problem.

Vertical Multiplication: The idea here is very much like multiplying a three-digit number by a two-digit number.

$$\begin{array}{rrrrl} & x^2 & - x & + 2 & \\ & & 2x & + 5 & \\ \hline 2x^3 & - 2x^2 & + 4x & & \text{This line is } 2x(x^2 - x + 2). \\ (+) & 5x^2 & - 5x & + 10 & \text{This line is } 5(x^2 - x + 2)\text{. Line up like terms as shown.} \\ \hline 2x^3 & + 3x^2 & - x & + 10 & \text{The sum of the above two lines.} \end{array}$$

As this example illustrates, when multiplying two polynomials, each term of the first factor is multiplied by each term of the second factor.

Practice Exercise 2 *In Problems 1–3, find each product.*

1. $(3x + 5)(x^2 - x + 2)$ **2.** $(3x - 4)(2x^2 + x - 3)$
3. $(x^2 + x + 1)(x^2 - x + 1)$

4 The next example illustrates the special case of multiplying a binomial by a binomial.

EXAMPLE 4 Multiplying Two Binomials

Find the product: $(3x - 8)(x + 2)$

Solution

$$\begin{aligned}(3x - 8)(x + 2) &= 3x \cdot (x + 2) - 8 \cdot (x + 2)\\ &= (3x \cdot x) + (3x \cdot 2) - (8 \cdot x) - (8 \cdot 2)\\ &= 3x^2 + 6x - 8x - 16\\ &= 3x^2 - 2x - 16\end{aligned}$$

◀

Notice in Example 4 that when multiplying a binomial times a binomial, the result consists of four multiplications: the products using the

- First terms in each binomial
- Outer terms in each binomial
- Inner terms in each binomial
- Last terms in each binomial

First, Outer, Inner, Last is often abbreviated as FOIL and is called the FOIL method for multiplying two binomials.

EXAMPLE 5 Using FOIL

Find each of the following products by using the FOIL method.

(a) $(3x - 2)(x + 5)$ (b) $(2x + 3)(5x + 1)$
(c) $(x + a)(x + b)$ (d) $(ax + b)(cx + d)$

Solution

(a) Outer, First, Inner, Last

$$\begin{aligned}(3x - 2)(x + 5) &= 3x(x + 5) - 2(x + 5)\\ &\quad\ \text{F} \qquad\ \text{O} \qquad\ \text{I} \qquad \text{L}\\ &= 3x \cdot x + 3x \cdot 5 - 2 \cdot x - 2 \cdot 5\\ &= 3x^2 + 15x - 2x - 10\\ &= 3x^2 + 13x - 10\end{aligned}$$

(b) Outer, First, Inner, Last

$$\begin{aligned}(2x + 3)(5x + 1) &\overset{\text{F} \quad \text{O} \quad \text{I} \quad \text{L}}{=} 2x \cdot 5x + 2x \cdot 1 + 3 \cdot 5x + 3 \cdot 1\\ &= 10x^2 + 2x + 15x + 3\\ &= 10x^2 + 17x + 3\end{aligned}$$

(c) Outer, First, Inner, Last

$$\begin{aligned}(x + a)(x + b) &\overset{\text{F} \quad \text{O} \quad \text{I} \quad \text{L}}{=} x \cdot x + b \cdot x + a \cdot x + ab\\ &= x^2 + bx + ax + ab\\ &= x^2 + (b + a)x + ab\\ &= x^2 + (a + b)x + ab\end{aligned}$$

(d) Outer, First, Inner, Last

$$\begin{aligned}(ax + b)(cx + d) &\overset{\text{F} \quad \text{O} \quad \text{I} \quad \text{L}}{=} (ax)(cx) + (ax)(d) + (cx)(b) + (b)(d)\\ &= acx^2 + adx + bcx + bd\\ &= acx^2 + (ad + bc)x + bd\end{aligned}$$

◀

Examples 5(c) and 5(d) show general results for multiplying any two binomials of the forms $(x + a)(x + b)$ or $(ax + b)(cx + d)$. These products are called **second-degree trinomials** because of the form of the answers. In general:

$$(x + a)(x + b) = x^2 + (a + b)x + ab \qquad \textbf{(1a)}$$

$$(ax + b)(cx + d) = acx^2 + (ad + bc)x + bd \qquad \textbf{(1b)}$$

Practice Exercise 3 *In Problems 1–4, find each product using the FOIL method.*

1. $(5x + 2)(3x + 1)$ **2.** $(x - 5)(x - 7)$

3. $(x - 3)(x + 2)$ **4.** $(2x - 3)(7x + 4)$

Special Products

5 Certain products, which we call **special products,** occur frequently in algebra. We can calculate them easily by performing the indicated multiplication and observing the results. Let's look at some examples.

EXAMPLE 6 Forming the Difference of Two Squares

Find each of the following products.

(a) $(x - 2)(x + 2)$ (b) $(x + 3)(x - 3)$ (c) $(x - a)(x + a)$

Solution

(a)
$$\begin{aligned}(x - 2)(x + 2) &= \overset{F}{x \cdot x} + \overset{O}{2 \cdot x} - \overset{I}{2 \cdot x} - \overset{L}{2 \cdot 2} \\ &= x^2 + 2x - 2x - 4 \\ &= x^2 - 4\end{aligned}$$

(b)
$$\begin{aligned}(x + 3)(x - 3) &= \overset{F}{x \cdot x} - \overset{O}{3 \cdot x} + \overset{I}{3 \cdot x} - \overset{L}{3 \cdot 3} \\ &= x^2 - 3x + 3x - 9 \\ &= x^2 - 9\end{aligned}$$

(c)
$$\begin{aligned}(x - a)(x + a) &= \overset{F}{x \cdot x} + \overset{O}{a \cdot x} - \overset{I}{a \cdot x} - \overset{L}{a \cdot a} \\ &= x^2 + ax - ax - a^2 \\ &= x^2 - a^2\end{aligned}$$

The general result given in Example 6(c) is called the **difference of two squares** because of the form of the answer.

Difference of Two Squares

$$(x - a)(x + a) = x^2 - a^2 \qquad \textbf{(2)}$$

Since multiplication is commutative,

$$(x - a)(x + a) = (x + a)(x - a) = x^2 - a^2$$

EXAMPLE 7 **Forming the Difference of Two Squares**

Find the following products by applying the formula for the difference of two squares:

(a) $(x - 9)(x + 9)$ (b) $(4x - 5)(4x + 5)$

Solution (a) $(x - 9)(x + 9) = x^2 - 9^2 = x^2 - 81$

(b) $(4x - 5)(4x + 5) = (4x)^2 - 5^2 = 16x^2 - 25$ Use $4x$ in place of x in equation (2) ◀

Practice Exercise 4 *In Problems 1–3, use equation (2) to find each product.*

1. $(x - 5)(x + 5)$ **2.** $(x + 4)(x - 4)$ **3.** $(2x + 1)(2x - 1)$ ◀

The next special product is the *square of a binomial*. Let's look at a few examples first.

EXAMPLE 8 **Squaring a Binomial**

Find each of the following products.

(a) $(x + 2)^2$ (b) $(x - 3)^2$ (c) $(x + a)^2$

Solution

(a)
$$\begin{aligned} (x + 2)^2 &= (x + 2)(x + 2) \\ &= x \cdot x + 2 \cdot x + 2 \cdot x + 2 \cdot 2 && \text{FOIL Method} \\ &= x^2 + 2x + 2x + 4 \\ &= x^2 + 4x + 4 \end{aligned}$$

(b)
$$\begin{aligned} (x - 3)^2 &= (x - 3)(x - 3) \\ &= x \cdot x - 3 \cdot x - 3 \cdot x + 3 \cdot 3 && \text{FOIL Method} \\ &= x^2 - 3x - 3x + 9 \\ &= x^2 - 6x + 9 \end{aligned}$$

(c)
$$\begin{aligned} (x + a)^2 &= (x + a)(x + a) \\ &= x \cdot x + a \cdot x + a \cdot x + a \cdot a && \text{FOIL Method} \\ &= x^2 + ax + ax + a^2 \\ &= x^2 + 2ax + a^2 \end{aligned}$$
◀

The general result given in Example 8(c) is called the **square of a binomial,** or **perfect square.** There are two forms of this formula:

Squares of Binomials (Perfect Squares)

$$(x + a)^2 = x^2 + 2ax + a^2 \qquad \textbf{(3a)}$$
$$(x - a)^2 = x^2 - 2ax + a^2 \qquad \textbf{(3b)}$$

We ask you to derive the second form (3b) in Problem 101 at the end of this section.

Warning: When using these formulas, do not forget the middle term, which is twice the product of the terms in the binomial. Do not make a careless error by writing $(x + a)^2$ as $x^2 + a^2$. To help you remember the middle term, write $(x + a)^2 = (x + a)(x + a)$ as the first step. ■

EXAMPLE 9 Squaring a Binomial

Find each of the following products by applying the formula for the square of a binomial.

(a) $(x + 4)^2$ (b) $(2x - 3)^2$

Solution (a)
$$(x + 4)^2 = (x + 4)(x + 4) = x^2 + 2\cdot(4\cdot x) + 4^2 = x^2 + 8x + 16$$

(b)
$$(2x - 3)^2 = (2x - 3)(2x - 3) = (2x)^2 - 2\cdot(2x\cdot 3) + 3^2 = 4x^2 - 12x + 9$$

Use $2x$ in the place of x in equation (3b). ◀

Practice Exercise 5 *In Problems 1–3, use equations (3a) or (3b) to find each product.*

1. $(x + 6)^2$ **2.** $(x - 5)^2$ **3.** $(2x + 3)^2$ ◀

Our next special product results in the *difference of two cubes*. Again, let's look at some examples. Notice that in these examples we are not multiplying two binomials together, but rather we are multiplying a binomial with a trinomial. As a result, we cannot use the FOIL method, but must multiply every term of one polynomial by every term of the other polynomial.

EXAMPLE 10 Forming the Sum/Difference of Two Cubes

Find each of the following products.

(a) $(x - 2)(x^2 + 2x + 4)$ (b) $(x - a)(x^2 + ax + a^2)$

(c) $(x + 1)(x^2 - x + 1)$

Solution (a)
$$(x - 2)(x^2 + 2x + 4) = x(x^2 + 2x + 4) - 2(x^2 + 2x + 4) = x^3 + 2x^2 + 4x - 2x^2 - 4x - 8 = x^3 - 8$$

(b)
$$(x - a)(x^2 + ax + a^2) = x(x^2 + ax + a^2) - a(x^2 + ax + a^2) = x^3 + ax^2 + a^2x - ax^2 - a^2x - a^3 = x^3 - a^3$$

(c)
$$(x + 1)(x^2 - x + 1) = x(x^2 - x + 1) + 1\cdot(x^2 - x + 1) = x^3 - x^2 + x + x^2 - x + 1 = x^3 + 1$$ ◀

The products in Examples 10(a) and (b) are called the **difference of two cubes** and the product in Example 10(c) is called the **sum of two cubes,** because of the form of the answers. In general:

Difference of Two Cubes; Sum of Two Cubes

$$(x - a)(x^2 + ax + a^2) = x^3 - a^3 \quad \textbf{(4a)}$$
$$(x + a)(x^2 - ax + a^2) = x^3 + a^3 \quad \textbf{(4b)}$$

The general result (4a) was derived in Example 10(b). We ask you to derive the sum of two cubes formula (4b) in Problem 102 at the end of this section.

Formulas (4a) and (4b) usually are not used to find products. Their main use will be seen in Section 2.8.

Practice Exercise 6 *In Problems 1–4, find each product.*

1. $(x - 5)(x^2 + 5x + 25)$ **2.** $(x + 4)(x^2 - 4x + 16)$
3. $(x + 5)(x^2 - 5x + 25)$ **4.** $(x - 4)(x^2 + 4x + 16)$

Our last special product is the *cube of a binomial*. We show it first with some examples.

EXAMPLE 11 Cubing a Binomial

Find each of the following products.

(a) $(x + 2)^3$ (b) $(x + a)^3$ (c) $(x - 4)^3$

Solution (a)
$$\begin{aligned}(x + 2)^3 &= (x + 2)(x + 2)(x + 2)\\ &= (x + 2)[(x + 2)(x + 2)] && \text{Associative Property}\\ &= (x + 2)[x^2 + 2\cdot(2\cdot x) + 2^2] && \text{Formula for square of a binomial (3a)}\\ &= (x + 2)(x^2 + 4x + 4) && \text{Simplify.}\\ &= x(x^2 + 4x + 4) + 2(x^2 + 4x + 4) && \text{Distributive Property}\\ &= x^3 + 4x^2 + 4x + 2x^2 + 8x + 8 && \text{Distributive Property}\\ &= x^3 + 6x^2 + 12x + 8 && \text{Combine like terms.}\end{aligned}$$

(b)
$$\begin{aligned}(x + a)^3 &= (x + a)(x + a)(x + a)\\ &= (x + a)[(x + a)(x + a)] && \text{Associative Property}\\ &= (x + a)(x^2 + 2ax + a^2) && \text{Formula for square of a binomial (3a)}\\ &= x(x^2 + 2ax + a^2) + a(x^2 + 2ax + a^2) && \text{Distributive Property}\\ &= x^3 + 2ax^2 + a^2x + ax^2 + 2a^2x + a^3 && \text{Distributive Property}\\ &= x^3 + 3ax^2 + 3a^2x + a^3 && \text{Combine like terms.}\end{aligned}$$

(c)
$$\begin{aligned}(x - 4)^3 &= (x - 4)(x - 4)(x - 4)\\ &= (x - 4)[(x - 4)(x - 4)] && \text{Associative Property}\\ &= (x - 4)[x^2 - 2(4\cdot x) + 4^2] && \text{Formula for square of a binomial (3b)}\\ &= (x - 4)(x^2 - 8x + 16) && \text{Simplify.}\\ &= x(x^2 - 8x + 16) - 4(x^2 - 8x + 16) && \text{Distributive Property}\\ &= x^3 - 8x^2 + 16x - 4x^2 + 32x - 64 && \text{Distributive Property}\\ &= x^3 - 12x^2 + 48x - 64 && \text{Combine like terms.}\end{aligned}$$

These products are called the **cubes of binomials,** or **perfect cubes.** There are two forms of the formula:

Cubes of Binomials (Perfect Cubes)

$$(x + a)^3 = x^3 + 3ax^2 + 3a^2x + a^3 \qquad \textbf{(5a)}$$
$$(x - a)^3 = x^3 - 3ax^2 + 3a^2x - a^3 \qquad \textbf{(5b)}$$

The general result (5a) was derived in Example 11(b). We ask you to derive (5b) in Problem 103 at the end of this section.

EXAMPLE 12 **Cubing a Binomial**

Find each of the following products by applying the formula for the cube of a binomial.

(a) $(x - 2)^3$ (b) $(2x + 5)^3$ (c) $(3x - 4)^3$

Solution (a) Use equation (5b).

$$\begin{aligned}(x - 2)^3 &= x^3 - 3 \cdot 2 \cdot x^2 + 3 \cdot 2^2 \cdot x - 2^3 \\ &= x^3 - 6x^2 + 12x - 8\end{aligned}$$

(b) Use equation (5a) with $2x$ in the place of x.

$$\begin{aligned}(2x + 5)^3 &= (2x)^3 + 3 \cdot 5 \cdot (2x)^2 + 3 \cdot 5^2 \cdot (2x) + 5^3 \\ &= 8x^3 + (3 \cdot 5 \cdot 4x^2) + (3 \cdot 25 \cdot 2x) + 125 \\ &= 8x^3 + 60x^2 + 150x + 125\end{aligned}$$

(c) Use equation (5b) with $3x$ in the place of x.

$$\begin{aligned}(3x - 4)^3 &= (3x)^3 - 3 \cdot 4 \cdot (3x)^2 + 3 \cdot 4^2 \cdot (3x) - 4^3 \\ &= 27x^3 - 3 \cdot 4 \cdot 9x^2 + 3 \cdot 16 \cdot 3x - 64 \\ &= 27x^3 - 108x^2 + 144x - 64\end{aligned}$$

◀

Practice Exercise 7 *In Problems 1–3, use equations (5a) or (5b) to find each product.*

1. $(x + 4)^3$ **2.** $(x - 5)^3$ **3.** $(2x - 3)^3$

Polynomials in Two or More Variables

Multiplying polynomials in two or more variables is performed in the same way as for polynomials in one variable.

EXAMPLE 13 **Multiplying Polynomials in Two Variables**

Find each of the following products.

(a) $(xy - 2)(2x^2 - xy + y^2)$ (b) $(x + 2y)(x - 2y)$

Solution (a)
$$\begin{aligned}(xy - 2)(2x^2 - xy + y^2) &= xy(2x^2 - xy + y^2) - 2(2x^2 - xy + y^2) \\ &= 2x^3y - x^2y^2 + xy^3 - 4x^2 + 2xy - 2y^2\end{aligned}$$

(b) $(x + 2y)(x - 2y) = x^2 - (2y)^2 = x^2 - 4y^2$ ◀

Practice Exercise 8 *Find each of the following products.*

1. $(x - y)(x^2 + y^2)$ **2.** $(ax + by)(ax - by)$

3. $(x + 2y)^2$ **4.** $(2x + y)(x + 3y)$

Summary

We conclude this section with a list of **formulas** and **special products.** In the list, a, b, c, and d are real numbers.

Second-Degree Trinomials

$$(x + a)(x + b) = x^2 + (a + b)x + ab \quad (1a)$$

$$(ax + b)(cx + d) = acx^2 + (ad + bc)x + bd \quad (1b)$$

Difference of Two Squares

$$(x - a)(x + a) = x^2 - a^2 \quad (2)$$

Squares of Binomials, or Perfect Squares

$$(x + a)^2 = x^2 + 2ax + a^2 \quad (3a)$$

$$(x - a)^2 = x^2 - 2ax + a^2 \quad (3b)$$

Difference of Two Cubes; Sum of Two Cubes

$$(x - a)(x^2 + ax + a^2) = x^3 - a^3 \quad (4a)$$

$$(x + a)(x^2 - ax + a^2) = x^3 + a^3 \quad (4b)$$

Cubes of Binomials, or Perfect Cubes

$$(x + a)^3 = x^3 + 3ax^2 + 3a^2x + a^3 \quad (5a)$$

$$(x - a)^3 = x^3 - 3ax^2 + 3a^2x - a^3 \quad (5b)$$

The formulas in equations (1)–(5) are used often and should be committed to memory. But if you forget one or are unsure of its form, you can always work a problem by multiplying each term of the first polynomial by each term of the second polynomial, or, in the case of multiplying two binominals, use the FOIL method.

Answers to Practice Exercises

Practice Exercise 1 **1.** $15x^4$ **2.** $12x^4 + 6x^3 - 9x$ **3.** $5x^4 - x^3$

Practice Exercise 2 **1.** $3x^3 + 2x^2 + x + 10$ **2.** $6x^3 - 5x^2 - 13x + 12$ **3.** $x^4 + x^2 + 1$

Practice Exercise 3 **1.** $15x^2 + 11x + 2$ **2.** $x^2 - 12x + 35$ **3.** $x^2 - x - 6$ **4.** $14x^2 - 13x - 12$

Practice Exercise 4 **1.** $x^2 - 25$ **2.** $x^2 - 16$ **3.** $4x^2 - 1$

Practice Exercise 5 **1.** $x^2 + 12x + 36$ **2.** $x^2 - 10x + 25$ **3.** $4x^2 + 12x + 9$

Practice Exercise 6 **1.** $x^3 - 125$ **2.** $x^3 + 64$ **3.** $x^3 + 125$ **4.** $x^3 - 64$

Practice Exercise 7 **1.** $x^3 + 12x^2 + 48x + 64$ **2.** $x^3 - 15x^2 + 75x - 125$ **3.** $8x^3 - 36x^2 + 54x - 27$

Practice Exercise 8 **1.** $x^3 - x^2y + xy^2 - y^3$ **2.** $a^2x^2 - b^2y^2$ **3.** $x^2 + 4xy + 4y^2$ **4.** $2x^2 + 7xy + 3y^2$

2.4 Assess Your Understanding

'Are You Prepared?' *Answers are given at the end of these exercises. If you get a wrong answer, read the pages listed in parentheses.*

1. Simplify: $x^4 \cdot x^8$ (pp. 62–64)

2. Use the Distributive Property to remove parentheses in the expression: $3x(x + 5)$ (p. 28)

3. If $P = 2x - 7$, then by the Principle of Substitution, $P(4x + 1) =$ __________. (p. 4)

Concepts and Vocabulary

4. *True or False:* The degree of the product of two polynomials equals the sum of their degrees.

5. *True or False:* $(a + b)^2 = a^2 + b^2$

6. To multiply a monomial and a polynomial we use the __________ Property.

7. FOIL is an acronym for __________, __________, __________, __________.

8. When multiplying two monomials in one variable we __________ the exponents.

Exercises

In Problems 9–24, find each product.

9. $x^2 \cdot x^3$
10. $x^4 \cdot x^3$
11. $3x^2 \cdot 4x^4$
12. $5x^2 \cdot 6x^4$
13. $-7x \cdot 8x^2$
14. $8x^3 \cdot (-6x)$
15. $3x(2x^3)$
16. $(2xy^2)(-7x^3y)$
17. $(-3x^2y^4)(-2xy^2)$
18. $(-5x^3y)(3xy^4)$
19. $(3xy)(4xy^2)$
20. $-2x(5x - 1)$
21. $(1 - 2x) \cdot 3x^2$
22. $(4 - 3x)(2x^2)$
23. $3xy(2x^2 - xy^2)$
24. $2x^2y^3(4xy^2 + 6x^3y)$

In Problems 25–82, find each product. Write your answer as a polynomial in standard form.

25. $x(x^2 + x - 4)$
26. $4x^2(x^3 - x + 2)$
27. $(x + 4)(x^2 - 2x + 3)$
28. $(x - 2)(x^2 + 5x - 2)$
29. $(3x - 2)(x^2 - 2x + 3)$
30. $(4x + 1)(x^2 + 5x - 2)$
31. $(-2x + 3)(-x^2 - x - 1)$
32. $(3 - 2x)(1 - x^3)$
33. $(x + 4)(x - 3)$
34. $(x + 6)(x - 2)$
35. $(x + 8)(2x + 1)$
36. $(2x - 1)(x + 2)$
37. $(-3x + 1)(x + 4)$
38. $(-2x - 1)(x + 1)$
39. $(1 - 4x)(2 - 3x)$
40. $(2 - x)(2 - 3x)$
41. $(2x + 3)(x^2 - 2)$
42. $(3x - 4)(x^2 + 1)$
43. $(3x - 5)(2x + 3)$
44. $(2x + 3)(-5x + 1)$
45. $(x - 7)(x + 7)$
46. $(x - 1)(x + 1)$
47. $(2x + 3)(2x - 3)$
48. $(3x + 2)(3x - 2)$
49. $(3x + 4)(3x - 4)$
50. $(5x - 3)(5x + 3)$
51. $(x + 4)^2$
52. $(x + 5)^2$
53. $(x - 4)^2$
54. $(x - 5)^2$
55. $(2x - 3)^2$
56. $(3x - 4)^2$
57. $(1 + x)^2 - (1 - x)^2$
58. $(1 + x)^2 + (1 - x)^2$
59. $(x + a)^2 - x^2$
60. $(x - a)^2 - x^2$
61. $(x - 1)(x^2 + x + 1)$
62. $(x + 1)(x^2 - x + 1)$
63. $(x + 3)(x^2 - 3x + 9)$
64. $(x - 3)(x^2 + 3x + 9)$
65. $(x + 5)^3$
66. $(x - 5)^3$
67. $(x + a)^3 - x^3$
68. $(x - a)^3 - x^3$
69. $(2x + 5)(x^2 + x + 1)$
70. $(3x - 4)(x^2 - x + 1)$
71. $(3x - 1)(2x^2 - 3x + 2)$
72. $(2x + 3)(3x^2 - 2x + 4)$
73. $(x^2 + x - 1)(x^2 - x + 1)$
74. $(x^2 + 2x + 1)(x^2 - 3x + 4)$
75. $(x + 2)^2(x - 1)$
76. $(x - 3)^2(x + 2)$
77. $(x - 1)^2(2x + 3)$
78. $(x + 5)^2(3x - 1)$
79. $(x - 1)(x - 2)(x - 3)$
80. $(x + 1)(x - 2)(x + 4)$
81. $(x + 1)^3 - (x - 1)^3$
82. $(x + 1)^3 - (x + 2)^3$

In Problems 83–100, perform the indicated operations. Express your answer as a polynomial.

83. $(x + y)(x - 2y)$
84. $(x - 2y)(x - y)$
85. $(2x - y)(3x + 4y)$
86. $(2x + 3y)(x + 2y)$
87. $(x - 2y)(x^2 + 2xy + 4y^2)$
88. $(x + 2y)(x^2 - 2xy + 4y^2)$
89. $(x - y)^2 - (x + y)^2$
90. $(x - y)^2 - (x^2 + y^2)$
91. $(x + 2y)^2 + (x - 3y)^2$
92. $(2x - y)^2 - (x - y)^2$
93. $[(x + y)^2 + z^2] + [x^2 + (y + z)^2]$
94. $[(x - y)^2 + z^2] + [x^2 + (y - z)^2]$
95. $(x - y)(x^2 + xy + y^2)$
96. $(x + y)(x^2 - xy + y^2)$
97. $(2x^2 + xy + y^2)(3x^2 - xy + 2y^2)$
98. $(3x^2 - xy + y^2)(2x^2 + xy - 4y^2)$

99. $(x + y + z)(x - y - z)$

100. $(x + y - z)(x - y + z)$

101. Derive equation (3b).

102. Derive equation (4b).

103. Derive equation (5b).

104. Develop a formula for $(x + a)^4$.

105. Explain why the degree of the product of two polynomials equals the sum of their degrees.

106. If asked to multiply two polynomials, would you prefer to work the problem horizontally or vertically? Explain the reason for your choice.

107. Explain the FOIL method.

'Are You Prepared?' Answers

1. x^{12} **2.** $3x^2 + 15x$ **3.** $(2x - 7)(4x + 1)$

2.5 Division of Polynomials

PREPARING FOR THIS SECTION *Before getting started, review the following:*

- Laws of Exponents (pp. 65–67)
- Subtraction of Polynomials (pp. 79–81)

Now work the 'Are You Prepared?' problems on page 100.

OBJECTIVES

1 Divide Monomials
2 Divide a Polynomial by a Monomial
3 Divide Two Polynomials

Dividing a Monomial by a Monomial

1 To divide two monomials we perform the indicated division by using the Laws of Exponents, as illustrated in the following examples.

EXAMPLE 1 **Dividing Monomials**

(a) $\dfrac{x^5}{x^2} = x^{5-2} = x^3$ (b) $\dfrac{18x^6}{3x^2} = \dfrac{18}{3} \cdot \dfrac{x^6}{x^2} = 6x^{6-2} = 6x^4$

(c) $\dfrac{4x^3}{8x^2} = \dfrac{1}{2}x$ ◀

2 To divide a polynomial by a monomial, divide each term of the polynomial by the monomial.

EXAMPLE 2 **Dividing a Monomial by a Polynomial**

(a) $$\frac{5x^5 + 3x^4 + 10x^3}{x^2} = \frac{5x^5}{x^2} + \frac{3x^4}{x^2} + \frac{10x^3}{x^2}$$ Divide each term by x^2.

$$= 5x^3 + 3x^2 + 10x$$

(b) $$\frac{4x^3 - 3x^2 + x - 2}{x} = \frac{4x^3}{x} - \frac{3x^2}{x} + \frac{x}{x} - \frac{2}{x}$$ Divide each term by x.

$$= 4x^2 - 3x + 1 - \frac{2}{x}$$ ◀

As Example 2 illustrates, the result of dividing a polynomial by a monomial may or may not result in a polynomial. Notice that the answer to Example 2(a) is a polynomial while the answer to Example 2(b) is not a polynomial.

Practice Exercise 1 *In Problems 1–4, perform the indicated operation.*

1. $\dfrac{7^7}{7^5}$ 2. $\dfrac{9x^8}{3x^2}$ 3. $\dfrac{12x^3 + 6x^2 - 2x}{2x}$ 4. $\dfrac{8x^4 + 2x^2 - 1}{4x^2}$

3 The procedure for dividing two polynomials is similar to the procedure for dividing two integers. Dividing two integers should be familiar to you, but we review it briefly below.

EXAMPLE 3 Dividing Two Integers

Divide 842 by 15.

Solution

$$\begin{array}{rrl} & 56 & \leftarrow \text{Quotient} \\ \text{Divisor} \rightarrow & 15\overline{)842} & \leftarrow \text{Dividend} \\ & \underline{75} & \leftarrow 5 \cdot 15 \text{ (Subtract)} \\ & 92 & \\ & \underline{90} & \leftarrow 6 \cdot 15 \text{ (Subtract)} \\ & 2 & \leftarrow \text{Remainder} \end{array}$$

$$\frac{842}{15} = 56 + \frac{2}{15}$$

In the long division process detailed in Example 3, the number 15 is called the **divisor;** the number 842 is called the **dividend;** the number 56 is called the **quotient;** and the number 2 is called the **remainder.**

To check the answer obtained in a division problem, multiply the quotient by the divisor and add the remainder. The answer should be the dividend.

$$(\text{Quotient})(\text{Divisor}) + \text{Remainder} = \text{Dividend}$$

For example, we can check the results obtained in Example 3 as follows:

$$(56)(15) + 2 = 840 + 2 = 842$$

To divide two polynomials, we first must write each polynomial in standard form. The process then follows a pattern similar to that of Example 3. The next example illustrates the procedure.

EXAMPLE 4 Dividing Two Polynomials

Find the quotient and remainder when $3x^2 - 7x + 5$ is divided by $x - 2$.

Solution Each of the polynomials is in standard form. The dividend is $3x^2 - 7x + 5$ and the divisor is $x - 2$.

STEP 1: Divide the leading term of the dividend, $3x^2$, by the leading term of the divisor, x. Enter the result, $3x$, over the term $3x^2$, as follows:

$$\begin{array}{r} 3x \phantom{{}- 7x + 5} \\ x - 2\overline{)3x^2 - 7x + 5} \end{array}$$

STEP 2: Multiply $3x$ by $x - 2$ and enter the result below the dividend, lining up the like terms.

$$\begin{array}{r} 3x \phantom{{}- 7x + 5} \\ x - 2\overline{)3x^2 - 7x + 5} \\ \underline{3x^2 - 6x} \phantom{{}+ 5} \end{array}$$

Step 3: Subtract and bring down the remaining term.

$$\begin{array}{r} 3x \phantom{{}-7x+5} \\ x-2\overline{)3x^2-7x+5} \\ \underline{3x^2-6x} \phantom{{}+5} \\ -x+5 \end{array} \quad \text{Subtract}$$

Step 4: Repeat Steps 1–3 using $(-x + 5)$ as the dividend. Divide the leading term of the dividend, $-x$, by the leading term of the divisor, x. Enter the result, -1, over the term $-7x$.

$$\begin{array}{r} 3x-1 \phantom{{}+5} \\ x-2\overline{)3x^2-7x+5} \\ \underline{3x^2-6x} \phantom{{}+5} \\ -x+5 \\ \underline{-x+2} \\ 3 \end{array} \quad \text{Subtract}$$

The quotient is $3x - 1$ and the remainder is 3.

Check: (Quotient)(Divisor) + Remainder $= (3x - 1)(x - 2) + 3$

$$= 3x^2 - 6x - 1x + 2 + 3$$
$$= 3x^2 - 7x + 5 = \text{Dividend}$$

$$\frac{3x^2 - 7x + 5}{x - 2} = 3x - 1 + \frac{3}{x - 2}$$ ◀

We'll do one more example with the step-by-step process.

EXAMPLE 5 Dividing Two Polynomials

Find the quotient and the remainder when

$$3x^3 + 4x^2 + x + 7 \qquad \text{is divided by} \qquad x^2 + 1$$

Solution Each of the given polynomials is in standard form. The dividend is $3x^3 + 4x^2 + x + 7$, and the divisor is $x^2 + 1$.

Step 1: Divide the leading term of the dividend, $3x^3$, by the leading term of the divisor, x^2. Enter the result, $3x$, over the term $3x^3$, as follows:

$$\begin{array}{r} 3x \phantom{{}+4x^2+x+7} \\ x^2+1\overline{)3x^3+4x^2+x+7} \end{array}$$

Step 2: Multiply $3x$ by $x^2 + 1$ and enter the result below the dividend, lining up like terms.

$$\begin{array}{r} 3x \phantom{{}+4x^2+x+7} \\ x^2+1\overline{)3x^3+4x^2+x+7} \\ \underline{3x^3 \phantom{{}+4x^2} +3x \phantom{{}+7}} \end{array} \quad \leftarrow 3x\cdot(x^2+1) = 3x^3 + 3x$$

↑ Notice that we align the $3x$ term under the x to make the next step easier.

Step 3: Subtract and bring down the remaining terms.

$$\begin{array}{r} 3x \\ x^2 + 1 \overline{)3x^3 + 4x^2 + x + 7} \\ \underline{3x^3 + 3x } \\ 4x^2 - 2x + 7 \end{array}$$

← Subtract.

← Bring down the $4x^2$ and the 7.

Step 4: Repeat Steps 1–3 using $4x^2 - 2x + 7$ as the dividend.

$$\begin{array}{r} 3x + 4 \\ x^2 + 1 \overline{)3x^3 + 4x^2 + x + 7} \\ \underline{3x^3 + 3x } \\ 4x^2 - 2x + 7 \\ \underline{4x^2 + 4} \\ -2x + 3 \end{array}$$

← Divide $4x^2$ by x^2 to get 4.

← Multiply $x^2 + 1$ by 4; subtract.

Since x^2 does not divide $-2x$ evenly (that is, the result is not a monomial), the process ends. The quotient is $3x + 4$, and the remainder is $-2x + 3$.

Check: (Quotient)(Divisor) + Remainder

$$\begin{aligned} &= (3x + 4)(x^2 + 1) + (-2x + 3) \\ &= 3x^3 + 4x^2 + 3x + 4 + (-2x + 3) \\ &= 3x^3 + 4x^2 + x + 7 = \text{Dividend} \end{aligned}$$

$$\frac{3x^3 + 4x^2 + x + 7}{x^2 + 1} = 3x + 4 + \frac{-2x + 3}{x^2 + 1}$$

◀

The remaining examples combine the steps involved in long division.

EXAMPLE 6 Dividing Two Polynomials

Find the quotient and remainder when $4x^3 + 2x^2 - 10x - 5$ is divided by $x + 2$.

Solution

$$\begin{array}{r} 4x^2 - 6x + 2 \\ x + 2 \overline{)4x^3 + 2x^2 - 10x - 5} \\ \underline{4x^3 + 8x^2 } \\ -6x^2 - 10x - 5 \\ \underline{-6x^2 - 12x } \\ 2x - 5 \\ \underline{2x + 4} \\ -9 \end{array}$$

Check: (Quotient)(Divisor) + Remainder

$$\begin{aligned} &= (4x^2 - 6x + 2)(x + 2) + (-9) \\ &= 4x^3 + 8x^2 - 6x^2 - 12x + 2x + 4 + (-9) \\ &= 4x^3 + 2x^2 - 10x - 5 = \text{Dividend} \end{aligned}$$

$$\frac{4x^3 + 2x^2 - 10x - 5}{x + 2} = 4x^2 - 6x + 2 + \frac{-9}{x + 2} = 4x^2 - 6x + 2 - \frac{9}{x + 2}$$

◀

EXAMPLE 7 Dividing Two Polynomials

Find the quotient and the remainder when

$$x^4 - 3x^3 + 2x - 5 \quad \text{is divided by} \quad x^2 - x + 1$$

Solution In setting up this division problem, it is necessary to either leave a space or insert $0x^2$ for the missing x^2 term in the dividend.

$$\begin{array}{rl}
 & x^2 - 2x - 3 \quad \leftarrow \text{Quotient} \\
\text{Divisor} \rightarrow \; x^2 - x + 1 & \overline{)x^4 - 3x^3 \qquad + 2x - 5} \quad \leftarrow \text{Dividend} \\
\text{Subtract} \rightarrow & \underline{x^4 - x^3 + x^2} \\
 & -2x^3 - x^2 + 2x - 5 \\
\text{Subtract} \rightarrow & \underline{-2x^3 + 2x^2 - 2x} \\
 & -3x^2 + 4x - 5 \\
\text{Subtract} \rightarrow & \underline{-3x^2 + 3x - 3} \\
 & x - 2 \quad \leftarrow \text{Remainder}
\end{array}$$

CHECK: (Quotient)(Divisor) + Remainder

$$= (x^2 - 2x - 3)(x^2 - x + 1) + x - 2$$
$$= x^4 - x^3 + x^2 - 2x^3 + 2x^2 - 2x - 3x^2 + 3x - 3 + x - 2$$
$$= x^4 - 3x^3 + 2x - 5 = \text{Dividend}$$

$$\frac{x^4 - 3x^3 + 2x - 5}{x^2 - x + 1} = x^2 - 2x - 3 + \frac{x - 2}{x^2 - x + 1}$$ ◀

EXAMPLE 8 Dividing Two Polynomials

Find the quotient and remainder when

$$2x^3 + 3x^2 - 2x - 3 \quad \text{is divided by} \quad 2x + 3$$

Solution

$$\begin{array}{rl}
 & x^2 - 1 \\
2x + 3 & \overline{)2x^3 + 3x^2 - 2x - 3} \\
 & \underline{2x^3 + 3x^2} \\
 & -2x - 3 \\
 & \underline{-2x - 3} \\
 & 0
\end{array}$$

CHECK: (Quotient)(Divisor) + Remainder

$$= (x^2 - 1)(2x + 3) + 0$$
$$= 2x^3 + 3x^2 - 2x - 3 = \text{Dividend}$$

$$\frac{2x^3 + 3x^2 - 2x - 3}{2x + 3} = x^2 - 1 + \frac{0}{2x + 3} = x^2 - 1$$ ◀

The process for dividing two polynomials leads to the following result:

Theorem Let Q be a polynomial of positive degree and let P be a polynomial whose degree is greater than the degree of Q. The remainder after dividing P by Q is either the zero polynomial or a polynomial whose degree is less than the degree of the divisor Q.

We conclude this section by revisiting the special products for the sum of two cubes and the difference of two cubes (Section 2.4). This time we will observe the result from a dividing perspective rather than from a multiplying perspective.

EXAMPLE 9 **Dividing the Difference of Two Cubes**

Find the quotient and remainder when $x^3 - 27$ is divided by $x - 3$

Solution

$$\begin{array}{r} x^2 + 3x + 9 \\ x - 3 \overline{) x^3 + 0x^2 + 0x - 27} \\ \underline{x^3 - 3x^2} \\ 3x^2 + 0x - 27 \\ \underline{3x^2 - 9x} \\ 9x - 27 \\ \underline{9x - 27} \\ 0 \end{array}$$

CHECK: (Quotient)(Divisor) + Remainder

$$\begin{aligned} &= (x^2 + 3x + 9)(x - 3) + 0 \\ &= x^3 - 3x^2 + 3x^2 - 9x + 9x - 27 \\ &= x^3 - 27 = \text{Dividend} \end{aligned}$$

$$\frac{x^3 - 27}{x - 3} = x^2 + 3x + 9$$

◀

EXAMPLE 10 **Dividing the Sum of Two Cubes**

Find the quotient and remainder when $x^3 + a^3$ is divided by $x + a$.

Solution

$$\begin{array}{r} x^2 - ax + a^2 \\ x + a \overline{) x^3 + 0x^2 + 0x + a^3} \\ \underline{x^3 + ax^2} \\ -ax^2 + 0x \\ \underline{-ax^2 - a^2x} \\ a^2x + a^3 \\ \underline{a^2x + a^3} \\ 0 \end{array}$$

CHECK: (Quotient)(Divisor) + Remainder

$$\begin{aligned} &= (x^2 - ax + a^2)(x + a) + 0 \\ &= (x^2 - ax + a^2)(x + a) \\ &= x^3 + ax^2 - ax^2 - a^2x + a^2x + a^3 \\ &= x^3 + a^3 = \text{Dividend} \end{aligned}$$

$$\frac{x^3 + a^3}{x + a} = x^2 - ax + a^2$$

◀

Example 10 shows that from the dividing perspective

$$x^3 + a^3 = (x + a)(x^2 - ax + a^2)$$

Similarly, you are asked to show $x^3 - a^3 = (x - a)(x^2 + ax + a^2)$ in Problem 53.

Practice Exercise 2 *Find the quotient and remainder when:*

1. $x^3 + 3x^2 - 7x + 4$ is divided by $x + 2$
2. $x^4 - 81$ is divided by $x^2 + 9$
3. $2x^3 - 2x^2 + 7x$ is divided by $2x - 2$

Answers to Practice Exercises

Practice Exercise 1 1. $7^2 = 49$ 2. $3x^6$ 3. $6x^2 + 3x - 1$ 4. $2x^2 + \frac{1}{2} - \frac{1}{4x^2}$

Practice Exercise 2 1. Quotient: $x^2 + x - 9$; remainder: 22
2. Quotient: $x^2 - 9$; remainder: 0
3. Quotient: $x^2 + \frac{7}{2}$; remainder: 7

2.5 Assess Your Understanding

'Are You Prepared?' *Answers are given at the end of these exercises. If you get a wrong answer, read the pages listed in parentheses.*

1. Simplify: $\frac{x^9}{x^5}$ (pp. 65–67)
2. Subtract: $(4x^3 + 5x^2 - 9x + 10) - (x^3 - 3x^2 - x + 6)$ (pp. 79–81)

Concepts and Vocabulary

3. The remainder after dividing two polynomials is either the ________ polynomial or a polynomial of degree ________ ________ the degree of the divisor.
4. To check division, the expression that is being divided, the dividend, should equal the product of the ________ and the ________ plus the ________.
5. *True or False:* When a polynomial is divided by a monomial, the result is always a polynomial.
6. The expression $\frac{45x^5}{9x^3}$ simplifies to ________.

Exercises

In Problems 7–16, simplify each expression.

7. $\frac{3^5}{3^2}$
8. $\frac{4^3}{4}$
9. $\frac{x^6}{x^2}$
10. $\frac{x^7}{x^5}$
11. $\frac{25x^4}{5x^2}$
12. $\frac{24x^5}{3x}$
13. $\frac{20y^4}{4y^3}$
14. $\frac{16z^3}{2z}$
15. $\frac{45x^2y^3}{9xy}$
16. $\frac{24x^3y}{3x^2y}$

In Problems 17–26, perform the indicated division.

17. $\frac{5x^3 - 3x^2 + x}{x}$
18. $\frac{4x^4 - 2x^3 + x^2}{x^2}$
19. $\frac{10x^5 - 5x^4 + 15x^2}{5x^2}$
20. $\frac{8x^3 + 16x^2 - 2x}{2x}$
21. $\frac{-21x^3 + x^2 - 3x + 4}{x}$
22. $\frac{-4x^4 + x^2 - 1}{x^2}$
23. $\frac{8x^3 - x^2 + 1}{x^2}$
24. $\frac{8x^4 + x^2 + 3}{x^3}$
25. $\frac{2x^3 - x^2 + 1}{x^3}$
26. $\frac{4x^2 + x + 5}{x^2}$

In Problems 27–56, find the quotient and the remainder. Check your work by verifying that

$$(Quotient)(Divisor) + Remainder = Dividend$$

27. $4x^3 - 3x^2 + x + 1$ divided by x

28. $3x^3 - x^2 + x - 2$ divided by x

29. $4x^3 - 3x^2 + x + 1$ divided by $x + 2$

30. $3x^3 - x^2 + x - 2$ divided by $x + 2$

31. $4x^3 - 3x^2 + x + 1$ divided by $x - 4$

32. $3x^3 - x^2 + x - 2$ divided by $x - 4$

33. $4x^3 - 3x^2 + x + 1$ divided by x^2

34. $3x^3 - x^2 + x - 2$ divided by x^2

35. $4x^3 - 3x^2 + x + 1$ divided by $x^2 + 2$

36. $3x^3 - x^2 + x - 2$ divided by $x^2 + 2$

37. $4x^3 - 3x^2 + x + 1$ divided by $x^3 - 1$

38. $3x^3 - x^2 + x - 2$ divided by $x^3 - 1$

39. $4x^3 - 3x^2 + x + 1$ divided by $x^2 + x + 1$

40. $3x^3 - x^2 + x - 1$ divided by $x^2 + x + 1$

41. $4x^3 - 3x^2 + x + 1$ divided by $x^2 - 3x - 4$

42. $3x^3 - x^2 + x - 2$ divided by $x^2 - 3x - 4$

43. $x^4 - 1$ divided by $x - 1$

44. $x^4 - 1$ divided by $x + 1$

45. $x^4 - 1$ divided by $x^2 - 1$

46. $x^4 - 1$ divided by $x^2 + 1$

47. $-4x^3 + x^2 - 4$ divided by $x - 1$

48. $-3x^4 - 2x - 1$ divided by $x - 1$

49. $1 - x^2 + x^4$ divided by $x^2 + x + 1$

50. $1 - x^2 + x^4$ divided by $x^2 - x + 1$

51. $1 - x^2 + x^4$ divided by $1 - x^2$

52. $1 - x^2 + x^4$ divided by $1 + x^2$

53. $x^3 - a^3$ divided by $x - a$

54. $x^3 + a^3$ divided by $x + a$

55. $x^4 - a^4$ divided by $x - a$

56. $x^5 - a^5$ divided by $x - a$

'Are You Prepared?' Answers

1. x^4 **2.** $3x^3 + 8x^2 - 8x + 4$

2.6 Factoring Polynomials: Greatest Common Factor; Factoring by Grouping

OBJECTIVES

1 Factor Out the Greatest Common Factor

2 Factor by Grouping

Consider the following products:

$$2x(x - 5) = 2x^2 - 10x$$
$$(2x + 3)(x - 4) = 2x^2 - 5x - 12$$
$$(x - 4)(x + 4) = x^2 - 16$$

In each example, the two polynomials on the left side are called **factors** of the polynomial on the right side. Expressing a given polynomial as a product of other polynomials—that is, finding factors of a polynomial—is called **factoring.** Factoring is the reverse of multiplying; think of it as "un-multiplying."

We shall restrict our discussion here to factoring polynomials in one variable into products of polynomials in one variable, where all coefficients are integers. We call this **factoring over the integers.**

In the remaining sections of Chapter 2, we discuss four types of factoring:

- factoring out the greatest common factor
- factoring by grouping
- factoring miscellaneous trinomials
- factoring special products

Any polynomial can be written as the product of 1 times itself or as -1 times its additive inverse. If a polynomial cannot be written as the product of two other polynomials (excluding 1 and -1), then the polynomial is said to be **prime.** When a polynomial has been written as a product consisting only of prime factors, it is said to be **factored completely.** Examples of prime polynomials are

$$2, \quad 3, \quad 5, \quad x, \quad x + 1, \quad x - 1, \quad 3x + 4$$

Factoring Out the Greatest Common Factor

1 The first factor to look for in factoring a polynomial is the presence of a common monomial factor in each term of the polynomial. If one is present, use the Distributive Property to factor it out.

EXAMPLE 1 Identifying Common Monomial Factors

Polynomial	Common Monomial Factor	Remaining Factor	Factored Form
(a) $2x + 4$	2	$x + 2$	$2x + 4 = 2(x + 2)$
(b) $3x - 6$	3	$x - 2$	$3x - 6 = 3(x - 2)$
(c) $2x^2 - 4x + 8$	2	$x^2 - 2x + 4$	$2x^2 - 4x + 8 = 2(x^2 - 2x + 4)$
(d) $8x - 12$	4	$2x - 3$	$8x - 12 = 4(2x - 3)$
(e) $x^2 + x$	x	$x + 1$	$x^2 + x = x(x + 1)$
(f) $x^3 - 3x^2$	x^2	$x - 3$	$x^3 - 3x^2 = x^2(x - 3)$
(g) $6x^2 + 9x$	$3x$	$2x + 3$	$6x^2 + 9x = 3x(2x + 3)$ ◀

Notice that, once all common monomial factors have been removed from a polynomial, the remaining factor is either a prime polynomial of degree 1 or a polynomial of degree 2 or higher. (Do you see why?)

Also notice that it is convenient to factor out the *greatest common factor*. For example, in Example 1(d) we could have factored out a 2, giving $8x - 12 = 2(4x - 6)$. But $4x - 6$ also has a factor of 2, so we factor again:

$$8x - 12 = 2(4x - 6) = 2 \cdot 2(2x - 3) = 4(2x - 3)$$

To avoid this extra step, try to factor out the greatest common factor from the start. For any factoring problem the result should be **factored completely** meaning the polynomial being factored has been written as a product consisting only of prime polynomial factors or as a product consisting of prime polynomial factors multiplied by a monomial factor, which may or may not be prime.

Practice Exercise 1 *Fill in the blanks in Problems 1–3 below.*

Polynomial	Common Monomial Factor	Remaining Factor	Factored Form
1. $7x + 21$	________	________	________
2. $9x^2 - 6x$	________	________	________
3. $x^4 - 2x^3$	________	________	________

EXAMPLE 2 **Factoring Out the Greatest Common Factor**

Factor out the greatest common factor:

(a) $6x^2 + 3x$ (b) $20x^3 - 12x^2$ (c) $14x^4 - 35x^2 - 7x$

Solution (a) $6x^2 + 3x = 3x(2x + 1)$

(b) $20x^3 - 12x^2 = 4x^2(5x - 3)$

(c) $14x^4 - 35x^2 - 7x = 7x(2x^3 - 5x - 1)$ ◀

WARNING: In Examples 2(a) and 2(c), the greatest common factor happened to be one of the terms within the polynomial being factored. When this is the case you must put a 1 in the place of that term in the factored polynomial. In Example 2(a), since $3x = 3x \cdot 1$, once the $3x$ is factored out, we are left with a 1 in the place of the factored polynomial where the $3x$ was. Do you see why the 1 must be there? (*Hint:* Check the factored answer by multiplying it out to get the original problem.) ■

When factoring, if the leading term is negative, we usually factor out the negative sign along with the greatest common factor.

EXAMPLE 3 **Factoring Out a Lead Negative**

Factor out the greatest common factor:

(a) $-10x^2 + 5x$ (b) $-7x^2 - 2x + 1$

Solution (a) $-10x^2 + 5x = -5x(2x - 1)$

(b) $-7x^2 - 2x + 1 = -1(7x^2 + 2x - 1)$ ◀

In Example 3(b), the only common factor was 1 or -1. Since the leading term is negative, we factored out -1 and as a result the sign of every term in the remaining factor changed from what it originally was.

Practice Exercise 2 *In Problems 1–6, factor out the greatest common factor.*

1. $9x^3 - 15x^2 + 12x$ **2.** $30x^4 - 6x^3$ **3.** $8x^4 + 12x + 4$

4. $-14x^2 - 35x$ **5.** $-39x^4 + 13x^3$ **6.** $-5x - 9y + z$ ◀

Factoring by Grouping

2 Sometimes a common factor does not occur in every term of the polynomial, but in each of several groups of terms that together makeup the polynomial. When this happens, the common factor can be factored out of each group by means of the Distributive Property. This technique is called **factoring by grouping.**

EXAMPLE 4 **Factoring by Grouping**

Factor completely by grouping: $x(x^2 + 2) + 3(x^2 + 2)$

Solution **METHOD 1:** Notice the common factor $x^2 + 2$. By factoring with the Distributive Property, we have

$$x(x^2 + 2) + 3(x^2 + 2) = (x + 3)(x^2 + 2)$$

Since $x^2 + 2$ and $x + 3$ are prime, the factorization is complete.

METHOD 2: Alternatively, we can think of $x^2 + 2$ as if it were a single term, for example, let's say $x^2 + 2 = P$. (Notice that this is similar to the substitution step in Example 3, Section 2.4.) With this substitution step, we have

$$\begin{aligned} x(x^2 + 2) + 3(x^2 + 2) &= x \cdot P + 3 \cdot P && \text{Substitute } P \text{ for } x^2 + 2. \\ &= (x + 3) \cdot P && \text{Factor out } P. \\ &= (x + 3)(x^2 + 2) && \text{Resubstitute } x^2 + 2 \text{ for } P. \end{aligned}$$

You can factor these problems with either method. If you initially prefer to use Method 2, work and practice enough of these problems so that eventually you are able to mentally think through the substitution step without writing it down.

EXAMPLE 5 Factoring by Grouping

Factor completely by grouping:

(a) $4x(x - 5) - 3(x - 5)$ (b) $3x(2x - 5) - (2x - 5)$
(c) $6x^2(x^2 + 1) + 2x(x^2 + 1)$

Solution

(a) $4x(x - 5) - 3(x - 5) = (4x - 3)(x - 5)$ — Factor out $(x - 5)$.

(b)
$$\begin{aligned} 3x(2x - 5) - (2x - 5) &= 3x(2x - 5) - 1(2x - 5) && \text{Multiplicative Identity} \\ &= (3x - 1)(2x - 5) && \text{Factor out } (2x - 5). \end{aligned}$$

(c)
$$\begin{aligned} 6x^2(x^2 + 1) + 2x(x^2 + 1) &= (6x^2 + 2x)(x^2 + 1) && \text{Factor out } (x^2 + 1). \\ &= 2x(3x + 1)(x^2 + 1) && \text{Factor out } 2x. \end{aligned}$$

In Example 5(b), we showed $2x - 5$ multiplied by 1, the multiplicative identity. This is to serve as a reminder to you that when we factor out $2x - 5$, we are left with a 1 in its place.

In Example 5(c), we initially factored out $x^2 + 1$. After the first step, we notice that $2x$ can still be factored out of $6x^2 + 2x$. After every factoring step you should check the problem to see if it is factored completely.

Example 5(c) could have been worked alternatively by factoring out the common factor of $2x$ at the beginning as follows:

$$6x^2(x^2 + 1) + 2x(x^2 + 1) = 2x[3x(x^2 + 1) + 1 \cdot (x^2 + 1)] = 2x(3x + 1)(x^2 + 1)$$

EXAMPLE 6 Factoring by Grouping

Factor completely by grouping: $x^3 - 4x^2 + 2x - 8$

Solution To see if factoring by grouping will work, group the first two terms and the last two terms. Then look for a common factor in each group. In this example, we can factor x^2 from $x^3 - 4x^2$ and 2 from $2x - 8$. The remaining factor in each case is the same, $x - 4$. This means factoring by grouping will work, as follows:

$$\begin{aligned} x^3 - 4x^2 + 2x - 8 &= (x^3 - 4x^2) + (2x - 8) \\ &= x^2(x - 4) + 2(x - 4) \\ &= (x^2 + 2)(x - 4) \end{aligned}$$

Since $x^2 + 2$ and $x - 4$ are prime, the factorization is complete.

Notice that in Example 6, by the Commutative Property

$$(x^2 + 2)(x - 4) = (x - 4)(x^2 + 2).$$

In other words, it doesn't matter which factor you list first in the answer.

EXAMPLE 7 **Factoring by Grouping**

Factor completely by grouping: $2x^3 + 6x^2 + x + 3$

Solution To see if factoring by grouping will work, group the first two terms and the last two terms. Then look for a common factor in each group. In this example, we can factor $2x^2$ out of the first group $2x^3 + 6x^2$ and we can factor a 1 out of the second group $x + 3$. The remaining factor in each case is $x + 3$. Since the remaining factor is the same in each group, factoring by grouping will work and we proceed as follows:

$$\begin{aligned} 2x^3 + 6x^2 + x + 3 &= (2x^3 + 6x^2) + (x + 3) \\ &= 2x^2(x + 3) + 1(x + 3) \\ &= (2x^2 + 1)(x + 3) \end{aligned}$$

Since $x + 3$ and $2x^2 + 1$ are prime, the factorization is complete. ◀

EXAMPLE 8 **Factoring by Grouping**

Factor completely by grouping: $3x^3 + 4x^2 - 6x - 8$

Solution **METHOD 1:** We begin by grouping the first two terms and the last two terms. For this problem the grouping is a little tricky because the third term is being subtracted. If we group the last two terms with the subtraction sign outside the parentheses, we have to remember to change the sign of every term inside the second set of parentheses. We proceed as follows:

$$\begin{aligned} 3x^3 + 4x^2 - 6x - 8 &= (3x^3 + 4x^2) - (6x + 8) && \text{Look closely at this step.} \\ &= x^2(3x + 4) - 2(3x + 4) && \text{Factor.} \\ &= (x^2 - 2)(3x + 4) && \text{Factor.} \end{aligned}$$

METHOD 2: We rewrite the subtraction as addition of the additive inverse:

$$\begin{aligned} 3x^3 + 4x^2 - 6x - 8 &= 3x^3 + 4x^2 + (-6x) + (-8) \\ &= (3x^3 + 4x^2) + [(-6x) + (-8)] \\ &= x^2(3x + 4) + (-2)(3x + 4) \\ &= [x^2 + (-2)](3x + 4) \\ &= (x^2 - 2)(3x + 4) \end{aligned}$$

◀

Either of the methods in Example 8 is acceptable for factoring this type of problem. Study them and pick the one that you find easier to understand. The important thing for you to do is to learn one of these methods and stick with it. Try not to switch back and forth between two methods when learning something new.

EXAMPLE 9 **Factoring by Grouping**

Factor completely by grouping: $2x^3 - 10x^2 - x + 5$

Solution METHOD 1:

$$\begin{aligned}2x^3 - 10x^2 - x + 5 &= (2x^3 - 10x^2) - (x - 5)\\ &= 2x^2(x - 5) - 1(x - 5)\\ &= (2x^2 - 1)(x - 5)\end{aligned}$$

METHOD 2:

$$\begin{aligned}2x^3 - 10x^2 - x + 5 &= 2x^3 + (-10x^2) + (-x) + 5\\ &= [2x^3 + (-10x^2)] + [(-x) + 5]\\ &= 2x^2[x + (-5)] + (-1)[x + (-5)]\\ &= [2x^2 + (-1)][x + (-5)]\\ &= (2x^2 - 1)(x - 5)\end{aligned}$$

◀

Practice Exercise 3 *In Problems 1–4, factor completely by grouping.*

1. $x(3x^2 - 5) + 2(3x^2 - 5)$
2. $4x^2(x + 7) + 6x(x + 7)$
3. $2x^3 - 8x^2 + 5x - 20$
4. $6x^3 - 30x^2 - x + 5$

Answers to Practice Exercises

Practice Exercise 1 1. $7; x + 3; 7(x + 3)$ 2. $3x; 3x - 2; 3x(3x - 2)$ 3. $x^3; x - 2; x^3(x - 2)$

Practice Exercise 2 1. $3x(3x^2 - 5x + 4)$ 2. $6x^3(5x - 1)$ 3. $4(2x^2 + 3x + 1)$ 4. $-7x(2x + 5)$ 5. $-13x^3(3x - 1)$ 6. $-1(5x + 9y - z)$

Practice Exercise 3 1. $(x + 2)(3x^2 - 5)$ 2. $2x(2x + 3)(x + 7)$ 3. $(2x^2 + 5)(x - 4)$ 4. $(6x^2 - 1)(x - 5)$

2.6 Assess Your Understanding

Concepts and Vocabulary

1. If a polynomial cannot be written as the product of two other polynomials (excluding 1 and −1), then the polynomial is said to be ________.
2. If factored completely, $5x^4 - 20x =$ ________.
3. Fill in the missing factor: $6x(4x + 1) - (4x + 1) = (\_\_\_\_\_\_\_\_)(4x + 1)$.
4. Fill in the missing factor: $-2x^3 - 5x^2 + 8 = -1(\_\_\_\_\_\_\_\_)$.

Exercises

In Problems 5–28, factor each polynomial by removing the common monomial factor. Factor out −1 if the leading coefficient is negative.

5. $3x + 6$
6. $7x - 14$
7. $ax^2 + a$
8. $ax - a$
9. $x^3 + x^2 + x$
10. $x^3 - x^2 + x$
11. $2x^2 + 2x + 2$
12. $3x^2 - 3x + 3$
13. $3x^2y - 6xy^2 + 12xy$
14. $60x^2y - 48xy^2 + 72x^3y$
15. $-5x^2y^3 - 45xy^2 - 50xy$
16. $-21x^3y^2 + 35x^2y^3 - 7x^2y^2$
17. $-3x^2 - 12x$
18. $-10x^3 - 25x^2$
19. $81x + 24y$
20. $42x + 30y$
21. $51x^3 + 17x^2$
22. $x^3 + x$
23. $18x^4 - 36x^2 + 63x$
24. $24x^4 + 64x^2 + 8x$
25. $-7x^2 - 7x$
26. $-3x + 12y - 18$
27. $x^4 + 81x^2$
28. $x^4 + 16x^2$

In Problems 29–68, factor completely each polynomial by grouping.

29. $x(x + 3) - 6(x + 3)$
30. $5(3x - 7) + x(3x - 7)$
31. $x(x^2 + 4) + 5(x^2 + 4)$
32. $2x(x^2 - 7) + 3(x^2 - 7)$
33. $9x(x + 5) - 2(x + 5)$
34. $8x(2x + 7) + 3(2x + 7)$
35. $3x(2x + 7) + (2x + 7)$
36. $x(x - 8) - (x - 8)$
37. $12x(x - 4) - (x - 4)$
38. $5x(2x - 1) + (2x - 1)$
39. $3x^2(4x + 5) - 6x(4x + 5)$
40. $8x^2(3x + 7) - 6x(3x + 7)$
41. $10x^2(3x + 11) - 5x(3x + 11)$
42. $6x^2(2x - 5) + 3x(2x - 5)$
43. $(x + 2)^2 - 5(x + 2)$
44. $(x - 1)^2 - 2(x - 1)$
45. $x^3 + x^2 + 2x + 2$
46. $x^3 + x^2 + 3x + 3$
47. $x^3 - x^2 + 3x - 3$
48. $x^3 - 3x^2 + 2x - 6$
49. $2x^3 - 10x^2 + 3x - 15$
50. $10x^3 - 15x^2 + 6x - 9$
51. $3x^3 + 12x^2 + 2x + 8$
52. $10x^3 + 5x^2 + 6x + 3$
53. $6x^3 - 15x^2 - 8x + 20$
54. $4x^3 - 3x^2 - 8x + 6$
55. $5x^3 + 2x^2 - 15x - 6$
56. $2x^3 + 8x^2 - 7x - 28$
57. $6x^3 - 12x^2 - 5x + 10$
58. $3x^3 - 24x^2 - 2x + 16$
59. $x^3 + 2x^2 + x + 2$
60. $2x^3 + 6x^2 + x + 3$
61. $3x^3 - 3x^2 + x - 1$
62. $5x^3 - 10x^2 + x - 2$
63. $3x^3 - 15x^2 - x + 5$
64. $4x^3 + 6x^2 - 2x - 3$
65. $3x^3 - 6x^2 - x + 2$
66. $6x^3 - 2x^2 - 3x + 1$
67. $x^2 + 2x - x - 2$
68. $x^2 - 3x - x + 3$

2.7 Factoring Second-Degree Trinomials

PREPARING FOR THIS SECTION *Before getting started, review the following:*

- Factor by Grouping (pp. 103–106)

Now work the 'Are You Prepared?' problems on page 116.

OBJECTIVES

1 Factor a Second-Degree Trinomial: $x^2 + Bx + C$
2 Factor a Second-Degree Trinomial: $Ax^2 + Bx + C$

In this section, we study techniques for factoring second-degree trinomials. Factoring trinomials is used extensively in algebra and as a result it is important that you master this topic. There are two types of trinomials we'll consider:

- $x^2 + Bx + C$
- $Ax^2 + Bx + C$

Let's start with the first type.

Factoring a Second-Degree Trinomial: $x^2 + Bx + C$

1 The idea behind factoring a second-degree polynomial like $x^2 + Bx + C$ is to see whether it can be made equal to the product of two, possibly equal, first-degree polynomials.

For example, we know that

$$(x + 3)(x + 4) = x^2 + 7x + 12$$

The factors of $x^2 + 7x + 12$ are $x + 3$ and $x + 4$. Notice the following:

$$x^2 + 7x + 12 = (x + 3)(x + 4)$$

12 is the product of 3 and 4
7 is the sum of 3 and 4

In general, if $x^2 + Bx + C = (x + a)(x + b)$, then $ab = C$ and $a + b = B$.

To factor a second-degree polynomial $x^2 + Bx + C$, find integers whose product is C and whose sum is B. That is, if there are numbers a, b, where $ab = C$ and $a + b = B$, then

$$x^2 + Bx + C = (x + a)(x + b)$$

EXAMPLE 1 Factoring a Trinomial

Factor completely: $x^2 + 7x + 10$

Solution First, determine all integers whose product is 10 and then compute each sum.

Integers whose product is 10	1, 10	−1, −10	2, 5	−2, −5
Sum	11	−11	7	−7

The integers 2 and 5 have a product of 10 and add up to 7, the coefficient of the middle term. As a result, we have

$$x^2 + 7x + 10 = (x + 2)(x + 5)$$

CHECK WITH THE FOIL METHOD:

$$(x + 2)(x + 5) = x^2 + 5x + 2x + 2 \cdot 5 = x^2 + 7x + 10$$ ◀

EXAMPLE 2 Factoring a Trinomial

Factor completely: $x^2 - 6x + 8$

Solution First, determine all integers whose product is 8 and then compute each sum.

Integers whose product is 8	1, 8	−1, −8	2, 4	−2, −4
Sum	9	−9	6	−6

The integers −2 and −4 have a product of 8 and a sum of −6, the coefficient of the middle term. As a result,

$$x^2 - 6x + 8 = (x - 2)(x - 4)$$

CHECK WITH THE FOIL METHOD:

$$(x - 2)(x - 4) = x^2 - 4x - 2x + (-2)(-4) = x^2 - 6x + 8$$ ◀

EXAMPLE 3 Factoring a Trinomial

Factor completely: $x^2 - x - 12$

Solution First, determine all integers whose product is −12 and then compute each sum.

Integers whose product is −12	1, −12	−1, 12	2, −6	−2, 6	3, −4	−3, 4
Sum	−11	11	−4	4	−1	1

The integers 3 and −4 have a product of −12 and a sum of −1, the coefficient of the middle term. Then,

$$x^2 - x - 12 = (x + 3)(x - 4)$$

CHECK: $(x + 3)(x - 4) = x^2 - 4x + 3x + 3 \cdot (-4) = x^2 - x - 12$ ◀

EXAMPLE 4 Factoring a Trinomial

Factor completely: $x^2 + 4x - 12$

Solution Referring to Example 3 above, the integers −2 and 6 have a product of −12 and a sum of 4. Then,

$$x^2 + 4x - 12 = (x - 2)(x + 6)$$

CHECK: $(x - 2)(x + 6) = x^2 + 6x - 2x + (-2)(6) = x^2 + 4x - 12$ ◀

REMEMBER: To avoid errors in factoring, always check your answer by multiplying it out to see if the result equals the original expression. ■

Practice Exercise 1 *Factor completely:*

1. $x^2 + 8x + 15$ **2.** $x^2 - 6x + 5$ **3.** $x^2 + x - 12$

EXAMPLE 5 Factoring a Trinomial

Factor completely: $x^2 - 2x + 12$

Solution First, determine all integers whose product is 12 and then compute each sum.

Integers whose product is 12	1, 12	−1, −12	2, 6	−2, −6	3, 4	−3, −4
Sum	13	−13	8	−8	7	−7

Since the coefficient of the middle term is −2 and none of the sums above equals −2, we conclude that $x^2 - 2x + 12$ cannot be factored. That is, $x^2 - 2x + 12$ is prime. ◀

Example 5 shows an important result:

In factoring $x^2 + Bx + C$, if none of the integers whose product is C adds to a sum equal to B, the polynomial is prime.

EXAMPLE 6 Identifying a Prime Polynomial

Show that each of the following polynomials is prime.

(a) $x^2 + 3x + 5$ (b) $x^2 + 9$

Solution (a) First, list the integers whose product is 5 and then compute each sum.

Integers whose product is 5	1, 5	−1, −5
Sum	6	−6

Since the coefficient of the middle term in $x^2 + 3x + 5$ is 3 and none of the sums equals 3, we conclude that $x^2 + 3x + 5$ is prime.

(b) First, list the integers whose product is 9 and then compute their sum.

Integers whose product is 9	1, 9	−1, −9	3, 3	−3, −3
Sum	10	−10	6	−6

Since the coefficient of the middle term in $x^2 + 9 = x^2 + 0x + 9$ is 0 and none of the sums equals 0, we conclude that $x^2 + 9$ is prime. ◀

Example 6(b) demonstrates a more general result:

Any polynomial of the form $x^2 + a^2$, a real, is prime.

Practice Exercise 2 *In Problems 1–6, factor completely each polynomial. If a polynomial cannot be factored, say that it is prime.*

1. $x^2 - 9x + 18$ **2.** $x^2 + 2x - 15$ **3.** $x^2 - 3x - 15$

4. $x^2 - 2x - 24$ **5.** $x^2 + 8x - 12$ **6.** $x^2 + 2$

Factoring a Second-Degree Trinomial: $Ax^2 + Bx + C$

2 To factor a second-degree polynomial $Ax^2 + Bx + C$, when $A \neq 1$ and A, B, and C have no common factors, follow these steps:

Steps for Factoring $Ax^2 + Bx + C$, Where $A \neq 1$, and A, B, and C Have No Common Factors

STEP 1: Find the value of AC.

STEP 2: Find integers whose product is AC that add up to B. That is, find a and b so that $ab = AC$ and $a + b = B$.

STEP 3: Write $Ax^2 + Bx + C = Ax^2 + ax + bx + C$.

STEP 4: Factor this last expression by grouping.

EXAMPLE 7 Factoring Trinomials

Factor completely: $2x^2 + 5x + 3$

Solution Comparing $2x^2 + 5x + 3$ to $Ax^2 + Bx + C$, we find that $A = 2$, $B = 5$, and $C = 3$.

STEP 1: The value of AC is $2 \cdot 3 = 6$.

STEP 2: Determine the integers whose product is $AC = 6$ and compute each sum.

Integers whose product is 6	1, 6	−1, −6	2, 3	−2, −3
Sum	7	−7	5	−5

The integers whose product is 6 that add up to $B = 5$ are 2 and 3.

STEP 3: $2x^2 + 5x + 3 = 2x^2 + 2x + 3x + 3$

STEP 4: Factor by grouping.

$$\begin{aligned} 2x^2 + 2x + 3x + 3 &= (2x^2 + 2x) + (3x + 3) \\ &= 2x(x + 1) + 3(x + 1) \\ &= (2x + 3)(x + 1) \end{aligned}$$

That is,

$$2x^2 + 5x + 3 = (2x + 3)(x + 1)$$

Alternatively, **STEP 3** could have been written as $2x^2 + 5x + 3 = 2x^2 + 3x + 2x + 3$.
Then

STEP 4:
$$\begin{aligned} 2x^2 + 3x + 2x + 3 &= (2x^2 + 3x) + (2x + 3) \\ &= x(2x + 3) + 1(2x + 3) \\ &= (x + 1)(2x + 3) \end{aligned}$$

Then,

$$2x^2 + 5x + 3 = (x + 1)(2x + 3)$$

CHECK: $(x + 1)(2x + 3) = 2x^2 + 3x + 2x + 3 = 2x^2 + 5x + 3$ ◀

For a problem like Example 7, students often ask, "In **STEP 3,** does it matter which term you write first, the $2x$ or the $3x$?" By working Example 7 both ways in **STEP 3,** you should realize that you can do it either way.

EXAMPLE 8 Factoring Trinomials

Factor completely: $2x^2 - x - 6$

Solution Comparing $2x^2 - x - 6$ to $Ax^2 + Bx + C$, we find that $A = 2, B = -1$, and $C = -6$.

STEP 1: The value of AC is $2 \cdot (-6) = -12$.

STEP 2: Determine the integers whose product is $AC = -12$ and compute each sum.

Integers whose product is −12	1, −12	−1, 12	2, −6	−2, 6	3, −4	−3, 4
Sum	−11	11	−4	4	−1	1

The integers whose product is −12 that add up to $B = -1$ are -4 and 3.

STEP 3: $2x^2 - x - 6 = 2x^2 - 4x + 3x - 6$

STEP 4: Factor by grouping.

$$\begin{aligned} 2x^2 - 4x + 3x - 6 &= (2x^2 - 4x) + (3x - 6) \\ &= 2x(x - 2) + 3(x - 2) \\ &= (2x + 3)(x - 2) \end{aligned}$$

As a result,

$$2x^2 - x - 6 = (2x + 3)(x - 2)$$

Alternatively, **Step 3** could have been written as $2x^2 - x - 6 = 2x^2 + 3x - 4x - 6$.
Then

Step 4:
$$\begin{aligned} 2x^2 + 3x - 4x - 6 &= (2x^2 + 3x) - (4x + 6) \\ &= x(2x + 3) - 2(2x + 3) \\ &= (x - 2)(2x + 3) \end{aligned}$$

That is,

$$2x^2 - x - 6 = (x - 2)(2x + 3)$$

Check: $(x - 2)(2x + 3) = 2x^2 + 3x - 4x - 6 = 2x^2 - x - 6$ ◀

As mentioned earlier, when splitting the middle term to factor a trinomial by grouping, you can write either term first, but notice what happened in Example 8. We rewrote $-x$ as the sum of $-4x$ and $3x$, one negative and one positive term. If you write $2x^2 - 4x + 3x - 6$, as it is written in the first solution, then you do not have to change the signs of the terms in the second grouping. But if you write $2x^2 + 3x - 4x - 6$, as in the alternative solution, then you must change the signs of the terms in the second grouping. Whenever you are factoring a trinomial by grouping and the middle term splits into a positive term and a negative term, write the negative term first to avoid a careless error from changing signs.

EXAMPLE 9 Factoring a Trinomial

Factor completely: $6x^2 + 11x - 10$

Solution Comparing $6x^2 + 11x - 10$ to $Ax^2 + Bx + C$, we find that $A = 6$, $B = 11$, and $C = -10$.

Step 1: The value of AC is $6 \cdot (-10) = -60$.

Step 2: Determine the integers whose product is $AC = -60$ and compute each sum.

Integers whose product is −60	−1, 60	1, −60	−2, 30	2, −30	−3, 20	3, −20	−4, 15	4, −15	−5, 12	5, −12	−6, 10	6, −10
Sum	59	−59	28	−28	17	−17	11	−11	7	−7	4	−4

The integers whose product is -60 that add up to $B = 11$ are -4 and 15.

Step 3: $6x^2 + 11x - 10 = 6x^2 - 4x + 15x - 10$ Write $-4x$ before $15x$.

Step 4: Factor by grouping.

$$\begin{aligned} 6x^2 - 4x + 15x - 10 &= (6x^2 - 4x) + (15x - 10) \\ &= 2x(3x - 2) + 5(3x - 2) \\ &= (2x + 5)(3x - 2) \end{aligned}$$

As a result,

$$6x^2 + 11x - 10 = (2x + 5)(3x - 2)$$

Check: $(2x + 5)(3x - 2) = 6x^2 - 4x + 15x - 10 = 6x^2 + 11x - 10$ ◀

EXAMPLE 10 Factoring a Trinomial

Factor completely: $-2x^2 + 11x + 6$

Solution Since the leading coefficient, -2, is negative, we begin by factoring out -1:

$$-2x^2 + 11x + 6 = (-1)(2x^2 - 11x - 6)$$

Now we proceed as before to factor $2x^2 - 11x - 6$ with $A = 2$, $B = -11$, and $C = -6$.

STEP 1: The value of AC is $2 \cdot (-6) = -12$.

STEP 2: Determine the integers whose product is $AC = -12$ and compute each sum.

Integers whose product is −12	−1, 12	1, −12	−2, 6	2, −6	−3, 4	3, −4
Sum	11	−11	4	−4	1	−1

The integers whose product is -12 that add up to $B = -11$ are 1 and -12.

STEP 3:
$$\begin{aligned} -2x^2 + 11x + 6 &= (-1)(2x^2 - 11x - 6) \\ &= (-1)[2x^2 - 12x + 1x - 6] \end{aligned}$$

STEP 4: Factor by grouping.

$$\begin{aligned} (-1)[2x^2 - 12x + 1x - 6] &= (-1)[(2x^2 - 12x) + (1x - 6)] \\ &= (-1)[2x(x - 6) + 1(x - 6)] \\ &= (-1)[(2x + 1)(x - 6)] \\ &= (-1)(2x + 1)(x - 6) \end{aligned}$$

As a result,

$$-2x^2 + 11x + 6 = (-1)(2x + 1)(x - 6)$$

CHECK:
$$\begin{aligned} (-1)(2x + 1)(x - 6) &= -1 \cdot [2x^2 - 12x + x - 6] \\ &= -1 \cdot [2x^2 - 11x - 6] = -2x^2 + 11x + 6 \end{aligned}$$

◀

EXAMPLE 11 Factoring a Trinomial

Factor completely: $2x^2 + x + 3$

Solution **STEP 1:** The value of AC is $2 \cdot 3 = 6$.

STEP 2: Determine the integers whose product is $AC = 6$ and compute each sum.

Integers whose product is 6	1, 6	−1, −6	2, 3	−2, −3
Sum	7	−7	5	−5

Since none of the sums equals $B = 1$, $2x^2 + x + 3$ is prime. ◀

Practice Exercise 3 *In Problems 1–5, factor completely each polynomial.*

1. $3x^2 + 16x - 12$ **2.** $6x^2 - 7x - 5$ **3.** $3x^2 + 7x - 20$

4. $4x^2 - 12x + 9$ **5.** $2x^2 + x + 1$

◀

Sometimes we need to factor out a common monomial factor first before proceeding with factoring the trinomial.

EXAMPLE 12 Factoring a Trinomial

Factor completely: $3x^3 - 15x^2 + 12x$

Solution The first thing to look for in any factoring problem is a common monomial factor. In this problem, each term has a common factor: $3x$.

Factor out $3x$: $3x^3 - 15x^2 + 12x = 3x(x^2 - 5x + 4)$

Check to see if the resulting polynomial, $x^2 - 5x + 4$, is factorable.

Integers whose product is 4	1, 4	−1, −4	2, 2	−2, −2
Sum	5	−5	4	−4

The integers -1 and -4 have a product of 4 and a sum of -5, the coefficient of the middle term. As a result we factor as follows:

$$\begin{aligned} 3x^3 - 15x^2 + 12x &= 3x(x^2 - 5x + 4) \\ &= 3x(x - 1)(x - 4) \end{aligned}$$

CHECK: $3x(x - 1)(x - 4) = 3x(x^2 - 4x - x + 4) = 3x(x^2 - 5x + 4)$
$= 3x^3 - 15x^2 + 12x$

◀

Some trinomials have a degree higher than 2, but can still be factored with the factoring techniques developed for trinomials of degree 2.

EXAMPLE 13 Factoring a Trinomial

Factor completely: $x^4 - 4x^2 - 5$

Solution First, determine all integers whose product is -5 and then compute each sum.

Integers whose product is −5	1, −5	−1, 5
Sum	−4	4

Since -4 is the coefficient of the middle term,

$$x^4 - 4x^2 - 5 = (x^2 - 5)(x^2 + 1)$$

CHECK: $(x^2 - 5)(x^2 + 1) = x^2 \cdot x^2 + 1x^2 - 5x^2 - 5 \cdot 1 = x^4 - 4x^2 - 5$

◀

EXAMPLE 14 Factoring a Trinomial

Factor completely: $3x^4 + 8x^2 + 4$

Solution Comparing $3x^4 + 8x^2 + 4$ to $Ax^4 + Bx^2 + C$, we find that $A = 3$, $B = 8$, and $C = 4$.

STEP 1: The value of AC is $3 \cdot 4 = 12$.

STEP 2: Determine the integers whose product is 12 and compute each sum.

Integers whose product is 12	1, 12	−1, −12	2, 6	−2, −6	3, 4	−3, −4
Sum	13	−13	8	−8	7	−7

The integers whose product is 12 that add up to $B = 8$ are 2 and 6.

STEP 3: $3x^4 + 8x^2 + 4 = 3x^4 + 6x^2 + 2x^2 + 4$

STEP 4: Factor by grouping.

$$\begin{aligned} 3x^4 + 6x^2 + 2x^2 + 4 &= (3x^4 + 6x^2) + (2x^2 + 4) \\ &= 3x^2(x^2 + 2) + 2(x^2 + 2) \\ &= (3x^2 + 2)(x^2 + 2) \end{aligned}$$

Then,

$$3x^4 + 8x^2 + 4 = (3x^2 + 2)(x^2 + 2)$$

CHECK: $(3x^2 + 2)(x^2 + 2) = 3x^4 + 6x^2 + 2x^2 + 4 = 3x^4 + 8x^2 + 4$ ◀

EXAMPLE 15 Factoring a Trinomial

Factor completely: $x^6 - 3x^3 - 10$

Solution First find all integers whose product is −10 and then compute each sum.

Integers whose product is −10	1, −10	−1, 10	2, −5	−2, 5
Sum	−9	9	−3	3

Since −3 is the coefficient of the middle term,

$$x^6 - 3x^3 - 10 = (x^3 + 2)(x^3 - 5)$$

CHECK: $(x^3 + 2)(x^3 - 5) = x^6 + 2x^3 - 5x^3 - 10 = x^6 - 3x^3 - 10$ ◀

Practice Exercise 4 *In Problems 1–6, factor completely each trinomial.*

1. $4x^2 + 28x + 24$ **2.** $x^3 - 2x^2 - 15x$ **3.** $6x^3 + 20x^2 + 16x$

4. $2x^3 - 6x^2 + 10x$ **5.** $-2x^2 + 2x + 12$ **6.** $3x^4 - 5x^2 - 2$

Answers to Practice Exercises

Practice Exercise 1 **1.** $(x + 5)(x + 3)$ **2.** $(x - 5)(x - 1)$ **3.** $(x + 4)(x - 3)$

Practice Exercise 2 **1.** $(x - 6)(x - 3)$ **2.** $(x + 5)(x - 3)$ **3.** prime **4.** $(x - 6)(x + 4)$ **5.** prime **6.** prime

Practice Exercise 3 **1.** $(3x - 2)(x + 6)$ **2.** $(3x - 5)(2x + 1)$ **3.** $(3x - 5)(x + 4)$ **4.** $(2x - 3)(2x - 3) = (2x - 3)^2$ **5.** prime

Practice Exercise 4 **1.** $4(x + 6)(x + 1)$ **2.** $x(x - 5)(x + 3)$ **3.** $2x(3x + 4)(x + 2)$ **4.** $2x(x^2 - 3x + 5)$ **5.** $-2(x - 3)(x + 2)$ **6.** $(3x^2 + 1)(x^2 - 2)$

2.7 Assess Your Understanding

'Are You Prepared?' *Answers are given at the end of these exercises. If you get a wrong answer, read the pages listed in parentheses.*

1. Factor completely: $9x^2 + 45x - 2x - 10$ (pp. 103–106)

2. Factor completely: $3x^2 - 7x + 15x - 35$ (pp. 103–106)

Concepts and Vocabulary

3. Fill in the missing factor:
$6x^2 + 5x - 4 = (3x + 4)$ (________)

4. *True or False:* The polynomial $x^2 + 25$ is prime.

Exercises

In Problems 5–102, factor completely each polynomial (over the integers). If the polynomial cannot be factored, say that it is prime.

5. $x^2 + 5x + 6$
6. $x^2 + 6x + 8$
7. $x^2 + 7x + 6$
8. $x^2 + 9x + 8$
9. $x^2 + 7x + 10$
10. $x^2 + 11x + 10$
11. $x^2 + 17x + 16$
12. $x^2 + 10x + 16$
13. $x^2 + 12x + 20$
14. $x^2 + 9x + 20$
15. $x^2 + 12x + 11$
16. $x^2 + 21x + 20$
17. $x^2 - 10x + 21$
18. $x^2 - 22x + 21$
19. $x^2 - 11x + 10$
20. $x^2 - 8x + 10$
21. $x^2 - 9x + 20$
22. $x^2 - 12x + 20$
23. $x^2 - 10x + 16$
24. $x^2 - 17x + 16$
25. $x^2 - 7x - 8$
26. $x^2 - 2x - 8$
27. $x^2 + 7x - 8$
28. $x^2 + 2x - 8$
29. $x^2 - 6x + 5$
30. $x^2 + 6x + 5$
31. $x^2 - 4x - 5$
32. $x^2 + 4x - 5$
33. $x^2 - 2x - 15$
34. $x^2 + 8x + 15$
35. $x^2 - 8x + 15$
36. $x^2 + 2x - 15$
37. $2x^2 - 12x - 32$
38. $3x^2 + 18x - 48$
39. $x^3 + 10x^2 + 16x$
40. $2x^3 - 20x^2 + 32x$
41. $x^3 - x^2 - 6x$
42. $x^3 - x^2 - 12x$
43. $x^3 - x^2 - 2x$
44. $x^3 - x^2 - 20x$
45. $x^3 + x^2 - 6x$
46. $x^3 + x^2 - 12x$
47. $x^3 + x^2 - 2x$
48. $x^3 + x^2 - 20x$
49. $3x^2 + 4x + 1$
50. $2x^2 + 3x + 1$
51. $2z^2 + 5z + 3$
52. $6z^2 + 5z + 1$
53. $3x^2 + 2x - 8$
54. $3x^2 + 10x + 8$
55. $3x^2 - 2x - 8$
56. $3x^2 - 10x + 8$
57. $3x^2 + 14x + 8$
58. $3x^2 - 14x + 8$
59. $3x^2 + 10x - 8$
60. $3x^2 - 10x - 8$
61. $12x^2 - x - 1$
62. $4x^2 - 9x + 2$
63. $2x^2 - 7x + 3$
64. $-6x^2 - 7x - 2$
65. $-4x^2 - 7x - 5$
66. $-10x^2 + 3x + 1$
67. $-2x^2 + 3x + 2$
68. $10x^2 + 21x - 10$
69. $10x^2 - 13x - 3$
70. $16x^2 - 14x + 3$
71. $16x^2 + 16x + 3$
72. $5x^2 + 13x + 6$
73. $12x^2 - 5x - 2$
74. $6x^2 + 19x + 3$
75. $2x^2 + 3x - 2$
76. $4x^2 + 8x + 3$
77. $8x^2 - 2x - 15$
78. $-3x^2 + 13x + 10$
79. $8x^2 - 26x + 15$
80. $25x^2 - 40x + 16$
81. $-3x^2 + x + 14$
82. $3x^2 + 5x + 2$
83. $12x^2 - 11x + 2$
84. $25x^2 + 5x - 2$
85. $3x^2 + 5x - 2$
86. $3x^2 + x + 6$
87. $8x^2 + 23x + 15$
88. $4x^2 - 3x + 6$
89. $10x^2 + 9x - 2$
90. $2x^2 + 5x - 12$
91. $18x^2 - 9x - 27$
92. $8x^2 - 6x - 2$
93. $8x^2 + 2x + 6$
94. $9x^2 - 3x + 3$
95. $x^2 - x + 4$
96. $x^2 + 6x + 9$
97. $4x^3 - 10x^2 - 6x$
98. $27x^3 - 9x^2 - 6x$
99. $x^4 + 2x^2 + 1$
100. $x^4 + 5x^2 + 4$
101. $x^4 + 8x^2 + 7$
102. $2x^4 - 3x^2 - 2$
103. Show that $x^2 + 4$ is prime.
104. Show that $x^2 + x + 1$ is prime.

'Are You Prepared?' Answers

1. $(9x - 2)(x + 5)$ **2.** $(x + 5)(3x - 7)$

2.8 Special Factoring Formulas

PREPARING FOR THIS SECTION *Before getting started, review the following:*

- Special Products (pp. 87–91)

Now work the 'Are You Prepared?' problems on page 120.

OBJECTIVES

1. Factor the Difference of Two Squares
2. Factor the Sum of Two Cubes
3. Factor the Difference of Two Cubes
4. Factor Perfect Square Trinomials

When you factor a polynomial, first check for common monomial factors. Then check for other types of factoring like factoring by grouping and factoring a second-degree trinomial. You can also check to see whether you can use one of the special product formulas from Section 2.4.

Difference of Two Squares	$x^2 - a^2 = (x - a)(x + a)$	**(1)**
Perfect Squares	$x^2 + 2ax + a^2 = (x + a)^2$	**(2)**
	$x^2 - 2ax + a^2 = (x - a)^2$	**(3)**
Sum of Two Cubes	$x^3 + a^3 = (x + a)(x^2 - ax + a^2)$	**(4)**
Difference of Two Cubes	$x^3 - a^3 = (x - a)(x^2 + ax + a^2)$	**(5)**

1 **EXAMPLE 1** **Factoring the Difference of Two Squares**

Factor completely each polynomial:

(a) $x^2 - 4$ (b) $x^4 - 16$ (c) $5x^2 - 45$

Solution (a) We notice that $x^2 - 4$ is the difference of two squares, x^2 and 2^2. Using equation (1), we find

$$x^2 - 4 = (x - 2)(x + 2)$$

(b) Again using equation (1), with $x^4 = (x^2)^2$ and $16 = 4^2$, we have

$$x^4 - 16 = (x^2 - 4)(x^2 + 4)$$

But $x^2 - 4$ is also the difference of two squares. The complete factorization is

$$x^4 - 16 = (x^2 - 4)(x^2 + 4) = (x - 2)(x + 2)(x^2 + 4)$$

(c) Do not forget to look for common monomial factors first.

$$5x^2 - 45 = 5(x^2 - 9) \quad \text{Factor 5 out first.}$$
$$= 5(x - 3)(x + 3) \quad \text{Factor the difference of two squares.}$$

◄

Remember, because of the Commutative Property, the order in which we write the factors down does not matter. For Example 1(a), we can also write the answer as

$$x^2 - 4 = (x - 2)(x + 2) = (x + 2)(x - 2)$$

Also remember that to avoid errors in factoring, always check your answer by multiplying it out to see if the result equals the original problem.

Practice Exercise 1 *In Problems 1–4, factor completely each polynomial.*

1. $x^2 - 36$ **2.** $25x^2 - 9$ **3.** $18x^2 - 2$ **4.** $x^4 - 81$

The next example illustrates how to factor the sum of two cubes.

2 **EXAMPLE 2** **Factoring the Sum of Two Cubes**

Factor completely:

(a) $x^3 + 8$ (b) $27x^3 + 1$

Solution (a) Equation (4) tells us that the sum of two cubes, $x^3 + a^3$, can be factored as $(x + a)(x^2 - ax + a^2)$. Because $x^3 + 8$ is the sum of two cubes, x^3 and 2^3, we have

$$x^3 + 8 = (x + 2)(x^2 - 2x + 4)$$

Since $x + 2$ and $x^2 - 2x + 4$ are prime polynomials, the factorization is complete.

(b) Again, equation (4) tells us that the sum of two cubes, $x^3 + a^3$, can be factored as $(x + a)(x^2 - ax + a^2)$. Because $27x^3 + 1$ is the sum of two cubes, $(3x)^3$ and 1^3, we have

$$27x^3 + 1 = (3x + 1)(9x^2 - 3x + 1)$$

Since $3x + 1$ and $9x^2 - 3x + 1$ are prime polynomials, the factorization is complete.

The next example illustrates how to factor the difference of two cubes.

3 **EXAMPLE 3** **Factoring the Difference of Two Cubes**

Factor completely:

(a) $x^3 - 1$ (b) $x^3 - 125$

Solution (a) Equation (5) tells us that the difference of two cubes, $x^3 - a^3$, can be factored as $(x - a)(x^2 + ax + a^2)$. Because $x^3 - 1$ is the difference of two cubes, x^3 and 1^3, we find

$$x^3 - 1 = (x - 1)(x^2 + x + 1)$$

Since $x - 1$ and $x^2 + x + 1$ are prime polynomials, the factorization is complete.

(b) Again, Equation (5) tells us that the difference of two cubes, $x^3 - a^3$, can be factored as $(x - a)(x^2 + ax + a^2)$. Because $x^3 - 125$ is the difference of two cubes, x^3 and 5^3, we have

$$x^3 - 125 = (x - 5)(x^2 + 5x + 25)$$

Since $x - 5$ and $x^2 + 5x + 25$ are prime polynomials, the factorization is complete.

Warning: Students often try to factor the second-degree factor that results from factoring the sum or difference of two cubes. *This second-degree factor is always prime and cannot be factored.* Check this by trying to factor some of the second-degree factors from the last two examples; that is, try to factor $x^2 + 5x + 25$ or $9x^2 - 3x + 1$.

Practice Exercise 2 *In Problems 1–3, factor completely each polynomial.*

1. $x^3 - 27$ **2.** $x^3 + 64$ **3.** $125x^3 + 8$

4 The next example illustrates how equations (2) and (3) are used in factoring trinomials that are perfect squares. After factoring out common monomial factors, a trinomial can be checked to see if it is a perfect square by noting whether the first and third terms are perfect squares and positive, and by noting whether the middle term is twice the product of the two numbers you would square to get the first and third terms. Some examples will give you the idea.

EXAMPLE 4 **Factoring Perfect Square Trinomials**

Factor completely:

(a) $x^2 + 6x + 9$ (b) $9x^2 - 6x + 1$ (c) $25x^2 + 30x + 9$

Solution (a) The first term, x^2, and the third term, $9 = 3^2$, are perfect squares. Because the middle term $6x$ is twice the product of x and 3, we have a perfect square.

$$x^2 + 6x + 9 = (x + 3)^2$$

(b) The first term, $9x^2 = (3x)^2$, and the third term, $1 = 1^2$, are perfect squares. Because the middle term $-6x$ is -2 times the product of $3x$ and 1, we have a perfect square.

$$9x^2 - 6x + 1 = (3x - 1)^2$$

(c) The first term, $25x^2 = (5x)^2$, and the third term, $9 = 3^2$, are perfect squares. Because the middle term $30x$ is twice the product of $5x$ and 3, we have a perfect square.

$$25x^2 + 30x + 9 = (5x + 3)^2$$

Practice Exercise 3 *In Problems 1–3, factor completely each polynomial.*

1. $x^2 + 8x + 16$ **2.** $9x^2 - 6x + 1$ **3.** $9x^2 + 12x + 4$

Summary of Factoring Techniques

We close this section with a summary of factoring techniques.

Type of Polynomial	Method	Example
Any polynomial	Look for common monomial factors. (Always do this first!)	$6x^2 + 9x = 3x(2x + 3)$
Binomials of degree 2 or higher	Check for a special product:	
	Difference of two squares, $x^2 - a^2$	$x^2 - 16 = (x - 4)(x + 4)$
	Difference of two cubes, $x^3 - a^3$	$x^3 - 64 = (x - 4)(x^2 + 4x + 16)$
	Sum of two cubes, $x^3 + a^3$	$x^3 + 27 = (x + 3)(x^2 - 3x + 9)$
Trinomials of degree 2	Check for a perfect square, $(x \pm a)^2$.	$x^2 + 8x + 16 = (x + 4)^2$
	See page 107.	$x^2 - x - 2 = (x - 2)(x + 1)$
	Follow the steps on page 110.	$6x^2 + x - 1 = (2x + 1)(3x - 1)$
Four or more terms	Grouping	$2x^3 - 3x^2 + 4x - 6 = (x^2 + 2)(2x - 3)$

Answers to Practice Exercises

Practice Exercise 1 **1.** $(x - 6)(x + 6)$ **2.** $(5x - 3)(5x + 3)$ **3.** $2(3x - 1)(3x + 1)$ **4.** $(x - 3)(x + 3)(x^2 + 9)$

Practice Exercise 2 **1.** $(x - 3)(x^2 + 3x + 9)$ **2.** $(x + 4)(x^2 - 4x + 16)$ **3.** $(5x + 2)(25x^2 - 10x + 4)$

Practice Exercise 3 **1.** $(x + 4)^2$ **2.** $(3x - 1)^2$ **3.** $(3x + 2)^2$

2.8 Assess Your Understanding

'Are You Prepared?' *Answers are given at the end of these exercises. If you get a wrong answer, read the pages listed in parentheses.*

1. Multiply: $(x - 5)(x + 5)$ (pp. 87–91)

2. Multiply: $(3x + 2)^2$ (pp. 87–91)

3. Multiply: $(x - 3)(x^2 + 3x + 9)$ (pp. 87–91)

Concepts and Vocabulary

4. Fill in the missing factor:
$4x^2 - 9 = (2x + 3)$ (__________)

5. *True or False:* $x^3 - 8 = (x - 2)(x - 2)(x - 2)$

6. *True or False:* $4x^2 - 8x + 9$ is a perfect square trinomial.

7. *True or False:* $x^3 + 27 = (x + 3)(x^2 - 3x + 9)$

8. *True or False:* In the formula,
$x^3 - a^3 = (x - a)(x^2 + ax + a^2)$,
the polynomial $x^2 + ax + a^2$ is always prime.

Exercises

In Problems 9–58, factor each polynomial completely.

9. $x^2 - 1$ **10.** $x^2 - 4$ **11.** $4x^2 - 1$ **12.** $9x^2 - 1$ **13.** $x^2 - 16$

14. $x^2 - 25$ **15.** $25x^2 - 4$ **16.** $36x^2 - 9$ **17.** $9x^2 - 16$ **18.** $16x^2 - 9$

19. $x^2 + 2x + 1$ **20.** $x^2 - 4x + 4$ **21.** $x^2 + 4x + 4$ **22.** $x^2 - 2x + 1$ **23.** $x^2 - 10x + 25$

24. $x^2 + 10x + 25$ **25.** $x^2 + 6x + 9$ **26.** $x^2 - 6x + 9$ **27.** $4x^2 + 4x + 1$ **28.** $9x^2 + 6x + 1$

29. $16x^2 + 8x + 1$ **30.** $25x^2 + 10x + 1$ **31.** $4x^2 + 12x + 9$ **32.** $9x^2 - 12x + 4$ **33.** $x^3 - x$

34. $2x^2 - 2$ **35.** $x^3 - 27$ **36.** $x^3 + 27$ **37.** $8x^3 + 27$ **38.** $27 - 8x^3$

39. $x^3 + 6x^2 + 9x$ **40.** $3x^2 - 18x + 27$ **41.** $x^4 - 81$ **42.** $x^4 - 1$ **43.** $x^6 - 2x^3 + 1$

44. $x^6 + 2x^3 + 1$ **45.** $x^7 - x^5$ **46.** $x^8 - x^5$ **47.** $2z^3 + 8z^2 + 8z$ **48.** $3y^3 - 6y^2 + 3y$

49. $16x^2 - 24x + 9$ **50.** $9x^2 + 24x + 16$ **51.** $2x^4 - 2x$ **52.** $2x^4 + 2x$ **53.** $x^4 + 2x^2 + 1$

54. $x^4 - 2x^2 + 1$ **55.** $16x^4 - 1$ **56.** $9x^4 - 9x^2$ **57.** $x^4 + 81x^2$ **58.** $x^4 + 16x^2$

59. Write a paragraph or two outlining a procedure for factoring a polynomial.

60. Make up three factoring problems: one that is a perfect square, one that is prime, and one that requires factoring by grouping. Give them to a friend to solve.

'Are You Prepared?' Answers

1. $x^2 - 25$ **2.** $9x^2 + 12x + 4$ **3.** $x^3 - 27$

Chapter Review

Things to Know

Exponents

$$a^n = \underbrace{a \cdot a \cdot \cdots \cdot a}_{n \text{ factors}} \quad n \text{ a positive integer}$$

$$a^0 = 1 \text{ if } a \neq 0$$

$$a^{-n} = \frac{1}{a^n} \quad \text{if } a \neq 0 \text{ and } n \text{ is a positive integer}$$

Laws of exponents

$$a^m \cdot a^n = a^{m+n}$$
$$(a^m)^n = a^{mn}$$
$$(a \cdot b)^n = a^n \cdot b^n$$
$$\frac{a^m}{a^n} = a^{m-n} \quad \text{if } a \neq 0$$
$$\frac{a^m}{a^n} = \frac{1}{a^{n-m}} \quad \text{if } a \neq 0$$
$$\left(\frac{a}{b}\right)^n = \frac{a^n}{b^n}$$

Monomial: Expression of the form ax^k, a a constant, x a variable, $k \geq 0$ an integer

Polynomial: Algebraic expression of the form $a_n x^n + a_{n-1}x^{n-1} + \cdots + a_1 x + a_0$, n a positive integer

Special products/factoring formulas

$$(x + a)(x + b) = x^2 + (a + b)x + ab$$
$$(ax + b)(cx + d) = acx^2 + (ad + bc)x + bd$$
$$(x - a)(x + a) = x^2 - a^2$$
$$(x + a)^2 = x^2 + 2ax + a^2, \qquad (x - a)^2 = x^2 - 2ax + a^2$$
$$(x - a)(x^2 + ax + a^2) = x^3 - a^3, \qquad (x + a)(x^2 - ax + a^2) = x^3 + a^3$$
$$(x + a)^3 = x^3 + 3ax^3 + 3a^2x + a^3$$
$$(x - a)^3 = x^3 - 3ax^2 + 3a^2x - a^3$$

Objectives

Section		You should be able to	Review Exercises
2.1	✓1	Use the laws of exponents (p. 62)	1–8
2.2	✓1	Recognize monomials (p. 70)	10, 13
	✓2	Combine like terms (p. 71)	9–14
	✓3	Recognize binomials and trinomials (p. 72)	9, 11, 12, 14
	✓4	Recognize polynomials (p. 72)	9–14
	✓5	Evaluate polynomials (p. 74)	15–24
2.3	✓1	Add polynomials (p. 78)	25–28, 33–36, 39
	✓2	Subtract polynomials (p. 79)	29–34, 37, 38, 40

2.4	✓1	Multiply monomials (p. 84)	41–44
	✓2	Multiply a monomial and a polynomial (p. 84)	49, 50, 55, 56
	✓3	Multiply polynomials (p. 85)	45–56
	✓4	Multiply two binomials with the FOIL method (p. 85)	53, 54
	✓5	Find special products (p. 87)	47–50
2.5	✓1	Divide monomials (p. 94)	57–60
	✓2	Divide a polynomial by a monomial (p. 94)	61–66
	✓3	Divide two polynomials (p. 95)	67–76
2.6	✓1	Factor out the greatest common factor (p. 102)	77–82
	✓2	Factor by grouping (p. 103)	95–98
2.7	✓1	Factor a second-degree trinomial: $x^2 + Bx + C$ (p. 107)	87–90, 99, 100
	✓2	Factor a second-degree trinomial: $Ax^2 + Bx + C$ (p. 110)	93, 94, 101–104
2.8	✓1	Factor the difference of two squares (p. 117)	79, 80
	✓2	Factor the sum of two cubes (p. 118)	83
	✓3	Factor the difference of two cubes (p. 118)	84
	✓4	Factor perfect square trinomials (p. 119)	85, 86, 91, 92

Review Exercises

In Problems 1–8, evaluate each expression.

1. $3^2 \cdot 3^0$ **2.** $2^0 \cdot 2^2$ **3.** $\dfrac{4^4}{4^2}$ **4.** $\dfrac{8^3}{8}$ **5.** $[(-2)^2]^3$ **6.** $(2^2)^3$ **7.** $(8^0)^4$ **8.** $(8^4)^0$

In Problems 9–14, combine all like terms. Tell whether the final expression is a monomial, binomial, or trinomial. List the leading coefficient and the degree.

9. $-5x^3 + 3x^2 - 9x^2 + 10$
10. $5x^4 - 19x^4 + 8x^4$
11. $9x^5 + x^5 + 12x^3$
12. $6x + 11$
13. $7x + x$
14. $7x^2 - 3x + 4$

In Problems 15–24, evaluate each polynomial for the given value(s) of the variable(s).

15. $3x^3 + x - 2$ for $x = 2$
16. $-2x^3 + x^2 - x$ for $x = -1$
17. $-x^2 + 1$ for $x = 5$
18. $x^2 + 2x + 4$ for $x = 2$
19. $5z^3 - z^2 + z$ for $z = -2$
20. $-y^3 + 3y^2 + 1$ for $y = 3$
21. $x^2y - xy^2 + 2$ for $x = -1, y = 1$
22. $x^3 - 3x^2y + 3xy^2 - y^3$ for $x = 2, y = 1$
23. $x^3 - x^2 + x - 1$ for $x = 1.2$
24. $x^3 + x^2 + x + 1$ for $x = -1.2$

In Problems 25–56, perform the indicated operations. Express your answer as a polynomial in standard form.

25. $(5x^2 - x + 2) + (-2x^2 + x + 1)$
26. $(4x^3 - x^2 + 1) + (4x^2 + x + 1)$
27. $3(x^2 + x - 2) + 2(3x^2 - x + 2)$
28. $4(-x^2 + 1) + 3(2x^2 + x + 2)$
29. $(3x^4 - x^2 + 1) - (2x^4 + x^3 - x + 2)$
30. $(2x^3 - x^2 + x - 4) - (-3x^3 + x^2 - 2x - 3)$
31. $2(-x^2 + x + 1) - 4(2x^2 + 3x - 2)$
32. $-4(2x^3 + x - 3) - 3(x^3 - x^2 + x - 1)$
33. $(4x^3 - x^2 - 2x + 3) + (-8x^3 + x^2 + 3x + 2) - (2x^3 - x^2 + 2x)$
34. $(8x^2 - x + 4) - (-2x^2 + 3x + 1) + (4x^2 + 1)$

35. $2(3x + 4y - 2) + 5(x + y - 1)$

36. $-2(2x - 3y) + 4(3x - 2y)$

37. $(2x^2 + 3xy + 4y^2) - (x^2 - y^2)$

38. $(-3x^2 + 2xy + 4y^2) - (x^2 - 2xy + y^2)$

39. $4x(x^2 + x + 2) + 8(x^3 - x^2 + 1)$

40. $2(x^3 - x^2 + x - 1) - x^2(2x - 3)$

41. $(9x^2)(-4x)$

42. $(7x)(8x^3)$

43. $(-5xy^3)(3x^2y^4)$

44. $(x^4y^2)(3x^0y^3)$

45. $(2x + 5)(x^2 - x - 1)$

46. $(x - 3)(x^3 + x^2 + x - 3)$

47. $(4x - 1)^2 \cdot (x + 2)$

48. $(x + 1)^2 \cdot (x + 2)$

49. $x(2x + 1)^3$

50. $x(x + 3)^3$

51. $(x + 1)^2 + x(x + 1)$

52. $(x^4 + x^2 + 1)(x^4 - x^2 + 1)$

53. $(2x + 3y)(3x - 4y)$

54. $(5x - 2y)(3x + y)$

55. $3x(2x - 1)^2 - 4x(x + 2) - 3(x + 2)(5x - 1)$

56. $4x^2(2x - 3) - 3(2x + 1)(3x - 2) + x(-x + 4)$

In Problems 57–66, perform the indicated division.

57. $\dfrac{45x^5}{-5x}$

58. $\dfrac{16x^7}{2x^4}$

59. $\dfrac{-x^4y^3}{xy^3}$

60. $\dfrac{28x^3y^5}{4xy^4}$

61. $\dfrac{3x^3 + x^2 - x + 4}{x}$

62. $\dfrac{-4x^3 - x^2 + 2x + 4}{x}$

63. $\dfrac{4x^3 - 8x^2 + 8}{2x^2}$

64. $\dfrac{3x^4 - 27x^2 + 9}{3x^3}$

65. $\dfrac{1 - x^2 + x^4}{x^4}$

66. $\dfrac{1 + x^2 + x^4}{x^2}$

In Problems 67–76, find the quotient and remainder. Check your work by verifying that

$$(Quotient)(Divisor) + Remainder = Dividend$$

67. $3x^3 - x^2 + x + 4$ divided by $x - 3$

68. $2x^3 - 3x^2 + x + 1$ divided by $x - 2$

69. $-3x^4 + x^2 + 2$ divided by $x^2 + 1$

70. $-4x^3 + x^2 - 2$ divided by $x^2 - 1$

71. $8x^4 - 2x^2 + 5x + 1$ divided by $x^2 - 3x + 1$

72. $3x^4 - x^3 - 8x + 4$ divided by $x^2 + 3x - 2$

73. $x^5 + 1$ divided by $x + 1$

74. $x^5 - 1$ divided by $x - 1$

75. $6x^5 + 3x^4 - 4x^3 - 2x^2 + 2x + 1$ divided by $2x + 1$

76. $6x^5 - 3x^4 - 4x^3 + 2x^2 + 2x - 1$ divided by $2x - 1$

In Problems 77–104, factor completely each polynomial (over the integers). If the polynomial cannot be factored, say that it is prime.

77. $3x^2 - 6x$

78. $2x^3 - 8x^2$

79. $9x^3 - x$

80. $x^3 - 4x$

81. $9x^3 + x$

82. $x^3 + 4x$

83. $8x^3 - 1$

84. $8x^3 + 1$

85. $x^3 - 6x^2 + 9x$

86. $3x^2 + 18x + 27$

87. $x^2 + 8x + 12$

88. $x^2 + 8x + 15$

89. $x^2 + 4x - 12$

90. $x^2 - 11x - 12$

91. $x^2 - 6x + 9$

92. $x^2 + 6x + 9$

93. $4x^2 - 8x - 5$

94. $2x^2 - 11x - 6$

95. $2x(3x + 1) - 3(3x + 1)$

96. $x^2(2x - 3) - 4(2x - 3)$

97. $x^3 - x^2 + x - 1$

98. $x^3 + x^2 + x + 1$

99. $x^2 + x + 1$

100. $x^2 - x + 1$

101. $12x^2 - 11x - 15$

102. $12x^2 + 11x - 15$

103. $12x^2 - 27x + 15$

104. $12x^2 - 6x - 15$

3 Rational Expressions

OUTLINE

After polynomials, rational expressions (quotients of polynomials) are the most basic algebraic expressions. In this chapter, we review the procedures used to add, subtract, multiply, and divide rational expressions.

3.1 Operations Using Fractions

PREPARING FOR THIS SECTION *Before getting started, review the following:*

- Express Rational Numbers as Decimals (pp. 15–16)
- Cancellation Property (p. 32)

Now work the 'Are You Prepared?' problems on page 139.

OBJECTIVES

1. Show Equivalency of Fractions
2. Reduce Fractions to Lowest Terms
3. Multiply Fractions
4. Divide Fractions
5. Add and Subtract Fractions with Equal Denominators
6. Add and Subtract Fractions with Unequal Denominators
7. Use the Least Common Multiple Method with Fractions

In Chapter 1 we defined a rational number as the quotient $\frac{a}{b}$ of two integers a and b, where $b \neq 0$. This form of a rational number is also called a **fraction.**

Figure 1

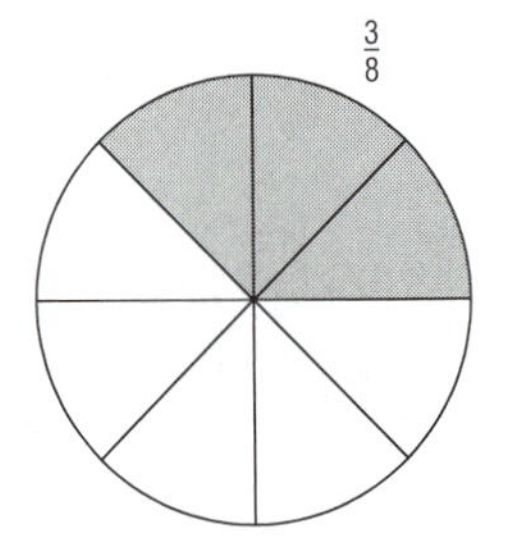

It is sometimes helpful to visualize fractions as "parts of a whole." For example, the fraction $\frac{3}{8}$ is 3 parts of a whole that has been divided into 8 equal parts. See Figure 1.

Two fractions that have different forms may be equal. Such fractions are called **equivalent.** For example, the fractions $\frac{3}{4}$ and $\frac{6}{8}$ are equivalent. To see why, imagine two pizzas illustrated in Figure 2. We divide the pizza in Figure 2(a) into 4 equal slices and the one in Figure 2(b) into 8 equal slices. If someone ate 3 slices from the pizza in Figure 2(a) ($\frac{3}{4}$ of the pizza) and someone else ate 6 slices from the pizza in Figure 2(b) ($\frac{6}{8}$ of the pizza), they have each eaten the same amount of pizza. That is, $\frac{3}{4}$ is equivalent to $\frac{6}{8}$; in other words, $\frac{3}{4} = \frac{6}{8}$.

Figure 2

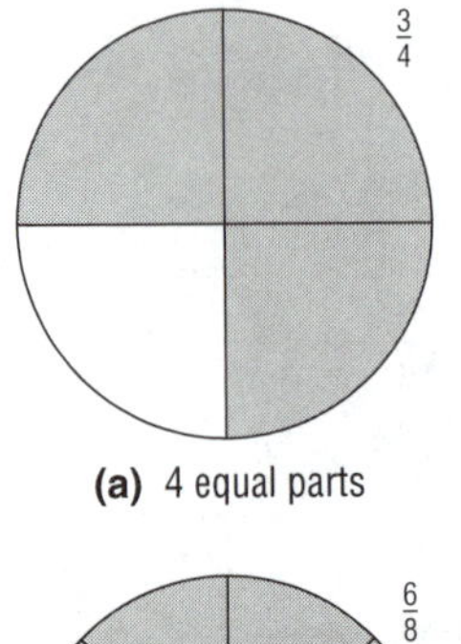

(a) 4 equal parts

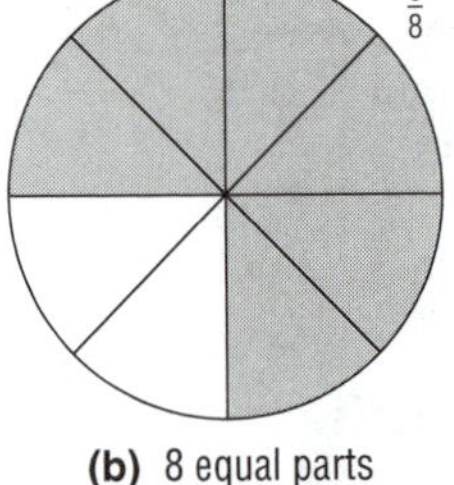

(b) 8 equal parts

We shall discuss three methods that can be used to check for equivalency of fractions.

METHOD 1: We can write each fraction as a decimal and then compare the decimal forms of the numbers. For example, the decimal forms of the fractions $\frac{3}{4}, \frac{6}{8}$, and $\frac{-9}{-12}$ are found as follows:

$$\begin{array}{r} 0.75 \\ 4\overline{)3.00} \\ \underline{2\,8} \\ 20 \\ \underline{20} \end{array} \qquad \begin{array}{r} 0.75 \\ 8\overline{)6.00} \\ \underline{5\,6} \\ 40 \\ \underline{40} \end{array} \qquad \begin{array}{r} 0.75 \\ 12\overline{)9.00} \\ \underline{8\,4} \\ 60 \\ \underline{60} \end{array} \qquad \frac{9}{12} = \frac{-9}{-12}$$

Or, on a calculator:

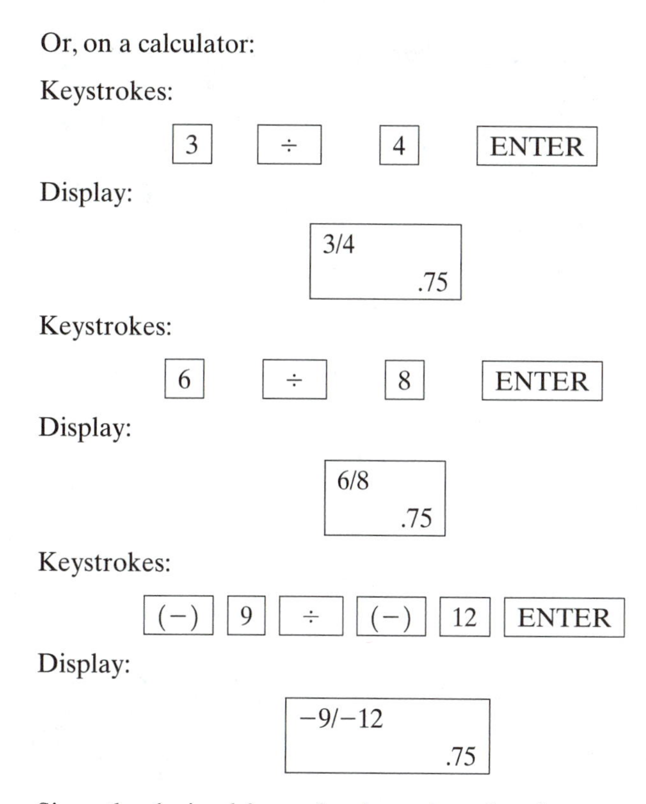

Since the decimal forms for these three fractions are each .75, the three fractions are equal to each other.

METHOD 2: To check for equivalency between two fractions, **cross multiply,** that is, multiply the numerator of each fraction by the denominator of the other fraction. If the resulting products are equal, the fractions are equivalent. For example,

$$\frac{3}{4} = \frac{6}{8} \qquad \text{since } 3 \cdot 8 = 24 \text{ and } 4 \cdot 6 = 24$$

$$\frac{3}{4} = \frac{-9}{-12} \qquad \text{since } 3 \cdot (-12) = -36 \text{ and } 4 \cdot (-9) = -36$$

$$\frac{8}{5} \neq \frac{31}{20} \qquad \text{since } 8 \cdot 20 = 160 \neq 5 \cdot 31 = 155$$

METHOD 3: We can use a Cancellation Property to reduce each fraction to the same fraction. For example,

$$\frac{6}{8} = \frac{3 \cdot \cancel{2}}{4 \cdot \cancel{2}} = \frac{3}{4} \qquad \frac{-9}{-12} = \frac{\cancel{-3} \cdot 3}{\cancel{-3} \cdot 4} = \frac{3}{4}$$ ◀

The following result is the basis for the cross multiplying technique used in Method 2.

If $b \neq 0$ and $d \neq 0$, then

$$\frac{a}{b} = \frac{c}{d} \quad \text{if and only if} \quad a \cdot d = b \cdot c \qquad \textbf{(1)}$$

EXAMPLE 1 **Showing Two Fractions Are Equivalent**

Show that the fractions $\frac{3}{8}$ and $\frac{18}{48}$ are equivalent by

(a) Showing that each one has the same decimal form.
(b) Demonstrating that property (1) is satisfied.
(c) Using the Cancellation Property.

Solution (a)

$$\begin{array}{r} 0.375 \\ 8\overline{)3.000} \\ \underline{2\,4} \\ 60 \\ \underline{56} \\ 40 \\ \underline{40} \end{array} \qquad \begin{array}{r} 0.375 \\ 48\overline{)18.000} \\ \underline{14\,4} \\ 3\,60 \\ \underline{3\,36} \\ 240 \\ \underline{240} \end{array}$$

Each fraction has the same decimal form, 0.375, so they are equivalent.

(b) $\frac{3}{8} = \frac{18}{48}$ since $3 \cdot 48 = 8 \cdot 18$ $(144 = 144)$

(c) $\frac{18}{48} = \frac{\cancel{6} \cdot 3}{\cancel{6} \cdot 8} = \frac{3}{8}$

↑ Cancellation Property

Practice Exercise 1 *In Problems 1–4, follow the directions given in Example 1 for each pair of fractions:*

1. $\frac{5}{8}, \frac{30}{48}$ **2.** $\frac{10}{3}, \frac{40}{12}$ **3.** $\frac{-2}{5}, \frac{6}{-15}$ **4.** $\frac{1215}{61}, \frac{51{,}030}{2562}$

The Cancellation Property (11d) given in Section 1.4 also enables us to form equivalent fractions. Consider, for example, the fraction $\frac{5}{3}$. By multiplying the numerator and denominator by the same nonzero integer, we can obtain other fractions equivalent to $\frac{5}{3}$. For example, each of the fractions below is equivalent to $\frac{5}{3}$:

$$\frac{5}{3} = \frac{5 \cdot 2}{3 \cdot 2} = \frac{10}{6} \qquad \frac{5}{3} = \frac{5 \cdot (-3)}{3 \cdot (-3)} = \frac{-15}{-9} \qquad \frac{5}{3} = \frac{5 \cdot 4}{3 \cdot 4} = \frac{20}{12}$$

2 In this collection of equivalent fractions, only one fraction has a numerator and a denominator containing no common factors (except 1 and −1), namely, $\frac{5}{3}$. When a fraction is written in the form $\frac{a}{b}$, where the integers a and b have no common factors except 1 and −1, we say the fraction is **reduced to lowest terms,** or **simplified.** To reduce a fraction to lowest terms,

we factor the numerator and the denominator and then cancel any common factors using the Cancellation Property from Section 1.4 repeated below:

$$\frac{a \cdot \cancel{c}}{b \cdot \cancel{c}} = \frac{a}{b} \qquad \text{if } b \neq 0, c \neq 0 \qquad \textbf{(2)}$$

We follow the common practice of using slash marks to indicate cancellation, as shown above and in the following example.

EXAMPLE 2 **Reducing a Fraction to Lowest Terms**

Reduce each fraction to lowest terms.

(a) $\frac{15}{35}$ (b) $\frac{12}{-9}$ (c) $\frac{-56}{12}$ (d) $\frac{-24}{-32}$ (e) $\frac{48}{24}$ (f) $\frac{24}{48}$

Solution (a) $\frac{15}{35} = \frac{\cancel{5} \cdot 3}{\cancel{5} \cdot 7} = \frac{3}{7}$ (b) $\frac{12}{-9} = \frac{\cancel{3} \cdot 4}{\cancel{3} \cdot (-3)} = \frac{4}{-3}$

(c) $\frac{-56}{12} = \frac{\cancel{4} \cdot (-14)}{\cancel{4} \cdot 3} = \frac{-14}{3}$ (d) $\frac{-24}{-32} = \frac{-\cancel{8} \cdot 3}{-\cancel{8} \cdot 4} = \frac{3}{4}$

(e) $\frac{48}{24} = \frac{\cancel{24} \cdot 2}{\cancel{24} \cdot 1} = \frac{2}{1} = 2$ (f) $\frac{24}{48} = \frac{\cancel{24} \cdot 1}{\cancel{24} \cdot 2} = \frac{1}{2}$ ◀

In writing a fraction, we will follow the usual practice of reducing it to lowest terms. In so doing, it is customary to write answers such as the one in Example 2(d) as $\frac{3}{4}$, as shown, rather than $\frac{-3}{-4}$. Look again at the solutions for Examples 2(e) and 2(f). When all the original factors in the denominator have been cancelled, as in 2(e), the factor 1 is understood to be present in the denominator, whether it is actually written or not. Writing the 1 in the denominator in a problem like 2(e) is optional and most of the time the 1 is simplified out of the problem. When all the original factors in the numerator are cancelled, as in 2(f), the factor 1 is understood to be present and must be written.

Practice Exercise 2 *In Problems 1–5, reduce each fraction to lowest terms.*

1. $\frac{18}{24}$ **2.** $\frac{63}{21}$ **3.** $\frac{-18}{-32}$ **4.** $\frac{-28}{12}$ **5.** $\frac{12}{36}$ ◀

Multiplication of Fractions

We begin with an example.

3 **EXAMPLE 3** **Multiplying Fractions**

Consider the following situation. After $\frac{1}{2}$ of a pizza has been finished, your friend Don appears and eats $\frac{1}{3}$ of the remaining pizza. How much of the original pizza did Don eat?

Solution Consider the illustration in Figure 3. We see that $\frac{1}{3}$ of $\frac{1}{2}$ is $\frac{1}{6}$ (shaded) of the whole pizza. That is,

Figure 3
6 equal slices

$\frac{1}{3}$ of $\frac{1}{2}$

$\frac{1}{6}$ of the whole

$$\frac{1}{3} \cdot \frac{1}{2} = \frac{1 \cdot 1}{3 \cdot 2} = \frac{1}{6}$$

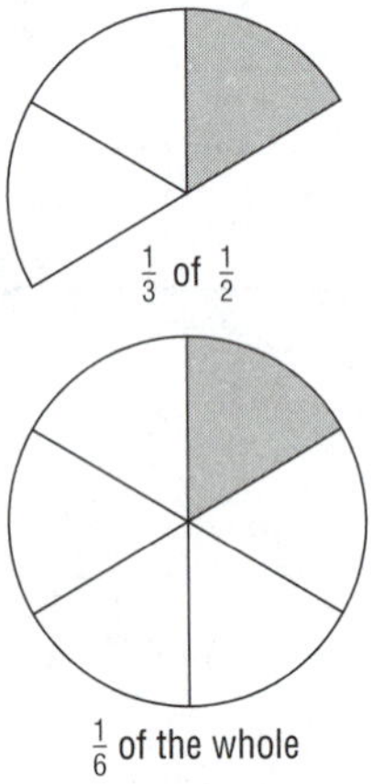

So, Don ate $\frac{1}{6}$ of the original pizza. ◀

The result from Example 3 shows that to multiply two fractions, multiply their numerators and their denominators. That is,

If $\frac{a}{b}$ and $\frac{c}{d}$, $b \neq 0, d \neq 0$, are two fractions, then

$$\frac{a}{b} \cdot \frac{c}{d} = \frac{a \cdot c}{b \cdot d} = \frac{ac}{bd} \quad \text{if } b \neq 0, d \neq 0 \quad \textbf{(3)}$$

EXAMPLE 4 **Multiplying Fractions**

Perform the indicated operation and simplify the result. Remember that the resulting fraction should be simplified (reduced to lowest terms).

(a) $\frac{8}{15} \cdot \frac{9}{4}$ (b) $\frac{-12}{25} \cdot \frac{15}{8}$

Solution (a) $\frac{8}{15} \cdot \frac{9}{4} = \frac{8 \cdot 9}{15 \cdot 4} = \frac{72}{60} = \frac{\cancel{12} \cdot 6}{\cancel{12} \cdot 5} = \frac{6}{5}$

(b) $\frac{-12}{25} \cdot \frac{15}{8} = \frac{-180}{200} = \frac{\cancel{20} \cdot (-9)}{\cancel{20} \cdot 10} = \frac{-9}{10}$ ◀

When multiplying fractions, it is usually simpler to first cancel all common factors and then multiply. For example, Example 4(a) may be worked as follows:

$$\frac{8}{15} \cdot \frac{9}{4} = \frac{8 \cdot 9}{15 \cdot 4} = \frac{\cancel{2} \cdot \cancel{2} \cdot 2 \cdot \cancel{3} \cdot 3}{\cancel{3} \cdot 5 \cdot \cancel{2} \cdot \cancel{2}} = \frac{6}{5}$$

Note: Slanting the cancellation marks in different directions for different factors, as we did above, is a good practice to follow, since it will help in checking for errors.

Practice Exercise 3 *In Problems 1–4, perform the indicated operation and simplify the result.*

1. $\frac{3}{5} \cdot \frac{2}{7}$ **2.** $\frac{-2}{9} \cdot \frac{12}{5}$ **3.** $\frac{18}{35} \cdot \frac{14}{9}$ **4.** $\frac{-25}{12} \cdot \frac{9}{-5}$ ◀

Division of Fractions

Again, we begin with an example.

4 **EXAMPLE 5** **Dividing Fractions**

Marsha has $\frac{2}{3}$ yard of ribbon. How many bows, each requiring $\frac{1}{6}$ yard of ribbon, can she make?

Solution We are looking for how many $\frac{1}{6}$'s there are in $\frac{2}{3}$, that is, we seek the quotient

$$\frac{\frac{2}{3}}{\frac{1}{6}}$$

Figure 4

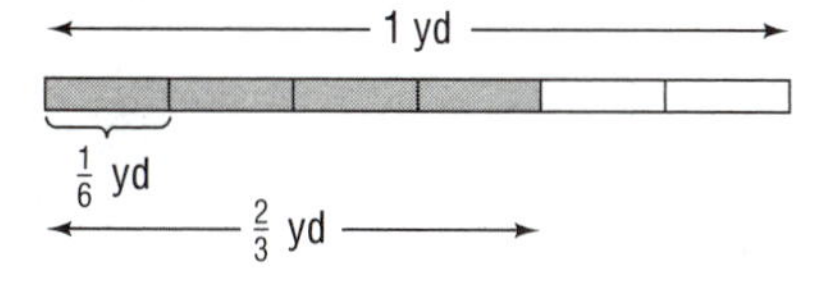

Figure 4 suggests that the answer is 4. To see why, notice that we are looking for how many $\frac{1}{6}$'s there are in $\frac{2}{3} = \frac{4}{6}$. Now we can see that the answer must be 4. That is,

$$\frac{\frac{2}{3}}{\frac{1}{6}} = \frac{2}{3} \div \frac{1}{6} = \frac{2}{3} \cdot \frac{6}{1} = \frac{12}{3} = 4$$

Marsha can make 4 bows. ◀

The result of Example 5 shows that to divide two fractions, we invert the fraction in the denominator and then multiply the two fractions. In other words, interchange the numerator and denominator of the fraction you are dividing by, and then multiply. That is,

If $\frac{a}{b}$ and $\frac{c}{d}$, $b \neq 0, d \neq 0$, are two fractions, then

$$\frac{\frac{a}{b}}{\frac{c}{d}} = \frac{a}{b} \div \frac{c}{d} = \frac{a}{b} \cdot \frac{d}{c} = \frac{ad}{bc} \quad \text{if } b \neq 0, c \neq 0, d \neq 0 \qquad \textbf{(4)}$$

EXAMPLE 6 Dividing Fractions

Perform the indicated operation and simplify the result.

(a) $\frac{\frac{8}{9}}{\frac{2}{3}}$ (b) $\frac{\frac{-24}{5}}{\frac{8}{25}}$

Solution (a) $\frac{\frac{8}{9}}{\frac{2}{3}} = \frac{8}{9} \div \frac{2}{3} = \frac{8}{9} \cdot \frac{3}{2} = \frac{8 \cdot 3}{9 \cdot 2} = \frac{\cancel{2} \cdot 2 \cdot 2 \cdot \cancel{3}}{\cancel{3} \cdot 3 \cdot \cancel{2}} = \frac{4}{3}$

↑ Invert $\frac{2}{3}$ and multiply

(b) $$\frac{\frac{-24}{5}}{\frac{8}{25}} = \frac{-24}{5} \div \frac{8}{25} = \frac{-24}{5} \cdot \frac{25}{8} = \frac{-24 \cdot 25}{5 \cdot 8} = \frac{-(\cancel{2} \cdot \cancel{2} \cdot \cancel{2} \cdot 3) \cdot 5 \cdot \cancel{5}}{\cancel{5} \cdot \cancel{2} \cdot \cancel{2} \cdot \cancel{2}}$$

Invert $\frac{8}{25}$ and multiply

$$= \frac{-15}{1} = -15$$ ◀

Practice Exercise 4 *In Problems 1–4, perform the indicated operation and simplify the result.*

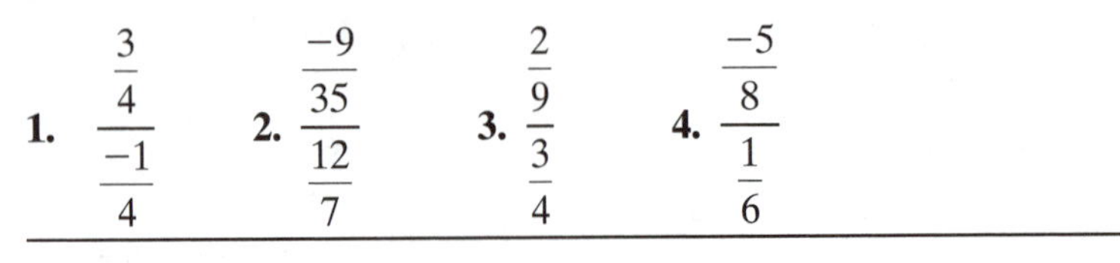

1. $\frac{\frac{3}{4}}{\frac{-1}{4}}$ **2.** $\frac{\frac{-9}{35}}{\frac{12}{7}}$ **3.** $\frac{\frac{2}{9}}{\frac{3}{4}}$ **4.** $\frac{\frac{-5}{8}}{\frac{1}{6}}$ ◀

Addition and Subtraction of Fractions

We begin with an example.

5 **EXAMPLE 7** **Adding Fractions**

If Mike eats $\frac{3}{8}$ of a pizza and Dan eats $\frac{2}{8}$ of the same pizza, how much of the pizza is eaten?

Solution Refer to Figure 5:

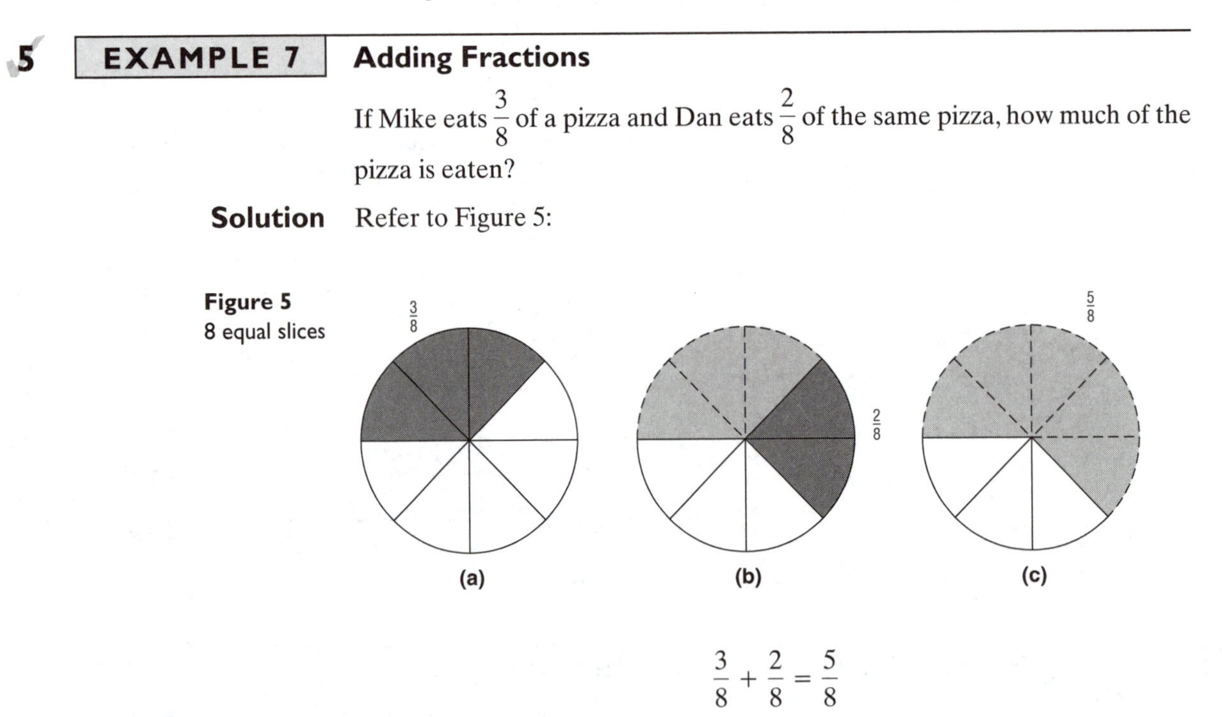

Figure 5 8 equal slices

$$\frac{3}{8} + \frac{2}{8} = \frac{5}{8}$$

That is, $\frac{5}{8}$ of the pizza is eaten. ◀

Example 7 shows that to add two fractions that have the same denominator, we simply add the numerators and keep the denominator intact.

If $\frac{a}{b}$ and $\frac{c}{b}$, $b \neq 0$, are two fractions, then

$$\frac{a}{b} + \frac{c}{b} = \frac{a+c}{b} \quad \text{if } b \neq 0 \qquad \textbf{(5)}$$

Notice in Equation (5) that each fraction has the same denominator b.

EXAMPLE 8 **Adding Fractions**

Perform the indicated operation and simplify the result.

(a) $\frac{1}{5} + \frac{3}{5}$ (b) $\frac{2}{9} + \frac{4}{9}$ (c) $\frac{-5}{12} + \frac{-1}{12}$

Solution (a) $\frac{1}{5} + \frac{3}{5} = \frac{1+3}{5} = \frac{4}{5}$

(b) $\frac{2}{9} + \frac{4}{9} = \frac{2+4}{9} = \frac{6}{9} = \frac{\cancel{3}\cdot 2}{\cancel{3}\cdot 3} = \frac{2}{3}$

(c) $\frac{-5}{12} + \frac{-1}{12} = \frac{-5+(-1)}{12} = \frac{-6}{12} = \frac{\cancel{6}\cdot(-1)}{\cancel{6}\cdot 2} = \frac{-1}{2}$ ◀

The rule for subtracting two fractions with the same denominator is obtained as follows:

$$\frac{a}{b} - \frac{c}{b} = \frac{a}{b} + \left(-\frac{c}{b}\right) = \frac{a}{b} + \frac{-c}{b} = \frac{a+(-c)}{b} = \frac{a-c}{b}$$

That is,

If $\frac{a}{b}$ and $\frac{c}{b}$, $b \neq 0$, are two fractions, then

$$\frac{a}{b} - \frac{c}{b} = \frac{a-c}{b} \quad \text{if } b \neq 0 \qquad \textbf{(6)}$$

Notice in equation (6) that each fraction has the same denominator b.

EXAMPLE 9 **Subtracting Fractions**

Perform the indicated operation and simplify the result.

(a) $\frac{9}{7} - \frac{2}{7}$ (b) $\frac{9}{14} - \frac{3}{14}$ (c) $\frac{-9}{10} - \frac{-13}{10}$

Solution (a) $\frac{9}{7} - \frac{2}{7} = \frac{9-2}{7} = \frac{7}{7} = 1$

(b) $\frac{9}{14} - \frac{3}{14} = \frac{9-3}{14} = \frac{6}{14} = \frac{\cancel{2}\cdot 3}{\cancel{2}\cdot 7} = \frac{3}{7}$

(c) $\frac{-9}{10} - \frac{-13}{10} = \frac{-9-(-13)}{10} = \frac{-9+13}{10} = \frac{4}{10} = \frac{\cancel{2}\cdot 2}{\cancel{2}\cdot 5} = \frac{2}{5}$ ◀

EXAMPLE 10 **Operations Involving Fractions**

Perform the indicated operations and simplify the result.

(a) $\frac{4}{9} - \frac{5}{9} + \frac{8}{9}$ (b) $\frac{2}{3}\cdot\frac{5}{4} - \frac{1}{6}$

Solution (a) $\frac{4}{9} - \frac{5}{9} + \frac{8}{9} = \frac{4-5}{9} + \frac{8}{9} = \frac{-1}{9} + \frac{8}{9} = \frac{-1+8}{9} = \frac{7}{9}$

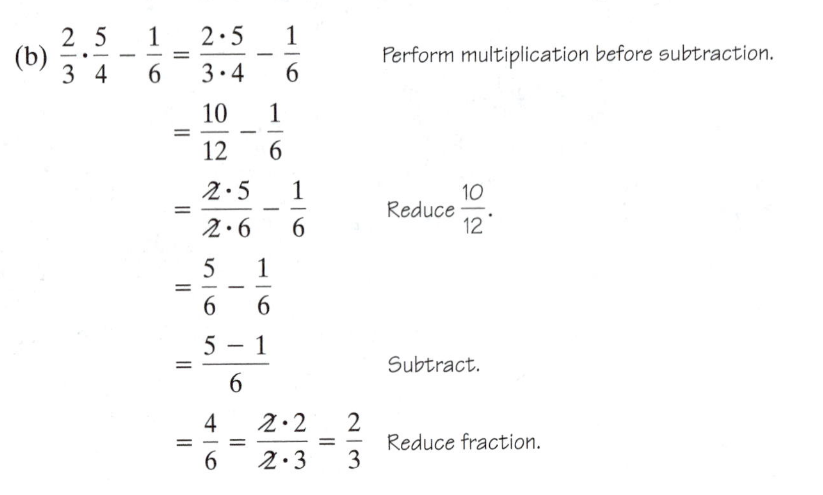

↑ Perform addition and subtraction left to right.

(b) $\frac{2}{3} \cdot \frac{5}{4} - \frac{1}{6} = \frac{2 \cdot 5}{3 \cdot 4} - \frac{1}{6}$ Perform multiplication before subtraction.

$= \frac{10}{12} - \frac{1}{6}$

$= \frac{\not{2} \cdot 5}{\not{2} \cdot 6} - \frac{1}{6}$ Reduce $\frac{10}{12}$.

$= \frac{5}{6} - \frac{1}{6}$

$= \frac{5-1}{6}$ Subtract.

$= \frac{4}{6} = \frac{\not{2} \cdot 2}{\not{2} \cdot 3} = \frac{2}{3}$ Reduce fraction. ◀

Practice Exercise 5 *In Problems 1–6, perform the indicated operation(s) and simplify the result.*

1. $\frac{3}{8} + \frac{5}{8}$ **2.** $\frac{2}{3} + \frac{2}{3}$ **3.** $\frac{8}{3} + \frac{13}{3}$ **4.** $\frac{9}{8} - \frac{5}{8}$ **5.** $\frac{5}{2} + \frac{3}{2} - \frac{1}{2}$

6. $\frac{4}{7} - \frac{3}{2} \cdot \frac{6}{7}$ ◀

6 The addition of fractions whose denominators are not equal requires a different approach. Consider the problem

$$\frac{2}{3} + \frac{1}{4}$$

Figure 6 illustrates the situation:

Figure 6

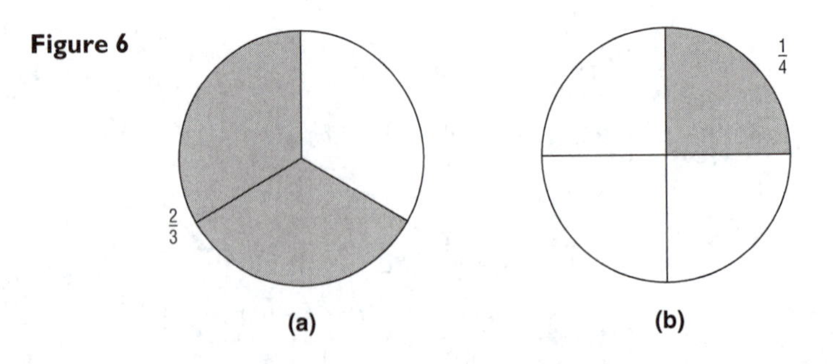

We need to divide each whole into smaller pieces, so that both $\frac{2}{3}$ and $\frac{1}{4}$ can be represented with the same size pieces. Since $\frac{2}{3} = \frac{2 \cdot 4}{3 \cdot 4} = \frac{8}{12}$ and $\frac{1}{4} = \frac{1 \cdot 3}{4 \cdot 3} = \frac{3}{12}$, we divide each whole into 12 equal parts, as shown in Figure 7.

Figure 7

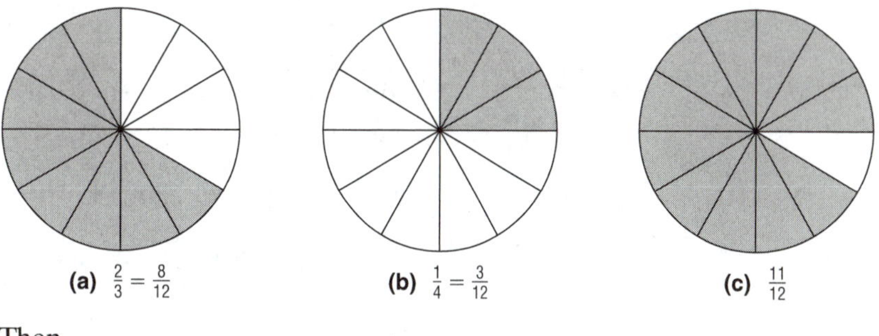

Then,

$$\frac{2}{3} + \frac{1}{4} = \frac{2 \cdot 4}{3 \cdot 4} + \frac{1 \cdot 3}{4 \cdot 3} = \frac{8}{12} + \frac{3}{12} = \frac{11}{12}$$

To add and subtract fractions whose denominators are not equal, we replace the given fractions by equivalent fractions that have the same denominator. The procedure for replacing a fraction with an equivalent fraction can be thought of in two ways:

- Use the Cancellation Property to obtain an equivalent fraction as in

$$\frac{2}{3} = \frac{2 \cdot 4}{3 \cdot 4} = \frac{8}{12}$$

- Multiply by 1, the multiplicative identity, as in

$$\frac{2}{3} = \frac{2}{3} \cdot 1 = \underset{\substack{\uparrow \\ \text{Since } 1 = \frac{4}{4}}}{\frac{2}{3} \cdot \frac{4}{4}} = \frac{2 \cdot 4}{3 \cdot 4} = \frac{8}{12}$$

EXAMPLE 11 **Adding and Subtracting Fractions with Unequal Denominators**

Perform the indicated operation and simplify the result:

(a) $\frac{3}{5} + \frac{5}{6}$ (b) $\frac{7}{9} - \frac{1}{2}$ (c) $\frac{-3}{8} + \frac{1}{5}$

Solution (a) $\frac{3}{5} + \frac{5}{6} = \frac{3 \cdot 6}{5 \cdot 6} + \frac{5 \cdot 5}{6 \cdot 5} = \frac{18}{30} + \frac{25}{30} = \frac{18 + 25}{30} = \frac{43}{30}$

(b) $\frac{7}{9} - \frac{1}{2} = \frac{7 \cdot 2}{9 \cdot 2} - \frac{1 \cdot 9}{2 \cdot 9} = \frac{14}{18} - \frac{9}{18} = \frac{5}{18}$

(c) $\frac{-3}{8} + \frac{1}{5} = \frac{-3 \cdot 5}{8 \cdot 5} + \frac{1 \cdot 8}{5 \cdot 8} = \frac{-15}{40} + \frac{8}{40} = \frac{-7}{40}$ ◀

In general, we have the following rules:

If $\frac{a}{b}$ and $\frac{c}{d}$, $b \neq 0, d \neq 0$, are two fractions, then

$$\frac{a}{b} + \frac{c}{d} = \frac{ad}{bd} + \frac{bc}{bd} = \frac{ad + bc}{bd} \quad \text{if } b \neq 0, d \neq 0 \qquad \textbf{(7)}$$

and

$$\frac{a}{b} - \frac{c}{d} = \frac{ad}{bd} - \frac{bc}{bd} = \frac{ad - bc}{bd} \quad \text{if } b \neq 0, d \neq 0 \qquad \textbf{(8)}$$

We will use these general rules in the next example.

EXAMPLE 12 **Adding and Subtracting Fractions with Unequal Denominators**

Perform the indicated operation and simplify the result.

(a) $\frac{1}{3} + \frac{2}{5}$ (b) $\frac{5}{2} - \frac{2}{5}$

Solution (a) $\frac{1}{3} + \frac{2}{5} = \frac{1}{3} \cdot \frac{5}{5} + \frac{3}{3} \cdot \frac{2}{5} = \frac{1 \cdot 5 + 3 \cdot 2}{3 \cdot 5} = \frac{5 + 6}{15} = \frac{11}{15}$

(b) $\frac{5}{2} - \frac{2}{5} = \frac{5}{2} \cdot \frac{5}{5} - \frac{2}{2} \cdot \frac{2}{5} = \frac{5 \cdot 5 - 2 \cdot 2}{2 \cdot 5} = \frac{25 - 4}{10} = \frac{21}{10}$ ◀

Practice Exercise 6 *In Problems 1–4, perform the indicated operation and simplify the result.*

1. $\frac{2}{3} + \frac{8}{5}$ **2.** $\frac{9}{5} - \frac{1}{2}$ **3.** $\frac{4}{9} + \frac{5}{7}$ **4.** $\frac{3}{10} + \frac{2}{3}$ ◀

Least Common Multiple (LCM)

7 If the denominators of two fractions to be added or subtracted have common factors, we do not have to use the general rules given by equations (7) and (8), since a more efficient method is available. Instead, we use the **least common multiple (LCM)** method of determining the smallest integer that is exactly divisible by each denominator. To find the least common multiple of two or more integers, first factor each integer completely into its prime factors. The LCM is the product of the different prime factors of each denominator, each factor appearing the greatest number of times it occurs in any one of the denominators. Some examples will give you the idea.

EXAMPLE 13 **Finding the Least Common Multiple**

Find the least common multiple of 12 and 32.

Solution First, we factor each integer completely:

$$12 = 2 \cdot 2 \cdot 3 \quad \text{and} \quad 32 = 2 \cdot 2 \cdot 2 \cdot 2 \cdot 2$$

Then to find the LCM, start by writing the factors of one of the integers, let's say 12 in this case. (Alternatively, you could start with the factors of 32.)

$$2 \cdot 2 \cdot 3$$

Next, look at the factored form of the other integer, 32. It contains the factor 2 repeated five times. Since our list contains 2 repeated twice, we insert three additional 2's to obtain the LCM of 12 and 32; that is, the LCM is

$$2 \cdot 2 \cdot 2 \cdot 2 \cdot 2 \cdot 3 = 96$$ ◀

EXAMPLE 14 **Finding the Least Common Multiple**

Find the least common multiple of 18 and 30.

Solution First, we factor each integer completely:

$$18 = 2 \cdot 3 \cdot 3 \quad \text{and} \quad 30 = 2 \cdot 3 \cdot 5$$

We start with the factors of 18:

$$2 \cdot 3 \cdot 3$$

Next, we examine the factors of 30. The factor 2 is already in our list, so we do not repeat it. The next factor 3 also is already in our list, so it is not repeated. Finally, the factor 5 is not in our list, so we insert it. The LCM of 18 and 30 is

$$2 \cdot 3 \cdot 3 \cdot 5 = 90$$ ◀

Practice Exercise 7 *Find the least common multiple of each pair of integers.*

1. 50 and 15 **2.** 12 and 42 **3.** 16 and 18

The next example illustrates how the LCM is used to add and subtract fractions.

EXAMPLE 15 **Using the Least Common Multiple with Fractions**

Perform the indicated operation and simplify the result.

(a) $\frac{5}{12} + \frac{1}{32}$ (b) $\frac{7}{18} - \frac{11}{30}$

Solution (a) From Example 13, we know the LCM of 12 and 32 is 96. We replace each fraction to be added by an equivalent fraction that has 96 as the denominator:

$$\frac{5}{12} + \frac{1}{32} = \frac{5 \cdot 8}{12 \cdot 8} + \frac{1 \cdot 3}{32 \cdot 3} = \frac{40}{96} + \frac{3}{96} = \frac{43}{96}$$

(b) From Example 14, we know the LCM of 18 and 30 is 90. We replace each fraction by an equivalent fraction that has 90 as the denominator:

$$\frac{7}{18} - \frac{11}{30} = \frac{7 \cdot 5}{18 \cdot 5} - \frac{11 \cdot 3}{30 \cdot 3} = \frac{35}{90} - \frac{33}{90} = \frac{2}{90} = \frac{\cancel{2} \cdot 1}{\cancel{2} \cdot 45} = \frac{1}{45}$$ ◀

If we had not used the LCM method to add the fractions in Example 15(a), but decided instead to use the general rule of equation (7), we would have made the problem more complicated than it needs to be, as the following demonstrates:

$$\frac{5}{12} + \frac{1}{32} = \frac{5 \cdot 32}{12 \cdot 32} + \frac{12 \cdot 1}{12 \cdot 32} = \frac{5 \cdot 32 + 12 \cdot 1}{12 \cdot 32} = \frac{160 + 12}{384} = \frac{172}{384} = \frac{\cancel{4} \cdot 43}{\cancel{4} \cdot 96} = \frac{43}{96}$$

↑ This part may prove difficult due to the size of the numbers.

Always look for common factors in the denominators of fractions to be added or subtracted and use the LCM method if any common factors are found.

Practice Exercise 8 *Perform the indicated operation and simplify the result.*

1. $\frac{1}{3} + \frac{2}{9}$ **2.** $\frac{3}{8} - \frac{1}{6}$ **3.** $\frac{-5}{4} + \frac{5}{14}$

The LCM method also works well when three or more fractions are to be added or subtracted.

EXAMPLE 16 Using the Least Common Multiple with Fractions

Perform the indicated operations and simplify the result.

$$\frac{1}{6} + \frac{2}{15} - \frac{3}{8}$$

Solution First, we factor completely each denominator to find the LCM:

$$6 = 2 \cdot 3 \qquad 15 = 3 \cdot 5 \qquad 8 = 2 \cdot 2 \cdot 2$$

We begin with $6 = 2 \cdot 3$. Since $15 = 3 \cdot 5$ has one 3, which is already in our list, and one 5, which is not, we insert a factor of 5 in our list to obtain

$$2 \cdot 3 \cdot 5$$

Now we look at $8 = 2 \cdot 2 \cdot 2$. It has three 2's and our list has only one. We insert two additional factors of 2 to obtain

$$2 \cdot 2 \cdot 2 \cdot 3 \cdot 5$$

The LCM of 6, 15, and 8 is $2 \cdot 2 \cdot 2 \cdot 3 \cdot 5 = 120$.

Next, we rewrite each fraction using 120 as the common denominator:

$$\begin{aligned}
\frac{1}{6} + \frac{2}{15} - \frac{3}{8} &= \frac{1 \cdot 20}{6 \cdot 20} + \frac{2 \cdot 8}{15 \cdot 8} - \frac{3 \cdot 15}{8 \cdot 15} \\
&= \frac{20}{120} + \frac{16}{120} - \frac{45}{120} \\
&= \frac{20 + 16 - 45}{120} \\
&= \frac{-9}{120} = \frac{-3}{40}
\end{aligned}$$

◀

Practice Exercise 9 *Perform the indicated operations and simplify the result.*

1. $\frac{1}{9} - \frac{7}{15} + \frac{4}{45}$ **2.** $\frac{5}{14} - \frac{3}{4} + \frac{9}{10}$ **3.** $1 + \frac{1}{2} + \frac{1}{3} + \frac{1}{4}$

◀

EXAMPLE 17 Using Fractions in An Application

Allysa and Ian order a large pizza that costs \$12. Allysa eats $\frac{2}{5}$ of the pizza while Ian eats the rest. Based on consumption, how much should each one pay to share the cost fairly?

Solution To find out how much each one pays, multiply the amount each person eats by the total cost of the pizza. Allysa pays

$$\frac{2}{5} \cdot \$12 = \frac{2}{5} \cdot \frac{12}{1} = \frac{24}{5} = \$4.80$$

Since Ian eats the rest of the pizza, he eats

$$1 - \frac{2}{5} = \frac{5}{5} - \frac{2}{5} = \frac{3}{5}$$

of the pizza. So Ian pays

$$\frac{3}{5} \cdot \$12 = \frac{3}{5} \cdot \frac{12}{1} = \frac{36}{5} = \$7.20$$

The total paid by Allysa and Ian is

$$\$4.80 + \$7.20 = \$12.00$$

as required.

Alternatively, we could have found Ian's cost by subtracting Allysa's cost from \$12:

$$\$12 - \$4.80 = \$7.20$$

◀

Answers to Practice Exercises

Practice Exercise 1

1. (a) 0.625 (b) $5 \cdot 48 = 8 \cdot 30 = 240$ (c) $\frac{30}{48} = \frac{\cancel{6} \cdot 5}{\cancel{6} \cdot 8} = \frac{5}{8}$

2. (a) 3.333... (b) $10 \cdot 12 = 3 \cdot 40 = 120$ (c) $\frac{40}{12} = \frac{\cancel{4} \cdot 10}{\cancel{4} \cdot 3} = \frac{10}{3}$

3. (a) -0.4 (b) $(-2)(-15) = 5 \cdot 6 = 30$

(c) $\frac{6}{-15} = \frac{\cancel{3} \cdot 2}{\cancel{3} \cdot (-5)} = \frac{2}{-5} = \frac{-2}{5}$

4. (a) 19.91803279 (b) $1215 \cdot 2562 = 61 \cdot 51{,}030 = 3{,}112{,}830$

(c) $\frac{51{,}030}{2562} = \frac{\cancel{42} \cdot 1215}{\cancel{42} \cdot 61} = \frac{1215}{61}$

Practice Exercise 2 **1.** $\frac{3}{4}$ **2.** 3 **3.** $\frac{9}{16}$ **4.** $\frac{-7}{3}$ **5.** $\frac{1}{3}$

Practice Exercise 3 **1.** $\frac{6}{35}$ **2.** $\frac{-8}{15}$ **3.** $\frac{4}{5}$ **4.** $\frac{15}{4}$

Practice Exercise 4 **1.** -3 **2.** $\frac{-3}{20}$ **3.** $\frac{8}{27}$ **4.** $\frac{-15}{4}$

Practice Exercise 5 **1.** 1 **2.** $\frac{4}{3}$ **3.** 7 **4.** $\frac{1}{2}$ **5.** $\frac{7}{2}$ **6.** $\frac{-5}{7}$

Practice Exercise 6 **1.** $\frac{34}{15}$ **2.** $\frac{13}{10}$ **3.** $\frac{73}{63}$ **4.** $\frac{29}{30}$

Practice Exercise 7 **1.** 150 **2.** 84 **3.** 144

Practice Exercise 8 **1.** $\frac{5}{9}$ **2.** $\frac{5}{24}$ **3.** $\frac{-25}{28}$

Practice Exercise 9 **1.** $\frac{-4}{15}$ **2.** $\frac{71}{140}$ **3.** $\frac{25}{12}$

3.1 Assess Your Understanding

'Are You Prepared?' *Answers are given at the end of these exercises. If you get a wrong answer, read the pages listed in parentheses.*

1. By the Cancellation Property, $\frac{5 \cdot a}{8 \cdot a} =$ __________. (p. 32)

2. The decimal form of $\frac{6}{11}$ is __________. (pp. 15–16)

Concepts and Vocabulary

3. *True or False:* Cross multiplying can be used to determine if two fractions are equivalent.

4. When reduced to lowest terms $\frac{-20}{36} =$ __________.

5. To multiply two fractions we multiply their __________ and their __________.

6. *True or False:* Multiplying fractions requires a common denominator.

7. The LCM of 60 and 45 is __________.

8. If the numerator and denominator of the fraction contain no common factors (except 1 and -1) the fraction is said to be in __________ __________.

9. *True or False:* Adding fractions requires a common denominator.

10. *True or False:* $\frac{1}{2}$ of $\frac{1}{3}$ is $\frac{2}{3}$

Exercises

In Problems 11–20, show that the fractions are equivalent by:

(a) *Showing that each one has the same decimal form*

(b) *Demonstrating that property (1) is satisfied*

(c) *Using the Cancellation Property (2)*

11. $\frac{5}{8}$ and $\frac{30}{48}$ **12.** $\frac{7}{8}$ and $\frac{28}{32}$ **13.** $\frac{10}{8}$ and $\frac{5}{4}$ **14.** $\frac{18}{5}$ and $\frac{108}{30}$ **15.** $\frac{-1}{8}$ and $\frac{-14}{112}$

16. $\frac{-2}{3}$ and $\frac{-16}{24}$ **17.** $\frac{48}{14}$ and $\frac{-72}{-21}$ **18.** $\frac{18}{12}$ and $\frac{-9}{-6}$ **19.** $\frac{-40}{32}$ and $\frac{-60}{48}$ **20.** $\frac{-9}{24}$ and $\frac{-6}{16}$

In Problems 21–28, reduce each fraction to lowest terms.

21. $\frac{30}{54}$ **22.** $\frac{21}{48}$ **23.** $\frac{82}{18}$ **24.** $\frac{84}{60}$

25. $\frac{-21}{36}$ **26.** $\frac{-14}{24}$ **27.** $\frac{-28}{-36}$ **28.** $\frac{-42}{-36}$

In Problems 29–64, perform the indicated operation(s) and simplify the result.

29. $\frac{5}{4}+\frac{1}{4}$ **30.** $\frac{2}{3}+\frac{5}{3}$ **31.** $\frac{13}{2}+\frac{9}{2}+\frac{11}{2}$ **32.** $\frac{6}{5}+\frac{7}{5}+\frac{2}{5}$

33. $\frac{13}{5}-\frac{8}{5}$ **34.** $\frac{9}{4}-\frac{6}{4}$ **35.** $\frac{9}{2}-\frac{7}{2}$ **36.** $\frac{8}{3}-\frac{1}{3}$

37. $\frac{8}{3}-\frac{1}{3}+\frac{2}{3}$ **38.** $\frac{3}{4}-\frac{7}{4}+\frac{1}{4}$ **39.** $\frac{9}{5}+\frac{6}{5}-\frac{4}{5}$ **40.** $\frac{7}{2}+\frac{3}{2}-\frac{5}{2}$

41. $\frac{3}{2}\cdot\frac{8}{9}$ **42.** $\frac{5}{6}\cdot\frac{18}{25}$ **43.** $\frac{16}{25}\cdot\frac{5}{8}$ **44.** $\frac{2}{3}\cdot\frac{9}{8}$

45. $\frac{25}{18}\cdot\frac{-6}{5}$ **46.** $\frac{5}{24}\cdot\frac{-12}{25}$ **47.** $\frac{8}{9}\cdot\frac{3}{2}\cdot\frac{18}{7}$ **48.** $\frac{9}{8}\cdot\frac{5}{18}\cdot\frac{9}{10}$

49. $\frac{1}{3}+\frac{2}{3}\cdot\frac{1}{2}$ **50.** $\frac{2}{3}\cdot\frac{3}{8}+\frac{3}{8}$ **51.** $\frac{4}{5}\cdot\frac{10}{3}-\frac{8}{3}$ **52.** $\frac{9}{8}\cdot\frac{10}{3}-\frac{3}{4}$

53. $\frac{-5}{8}\cdot\frac{12}{25}$ **54.** $\frac{-4}{9}\cdot\frac{33}{40}$ **55.** $\frac{-16}{-3}\cdot\frac{9}{20}$ **56.** $\frac{21}{4}\cdot\frac{-8}{-9}$

57. $\dfrac{\frac{15}{8}}{\frac{25}{2}}$ **58.** $\dfrac{\frac{14}{9}}{\frac{42}{5}}$ **59.** $\dfrac{\frac{-15}{32}}{\frac{25}{24}}$ **60.** $\dfrac{\frac{24}{35}}{\frac{-8}{15}}$

61. $\dfrac{\frac{12}{5}}{\frac{24}{15}}$ **62.** $\dfrac{\frac{5}{4}}{\frac{5}{12}}$ **63.** $\dfrac{\frac{22}{9}}{\frac{33}{12}}$ **64.** $\dfrac{\frac{6}{35}}{\frac{10}{25}}$

In Problems 65–74, find the LCM of the given integers.

65. 6 and 4 **66.** 18 and 27 **67.** 6 and 9 **68.** 8 and 20 **69.** 12 and 18
70. 18 and 24 **71.** 15 and 18 **72.** 15 and 24 **73.** 4, 6, and 15 **74.** 6, 9, and 12

In Problems 75–128, perform the indicated operation(s) and simplify the result.

75. $\frac{4}{3}+\frac{3}{2}$ **76.** $\frac{2}{3}+\frac{1}{2}$ **77.** $\frac{1}{5}+\frac{3}{4}$ **78.** $\frac{9}{10}+\frac{2}{3}$

79. $\frac{4}{9}+\frac{3}{8}$ **80.** $\frac{6}{7}+\frac{1}{2}$ **81.** $\frac{8}{3}-\frac{1}{2}$ **82.** $\frac{2}{3}-\frac{1}{2}$

83. $\frac{1}{5}-\frac{3}{4}$ **84.** $\frac{9}{10}-\frac{2}{3}$ **85.** $\frac{4}{9}-\frac{3}{8}$ **86.** $\frac{6}{7}-\frac{1}{2}$

87. $\frac{2}{7}+\frac{5}{14}$ **88.** $\frac{1}{6}+\frac{5}{12}$ **89.** $\frac{9}{13}+\frac{11}{26}$ **90.** $\frac{2}{9}+\frac{13}{18}$

91. $\frac{9}{10}-\frac{1}{5}$ **92.** $\frac{9}{13}-\frac{11}{26}$ **93.** $\frac{5}{12}-\frac{11}{24}$ **94.** $\frac{7}{8}-\frac{9}{16}$

95. $\frac{5}{9}+\frac{5}{6}$ **96.** $\frac{1}{4}+\frac{3}{14}$ **97.** $\frac{1}{6}+\frac{11}{15}$ **98.** $\frac{5}{12}+\frac{7}{20}$

99. $\frac{3}{8}+\frac{5}{12}$ **100.** $\frac{1}{6}+\frac{4}{9}$ **101.** $\frac{5}{6}+\frac{3}{4}$ **102.** $\frac{7}{8}+\frac{5}{6}$

103. $\frac{5}{9}-\frac{5}{6}$ **104.** $\frac{1}{4}-\frac{3}{14}$ **105.** $\frac{1}{6}-\frac{11}{15}$ **106.** $\frac{5}{12}-\frac{7}{20}$

107. $\frac{3}{8}-\frac{5}{12}$ **108.** $\frac{1}{6}-\frac{4}{9}$ **109.** $\frac{5}{6}-\frac{3}{4}$ **110.** $\frac{7}{8}-\frac{5}{6}$

111. $\frac{5}{2}-\frac{8}{3}+\frac{1}{6}$ **112.** $\frac{4}{5}+\frac{2}{3}-\frac{4}{15}$ **113.** $\frac{3}{4}+\frac{8}{3}-\frac{11}{12}$ **114.** $\frac{2}{7}-\frac{2}{3}+\frac{5}{21}$

115. $\frac{1}{12}+\frac{4}{3}+\frac{3}{4}$ **116.** $\frac{3}{20}+\frac{4}{5}+\frac{1}{4}$ **117.** $\frac{1}{2}+\frac{1}{3}+\frac{1}{4}$ **118.** $\frac{1}{2}+\frac{1}{3}-\frac{1}{4}$

119. $\frac{3}{8}+\frac{1}{4}-\frac{1}{6}$ **120.** $\frac{5}{8}-\frac{3}{4}+\frac{5}{6}$ **121.** $\frac{3}{4}\cdot\left(\frac{3}{2}+\frac{2}{3}\right)$ **122.** $\frac{3}{2}\cdot\left(\frac{1}{4}-\frac{3}{8}\right)$

123. $\frac{5}{6}+\frac{3}{4}-\left(\frac{1}{2}\right)^2$ **124.** $\frac{1}{6}-\frac{1}{4}+\left(\frac{1}{2}\right)^2$ **125.** $\frac{5}{12}-\frac{1}{18}+\frac{7}{30}$ **126.** $\frac{9}{10}-\frac{2}{25}+\frac{5}{12}$

127. $\frac{4}{5}\cdot\frac{25}{8}+\frac{5}{2}\cdot\frac{3}{25}$ **128.** $\frac{3}{4}\cdot\frac{8}{9}-\frac{5}{12}\cdot\frac{18}{5}$

---

129. Mike and Kathleen order a large pizza that costs \$10.00. Mike eats $\frac{5}{8}$ of the pizza and Kathleen eats the rest. Based on consumption, how much should each one pay to share the cost fairly?

130. Don ordered a jumbo pizza and ate $\frac{1}{3}$ of it. His friend Tom ate $\frac{3}{4}$ of what was left. How much of the original pizza did Tom eat? Who ate more, Don or Tom? How much pizza was left over?

131. Katy, Mike, and Danny agree to order an extra-large pizza for \$15 and divide the cost based on consumption. The pizza arrives cut into 8 equal slices. Katy eats 1 slice, Mike eats 2 slices, and Danny eats 3 slices; 2 slices remain uneaten. How much should each one pay?

132. Katy, Mike, and Danny again agree to order an extra-large pizza for \$15 and divide the cost based on consumption. This time, the pizza arrives cut into 7 equal slices. Katy eats 1 slice, Mike eats 2 slices, and Danny eats 3 slices; 1 slice remains uneaten. How much should each one pay?

133. Write a few paragraphs to explain the rule for adding two fractions. Be sure to include a justification for this rule.

134. Your friend just added $\frac{1}{2}+\frac{1}{3}$ as follows: $\frac{1}{2}+\frac{1}{3}=\frac{2}{5}$. Explain the mistake and convince your friend that $\frac{1}{2}+\frac{1}{3}=\frac{5}{6}$.

'Are You Prepared?' Answers

1. $\frac{5}{8}$ **2.** $0.5\overline{454}$

3.2 Reducing Rational Expressions to Lowest Terms

PREPARING FOR THIS SECTION *Before getting started, review the following:*

- Domain of a Variable (pp. 53–54)
- Evaluate Polynomials (pp. 74–75)
- Factoring Polynomials (Sections 2.6, 2.7, 2.8)
- Reduce Fractions to Lowest Terms (pp. 128–129)

Now work the 'Are You Prepared?' problems on page 146.

OBJECTIVES

1 Reduce Rational Expressions

2 Evaluate Rational Expressions

If we form the quotient of two polynomials, the result is called a **rational expression.** Some examples of rational expressions are

(a) $\dfrac{x^3 + 1}{x}$ (b) $\dfrac{3x^2 + x - 2}{x^2 + 5}$ (c) $\dfrac{x}{x^2 - 1}$ (d) $\dfrac{xy^2}{(x - y)^2}$

Expressions (a), (b), and (c) are rational expressions in one variable, x, whereas (d) is a rational expression in two variables, x and y.

1 Rational expressions are described in the same manner as fractions. In expression (a), the polynomial $x^3 + 1$ is called the **numerator,** and x is called the **denominator.** When the numerator and denominator of a rational expression contain no common factors (except 1 and -1), we say the rational expression is **reduced to lowest terms,** or **simplified.**

As mentioned in Section 1.8, division by 0 is not defined. Because of this, the polynomial in the denominator of a rational expression cannot be equal to 0. For example, in the expression $\dfrac{x^3 + 1}{x}$, x cannot take on the value of 0. The domain of the variable x is $\{x | x \neq 0\}$.

A rational expression is reduced to lowest terms by completely factoring the numerator and the denominator and canceling any common factors by using the Cancellation Property,

$$\frac{ac}{bc} = \frac{a}{b} \qquad \text{if } b \neq 0, c \neq 0 \qquad \textbf{(1)}$$

We shall follow the previous practice of using a slash mark to indicate cancellation. For example,

$$\frac{x^2 - 1}{x^2 - 2x - 3} = \frac{(x - 1)\cancel{(x + 1)}}{(x - 3)\cancel{(x + 1)}} = \frac{x - 1}{x - 3}$$

EXAMPLE 1 **Reducing a Rational Expression To Lowest Terms**

Reduce to lowest terms: $\dfrac{x^2 + 4x + 4}{x^2 + 3x + 2}$

Solution We begin by factoring the numerator and the denominator:

$$x^2 + 4x + 4 = (x + 2)(x + 2)$$
$$x^2 + 3x + 2 = (x + 2)(x + 1)$$

Since a common factor, $x + 2$, appears, the original expression is not in lowest terms. To reduce it to lowest terms, we use the Cancellation Property:

$$\frac{x^2 + 4x + 4}{x^2 + 3x + 2} = \frac{\cancel{(x+2)}(x + 2)}{\cancel{(x+2)}(x + 1)} = \frac{x + 2}{x + 1}$$ ◀

WARNING: Apply the Cancellation Property only to rational expressions *written in factored form.* Be sure to cancel only common factors! That is, you can only cancel factors that are exactly alike! ■

EXAMPLE 2 Reducing a Rational Expression To Lowest Terms

Reduce to lowest terms: $\dfrac{x^3 + 5x^2}{x^2 + x}$

Solution We rewrite the rational expression so that both the numerator and denominator are factored completely:

$$\frac{x^3 + 5x^2}{x^2 + x} = \frac{x^2(x + 5)}{x(x + 1)} = \frac{\cancel{x} \cdot x(x + 5)}{\cancel{x}(x + 1)} = \frac{x(x + 5)}{x + 1}$$ ◀

Once a rational expression has been reduced to lowest terms, it may be left in factored form or multiplied out. The simplified form of the solution to Example 2 may be written as

$$\frac{x(x + 5)}{x + 1} \quad \text{or as} \quad \frac{x^2 + 5x}{x + 1}$$

Although a rational expression can be left in factored form or multiplied out, the preferred form is to leave the rational expression in factored form.

EXAMPLE 3 Reducing a Rational Expression To Lowest Terms

Reduce to lowest terms: $\dfrac{x^4 - 8x}{x^2 - 2x}$

Solution

$$\frac{x^4 - 8x}{x^2 - 2x} = \frac{x(x^3 - 8)}{x(x - 2)}$$ Factor out monomial factors.

$$= \frac{\cancel{x(x-2)}(x^2 + 2x + 4)}{\cancel{x(x-2)}}$$ Factor the difference of two cubes.

$$= \frac{x^2 + 2x + 4}{1}$$ Cancel.

$$= x^2 + 2x + 4$$ Simplify. ◀

Practice Exercise 1 *Reduce each rational expression to lowest terms:*

1. $\dfrac{x^2 + 3x + 2}{x^2 + 2x}$ **2.** $\dfrac{x^3 + 8}{x^2 - 4}$ **3.** $\dfrac{6x^2 - 12x}{2x - 4}$

Recall from Section 2.7 that if the leading coefficient of a polynomial in standard form is negative, it is usually easier to factor out -1 first before factoring further. For example,

$$-x^2 - x + 6 = -1 \cdot (x^2 + x - 6) = (-1)(x + 3)(x - 2)$$

When factoring out -1, remember to *change the sign of each term* in the remaining polynomial factor.

The next example illustrates this procedure as it applies to reducing rational expressions.

EXAMPLE 4 **Reducing a Rational Expression To Lowest Terms**

Reduce to lowest terms: $\dfrac{6 - x - x^2}{x^2 - x - 12}$

Solution

$$\frac{6 - x - x^2}{x^2 - x - 12} = \frac{-x^2 - x + 6}{x^2 - x - 12} = \frac{(-1)(x^2 + x - 6)}{x^2 - x - 12}$$

Write numerator in standard form; Factor out -1

$$= \frac{(-1)\cancel{(x + 3)}(x - 2)}{(x - 4)\cancel{(x + 3)}} = \frac{(-1)(x - 2)}{x - 4}$$

Factor trinomials; Cancel

There are other ways to write the solution to Example 4:

$$\frac{-(x - 2)}{x - 4} \quad \text{or} \quad \frac{-x + 2}{x - 4} \quad \text{or} \quad \frac{2 - x}{x - 4} \quad \text{or} \quad -\frac{x - 2}{x - 4}$$

Can you find any other ways to write the solution to Example 4?

Practice Exercise 2 *Reduce each rational expression to lowest terms.*

1. $\dfrac{36 - x^2}{x^2 + 6x}$ **2.** $\dfrac{x^2 + 6x + 9}{9 - x^2}$ **3.** $\dfrac{x^2 - 5x + 6}{2x^2 - x^3}$

Evaluating Rational Expressions

2 To **evaluate** a rational expression means to evaluate the polynomial in the numerator and the polynomial in the denominator. However, the polynomial in the denominator of a rational expression cannot have a value equal to 0, since division by 0 is not defined. In other words, the domain of the variable of a rational expression must exclude any values that cause the polynomial in the denominator to have a value equal to 0.

EXAMPLE 5 **Evaluating a Rational Expression**

Evaluate the rational expression below for each value of x.

$$\frac{x^2 - 2x + 1}{x^2 - 9}$$

(a) $x = 2$ (b) $x = -1$ (c) $x = 0$ (d) $x = 1$

Solution (a) We substitute 2 for x in the rational expression to obtain

$$\frac{x^2 - 2x + 1}{x^2 - 9} \underset{x = 2}{\uparrow}= \frac{(2)^2 - 2(2) + 1}{(2)^2 - 9} = \frac{4 - 4 + 1}{4 - 9} = \frac{1}{-5} = -\frac{1}{5}$$

(b) $$\frac{x^2 - 2x + 1}{x^2 - 9} \underset{x = -1}{\uparrow}= \frac{(-1)^2 - 2(-1) + 1}{(-1)^2 - 9} = \frac{1 + 2 + 1}{1 - 9} = \frac{4}{-8} = -\frac{1}{2}$$

(c) $$\frac{x^2 - 2x + 1}{x^2 - 9} \underset{x = 0}{\uparrow}= \frac{(0)^2 - 2(0) + 1}{(0)^2 - 9} = \frac{1}{-9} = -\frac{1}{9}$$

(d) $$\frac{x^2 - 2x + 1}{x^2 - 9} \underset{x = 1}{\uparrow}= \frac{(1)^2 - 2(1) + 1}{(1)^2 - 9} = \frac{1 - 2 + 1}{1 - 9} = \frac{0}{-8} = 0$$

◀

The domain of the variable x in the rational expression given in Example 5 is any real number except 3 and -3, since these values cause the denominator, $x^2 - 9$, to equal 0.

Practice Exercise 3

1. Rework Example 5 using a calculator. Be sure to use the keys for parentheses as needed.
2. On your calculator, try to evaluate the rational expression of Example 5 for $x = 3$. What happens? Can you explain why?
3. Use a calculator to evaluate $\dfrac{2x^2 - 3x + 4}{x + 5}$ for

   (a) $x = -3$ (b) $x = 0$ (c) $x = \dfrac{1}{2}$

◀

EXAMPLE 6 Evaluating a Rational Expression

Evaluate the rational expression below for each value of x.

$$\frac{(x + 2)(x - 3)}{(x + 1)(x - 4)}$$

(a) $x = 0$ (b) $x = 1$ (c) $x = 3$

Solution (a) $$\frac{(x + 2)(x - 3)}{(x + 1)(x - 4)} \underset{x = 0}{\uparrow}= \frac{(0 + 2)(0 - 3)}{(0 + 1)(0 - 4)} = \frac{(2)(-3)}{(1)(-4)} = \frac{-6}{-4} = \frac{3}{2}$$

(b) $$\frac{(x + 2)(x - 3)}{(x + 1)(x - 4)} \underset{x = 1}{\uparrow}= \frac{(1 + 2)(1 - 3)}{(1 + 1)(1 - 4)} = \frac{(3)(-2)}{(2)(-3)} = \frac{-6}{-6} = 1$$

(c) $$\frac{(x + 2)(x - 3)}{(x + 1)(x - 4)} \underset{x = 3}{\uparrow}= \frac{(3 + 2)(3 - 3)}{(3 + 1)(3 - 4)} = \frac{(5)(0)}{(4)(-1)} = \frac{0}{-4} = 0$$

◀

For the rational expression given in Example 6, the values $x = -1$ and $x = 4$ must be excluded from the domain of the variable, since each of these values causes the denominator, $(x + 1)(x - 4)$, to equal 0. The domain of x is any real number except -1 and 4.

Answers to Practice Exercises

Practice Exercise 1 **1.** $\frac{x + 1}{x}$ **2.** $\frac{x^2 - 2x + 4}{x - 2}$ **3.** $3x$

Practice Exercise 2 **1.** $\frac{(-1)(x - 6)}{x}$ **2.** $-\frac{x + 3}{x - 3}$ **3.** $-\frac{x - 3}{x^2}$

Practice Exercise 3 **1.** Same answers as in Example 5

2. Error or E or ERR: DIVIDE BY 0 appears.

3. (a) $\frac{31}{2}$ (b) $\frac{4}{5}$ (c) $\frac{6}{11}$

3.2 Assess Your Understanding

'Are You Prepared?' *Answers are given at the end of these exercises. If you get a wrong answer, read the pages listed in parentheses.*

1. What is the domain of the variable x in $\frac{x + 3}{x - 5}$? (pp. 53–54)

2. Reduce to lowest terms: $\frac{12}{40}$ (pp. 128–129)

3. Factor completely: $4x^2 - 5x - 6$ (Section 2.7)

4. Factor completely: $2x^4 - 10x^2 + 8$ (Section 2.6, 2.8)

5. Evaluate $x^3 - 3x^2 + 4x$ when $x = -2$. (pp. 74–75)

6. Factor completely: $x^3 - 27$ (Section 2.8)

Concepts and Vocabulary

7. The domain of the variable in a rational expression must exclude any values that cause the ________ polynomial to have a value of 0.

8. *True or False:* The rational expression $\frac{x^2 - 4}{x^2 + 2x}$ is in lowest terms.

9. *True or False:* The domain of the variable x in the rational expression $\frac{x - 5}{x + 2}$ is $x \neq 0$.

10. When $x = 4$, $\frac{x + 7}{x - 2} =$ ________.

Exercises

In Problems 11–50, reduce each rational expression to lowest terms.

11. $\frac{2x - 4}{4x}$ **12.** $\frac{3x^2}{9x + 18}$ **13.** $\frac{x^2 - x}{x^2 + x}$ **14.** $\frac{x^3 - x}{x^3 + x^2}$

15. $\frac{x^2 + 4x + 4}{x^2 + 6x + 8}$ **16.** $\frac{x^2 + 3x + 2}{x^2 - 4}$ **17.** $\frac{x^2 - 9}{x^2 - 6x + 9}$ **18.** $\frac{x^2 - 4x + 4}{x^2 - 2x}$

19. $\frac{x^2 + x - 2}{x^3 + 4x^2 + 4x}$ **20.** $\frac{x^2 + 6x + 9}{3x^2 + 9x}$ **21.** $\frac{5x + 10}{x^2 - 4}$ **22.** $\frac{4x + 8}{12x + 24}$

23. $\frac{x^2 - 2x}{3x - 6}$ **24.** $\frac{15x^2 + 24x}{3x^2}$ **25.** $\frac{24x}{12x^2 - 6x}$ **26.** $\frac{3x - 12}{x^2 - 16}$

27. $\frac{y^2 - 25}{2y - 10}$ **28.** $\frac{3y + 2}{3y^2 + 5y + 2}$ **29.** $\frac{x^2 + 4x - 5}{x - 1}$ **30.** $\frac{x - x^2}{x^2 + x - 2}$

31. $\frac{x^2 - 4}{x^2 + 5x + 6}$ **32.** $\frac{x^2 + x - 6}{9x - x^3}$ **33.** $\frac{x^2 - x - 2}{2x^2 - 3x - 2}$ **34.** $\frac{2x^2 - x - 1}{4x^2 - 1}$

35. $\dfrac{x^3 + x^2 + x}{x^3 - 1}$ **36.** $\dfrac{x^3 + 1}{x^3 - x^2 + x}$ **37.** $\dfrac{3x^2 + 6x}{3x^2 + 5x - 2}$ **38.** $\dfrac{4x^2 - 12x}{3x^2 - 10x + 3}$

39. $\dfrac{x^4 - 1}{x^3 - x}$ **40.** $\dfrac{x^3 - 1}{x^3 - x^2}$

41. $\dfrac{(x^2 - 3x - 10)(x^2 + 4x - 21)}{(x^2 + 2x - 35)(x^2 + 9x + 14)}$ **42.** $\dfrac{(x^2 - x - 6)(x^2 - 25)}{(x^2 - 4x - 5)(x^2 + 2x - 15)}$

43. $\dfrac{x^2 + 5x - 14}{2 - x}$ **44.** $\dfrac{2x^2 + 5x - 3}{1 - 2x}$ **45.** $\dfrac{2x^3 - x^2 - 10x}{x^3 - 2x^2 - 8x}$

46. $\dfrac{4x^4 + 2x^3 - 6x^2}{4x^4 + 26x^3 + 30x^2}$ **47.** $\dfrac{(x - 4)^2 - 9}{(x + 3)^2 - 16}$ **48.** $\dfrac{(x + 2)^2 - 8x}{(x - 2)^2}$

49. $\dfrac{6x(x - 1) - 12}{x^3 - 8 - (x - 2)^2}$ **50.** $\dfrac{3(x - 2)^2 + 17(x - 2) + 10}{2(x - 2)^2 + 7(x - 2) - 15}$

In Problems 51–60, evaluate each rational expression for the given value of the variable.

51. $\dfrac{x^2 + 1}{3x}$ for $x = 3$ **52.** $\dfrac{6x}{x^2 - 1}$ for $x = 2$

53. $\dfrac{x^2 - 4x + 4}{x^2 - 25}$ for $x = -4$ **54.** $\dfrac{x^2 - 6x + 9}{x^2 - 16}$ for $x = -5$

55. $\dfrac{9x^2}{x^2 + 1}$ for $x = -1$ **56.** $\dfrac{9x^2}{x^2 + 1}$ for $x = 1$

57. $\dfrac{x^2 + x + 1}{x^2 - x + 1}$ for $x = 1$ **58.** $\dfrac{x^2 + x - 1}{x^2 - x + 1}$ for $x = -1$

59. $\dfrac{1 + x + x^2}{x^2}$ for $x = 3.21$ **60.** $\dfrac{5x^2}{1 - x^2}$ for $x = 4.23$

In Problems 61–70, determine which of the value(s) given below, if any, must be excluded from the domain of the variable in each rational expression:

(a) $x = 3$ (b) $x = 1$ (c) $x = 0$ (d) $x = -1$

61. $\dfrac{x^2 - 1}{x}$ **62.** $\dfrac{x^2 + 1}{x}$ **63.** $\dfrac{x}{x^2 - 9}$ **64.** $\dfrac{x}{x^2 + 9}$ **65.** $\dfrac{x^2}{x^2 + 1}$

66. $\dfrac{x^3}{x^2 - 1}$ **67.** $\dfrac{x^2 + 5x - 10}{x^3 - x}$ **68.** $\dfrac{-9x^2 - x + 1}{x^3 + x}$ **69.** $\dfrac{x^2 + x + 1}{x^2 - x + 1}$ **70.** $\dfrac{x^2 + x + 1}{x^4 + x^2 + 1}$

In Problems 71–74, find a rational expression equal to the one given that has a denominator of $x - 4$.

71. $\dfrac{5}{4 - x}$ **72.** $\dfrac{-4}{4 - x}$ **73.** $\dfrac{2x + 3}{4 - x}$ **74.** $\dfrac{3x - 4}{4 - x}$

75. Evaluate the rational expression $\dfrac{x}{x^2 - 1}$ for

(a) $x = 1.1$
(b) $x = 1.01$
(c) $x = 1.001$
(d) $x = 1.0001$
(e) Can $x = 1$? Explain.

76. Evaluate the rational expression $\dfrac{x}{x^2 - 1}$ for

(a) $x = 100$
(b) $x = 1000$
(c) $x = 10{,}000$
(d) $x = 100{,}000$
(e) As x gets larger, what is happening to the value of the rational expression?

'Are You Prepared?' Answers

1. $\{x | x \neq 5\}$; or any real number except 5. **2.** $\dfrac{3}{10}$ **3.** $(4x + 3)(x - 2)$ **4.** $2(x - 2)(x + 2)(x - 1)(x + 1)$ **5.** -28

6. $(x - 3)(x^2 + 3x + 9)$

3.3 Multiplication and Division of Rational Expressions

PREPARING FOR THIS SECTION *Before getting started, review the following:*

- Factoring Polynomials (Sections 2.6, 2.7, 2.8)
- Multiplying Fractions (pp. 129–130)
- Dividing Fractions (pp. 130–132)
- Reducing Rational Expressions (pp. 142–144)

Now work the 'Are You Prepared?' problems on page 152.

OBJECTIVES

1 Multiply Rational Expressions

2 Divide Rational Expressions

The rules for multiplying and dividing rational expressions are the same as the rules for multiplying and dividing fractions.

Multiplication of Rational Expressions

If $\frac{a}{b}$ and $\frac{c}{d}$, $b \neq 0, d \neq 0$, are two rational expressions, their product is given by the rule

$$\frac{a}{b} \cdot \frac{c}{d} = \frac{a \cdot c}{b \cdot d} \quad \text{if } b \neq 0, d \neq 0 \qquad \textbf{(1)}$$

In using equation (1), be sure to factor each polynomial completely so that common factors can be cancelled. Leave your answer in factored form.

1 EXAMPLE 1 Multiplying Rational Expressions

Perform the indicated operation and simplify the result. Leave your answer in factored form.

(a) $\frac{x-1}{6x} \cdot \frac{2x^3}{x^2-1}$ (b) $\frac{4x-8}{x^2+x} \cdot \frac{x^2-1}{3x+6}$

Solution (a)

$$\frac{x-1}{6x} \cdot \frac{2x^3}{x^2-1} = \frac{(x-1) \cdot 2x^3}{6x(x^2-1)} \qquad \text{Apply Equation (1).}$$

$$= \frac{\cancel{(x-1)} \cdot \cancel{2} \cdot \cancel{x} \cdot x \cdot x}{\cancel{2} \cdot 3 \cdot \cancel{x} \cdot \cancel{(x-1)}(x+1)} \qquad \text{Factor.}$$

$$= \frac{x^2}{3(x+1)} \qquad \text{Reduce.}$$

(b)

$$\frac{4x-8}{x^2+x} \cdot \frac{x^2-1}{3x+6} = \frac{(4x-8)(x^2-1)}{(x^2+x)(3x+6)} \qquad \text{Apply Equation (1).}$$

$$= \frac{4(x-2)(x-1)\cancel{(x+1)}}{x\cancel{(x+1)}\cdot 3(x+2)} \quad \text{Factor}$$

$$= \frac{4(x-2)(x-1)}{3x(x+2)} \quad \text{Reduce}$$

Notice in Example 1(b) that we wrote the factors of the answer so that the numerical factors appear first. In this way, numerical factors, like the 3 in the denominator of Example 1(b), will not be mistaken for exponents.

EXAMPLE 2 **Multiplying Rational Expressions**

Perform the indicated operation and simplify the result:

$$\frac{x^2 - 2x + 1}{x^3 + x} \cdot \frac{4x^2 + 4}{x^2 + x - 2}$$

Leave your answer in factored form.

Solution

$$\frac{x^2 - 2x + 1}{x^3 + x} \cdot \frac{4x^2 + 4}{x^2 + x - 2} = \frac{(x^2 - 2x + 1)(4x^2 + 4)}{(x^3 + x)(x^2 + x - 2)}$$

$$= \frac{\cancel{(x-1)}(x-1)\cdot 4\cancel{(x^2+1)}}{x\cancel{(x^2+1)}(x+2)\cancel{(x-1)}}$$

$$= \frac{4(x-1)}{x(x+2)}$$

Practice Exercise 1 *In Problems 1 and 2, perform the indicated operation and simplify the result. Leave your answer in factored form.*

1. $\dfrac{x^2 + 3x + 2}{x^3 + x} \cdot \dfrac{x^2 - 2x}{x^2 - 4}$ **2.** $\dfrac{6x - 3}{x^3 - 27} \cdot \dfrac{x^2 - 9}{4x^2 - 1}$

Division of Rational Expressions

If $\dfrac{a}{b}$ and $\dfrac{c}{d}$, $b \neq 0, c \neq 0, d \neq 0$, are two rational expressions, their quotient is given by the rule

$$\frac{\frac{a}{b}}{\frac{c}{d}} = \frac{a}{b} \div \frac{c}{d} = \frac{a}{b} \cdot \frac{d}{c} = \frac{ad}{bc} \quad \text{if } b \neq 0, c \neq 0, d \neq 0 \qquad \textbf{(2)}$$

2 **EXAMPLE 3** **Dividing Rational Expressions**

Perform the indicated operation and simplify the result. Leave your answer in factored form.

(a) $\dfrac{\dfrac{6x}{x+1}}{\dfrac{3x^2}{x^2-1}}$ (b) $\dfrac{\dfrac{x+3}{x^2-4}}{\dfrac{x^2-x-12}{x^3-8}}$

Solution (a) $\dfrac{\dfrac{6x}{x+1}}{\dfrac{3x^2}{x^2-1}} = \dfrac{6x}{x+1} \div \dfrac{3x^2}{x^2-1}$ — Rewrite as division of two fractions.

$= \dfrac{6x}{x+1} \cdot \dfrac{x^2-1}{3x^2}$ — Rule for division of fractions

$= \dfrac{6x(x^2-1)}{(x+1)\cdot 3x^2}$ — Rule for multiplication of fractions

$= \dfrac{2\cdot \cancel{3}\cdot \cancel{x}\cdot (x-1)\cancel{(x+1)}}{\cancel{3}\cdot \cancel{x}\cdot x\cdot \cancel{(x+1)}}$ — Factor.

$= \dfrac{2(x-1)}{x}$ — Reduce.

(b) $\dfrac{\dfrac{x+3}{x^2-4}}{\dfrac{x^2-x-12}{x^3-8}} = \dfrac{x+3}{x^2-4} \div \dfrac{x^2-x-12}{x^3-8}$ — Rewrite as division of two fractions.

$= \dfrac{x+3}{x^2-4} \cdot \dfrac{x^3-8}{x^2-x-12}$ — Rule for division of fractions

$= \dfrac{(x+3)(x^3-8)}{(x^2-4)(x^2-x-12)}$ — Rule for multiplication of fractions

$= \dfrac{\cancel{(x+3)}\cancel{(x-2)}(x^2+2x+4)}{\cancel{(x-2)}(x+2)(x-4)\cancel{(x+3)}}$ — Factor.

$= \dfrac{x^2+2x+4}{(x+2)(x-4)}$ — Reduce. ◀

In the next example we combine the operations of multiplication and division with fractions.

EXAMPLE 4 **Dividing Rational Expressions**

Perform the indicated operations and simplify the result. Leave your answer in factored form:

$$\frac{\dfrac{6x^2}{x^3-1} \cdot \dfrac{4x^2-4}{9}}{\dfrac{4x^3+7x^2-2x}{3x^2+3x+3}}$$

Solution $\dfrac{\dfrac{6x^2}{x^3-1} \cdot \dfrac{4x^2-4}{9}}{\dfrac{4x^3+7x^2-2x}{3x^2+3x+3}} = \dfrac{\dfrac{6x^2(4x^2-4)}{(x^3-1)\cdot 9}}{\dfrac{4x^3+7x^2-2x}{3x^2+3x+3}}$ — Perform the multiplication in the numerator.

$= \dfrac{6x^2(4x^2-4)}{(x^3-1)\cdot 9} \div \dfrac{4x^3+7x^2-2x}{3x^2+3x+3}$ — Rewrite as division of two fractions.

$$= \frac{6x^2(4x^2-4)}{(x^3-1)\cdot 9}\cdot\frac{3x^2+3x+3}{4x^3+7x^2-2x}$$ Rule for division of fractions.

$$= \frac{6x^2(4x^2-4)(3x^2+3x+3)}{(x^3-1)\cdot 9\cdot(4x^3+7x^2-2x)}$$ Rule for multiplication of fractions.

$$= \frac{2\cdot\cancel{3}\cdot\cancel{x}\cdot x\cdot 4\cancel{(x-1)}(x+1)\cdot\cancel{3(x^2+x+1)}}{\cancel{(x-1)(x^2+x+1)}\cdot\cancel{3}\cdot\cancel{3}\cdot\cancel{x}(x+2)(4x-1)}$$ Factor.

$$= \frac{8x(x+1)}{(x+2)(4x-1)}$$ Reduce. ◀

EXAMPLE 5 Dividing Rational Expressions

Perform the indicated operations and simplify the result. Leave your answer in factored form:

$$\frac{\frac{x^2-3x-10}{4x^2+8x}}{x-5}$$

Solution

$$\frac{\frac{x^2-3x-10}{4x^2+8x}}{x-5} = \frac{\frac{x^2-3x-10}{4x^2+8x}}{\frac{x-5}{1}} = \frac{x^2-3x-10}{4x^2+8x}\div\frac{x-5}{1} = \frac{x^2-3x-10}{4x^2+8x}\cdot\frac{1}{x-5}$$

(Rewrite $x-5$ as $\frac{x-5}{1}$; Rewrite as division of fractions; Rule for division of fractions)

$$= \frac{(x^2-3x-10)(1)}{(4x^2+8x)(x-5)} = \frac{\cancel{(x-5)}\cancel{(x+2)}(1)}{4x\cancel{(x+2)}\cancel{(x-5)}} = \frac{1}{4x}$$

(Rule for multiplication of fractions; Factor; Reduce) ◀

Practice Exercise 2 *In Problems 1–3, perform the indicated operation(s) and simplify the result. Leave your answer in factored form.*

1. $\frac{\frac{x^2+x-6}{2x}}{\frac{x+3}{x^3}}$ 2. $\frac{\frac{x^2-16}{2x^2+8x}}{x-4}$ 3. $\frac{\frac{x^2}{x^2+4}\cdot\frac{x^2-4}{3x}}{\frac{x^4-8x}{2x^2+8}}$ ◀

Answers to Practice Exercises

Practice Exercise 1 1. $\frac{x+1}{x^2+1}$ 2. $\frac{3(x+3)}{(x^2+3x+9)(2x+1)}$

Practice Exercise 2 1. $\frac{(x-2)x^2}{2}$ 2. $\frac{1}{2x}$ 3. $\frac{2(x+2)}{3(x^2+2x+4)}$

3.3 Assess Your Understanding

'Are You Prepared?' *Answers are given at the end of these exercises. If you get a wrong answer, read the pages listed in parentheses.*

1. Multiply: $\frac{12}{17} \cdot \frac{2}{3}$ (pp. 129–130)

2. Divide: $\frac{\frac{8}{9}}{\frac{4}{7}}$ (pp. 130–132)

3. Reduce: $\frac{5x^2 - 14x - 3}{x^2 - 9}$ (pp. 142–144)

4. Reduce: $\frac{x^3 + 8}{(x + 2)^3}$ (pp. 142–144)

5. Factor completely: $4x^2 + 12x + 9$ (Section 2.8)

6. Factor completely: $4x^2 + 15x + 9$ (Section 2.7)

Concepts and Vocabulary

7. To multiply two rational expressions, multiply their __________ and their __________.

8. Fill in the blank: $\frac{\frac{x - 5}{x}}{\frac{x^2 - 1}{x + 2}} = \frac{x - 5}{x} \cdot$ __________.

Exercises

In Problems 9–38, perform the indicated operation and simplify the result. Leave your answer in factored form.

9. $\frac{3x}{4} \cdot \frac{12}{x}$

10. $\frac{5x^2}{18} \cdot \frac{9}{2x}$

11. $\frac{8x^2}{x + 1} \cdot \frac{x^2 - 1}{2x}$

12. $\frac{x^2 - 4}{3x^2} \cdot \frac{9x}{x + 2}$

13. $\frac{x + 5}{6x} \cdot \frac{x^2}{2x + 10}$

14. $\frac{x - 6}{6x^2} \cdot \frac{2x}{3x - 18}$

15. $\frac{4x - 8}{3x + 6} \cdot \frac{3}{2x - 4}$

16. $\frac{5}{8x - 2} \cdot \frac{4x - 1}{10x}$

17. $\frac{3x + 9}{2x - 4} \cdot \frac{x^2 - 4}{6x}$

18. $\frac{2x + 8}{6x} \cdot \frac{3x^2}{x^2 - 16}$

19. $\frac{3x - 6}{5x} \cdot \frac{x^2 - x - 6}{x^2 - 4}$

20. $\frac{9x - 15}{2x - 2} \cdot \frac{1 - x^2}{6x - 10}$

21. $\frac{4x^2 - 1}{x^2 - 16} \cdot \frac{x^2 - 4x}{2x + 1}$

22. $\frac{12}{x^2 - x} \cdot \frac{x^2 - 1}{4x - 2}$

23. $\frac{4x - 8}{-3x} \cdot \frac{12}{12 - 6x}$

24. $\frac{6x - 27}{5x} \cdot \frac{2}{4x - 18}$

25. $\frac{x^2 - 3x - 10}{x^2 + 2x - 35} \cdot \frac{x^2 + 4x - 21}{x^2 + 9x + 14}$

26. $\frac{x^2 - x - 6}{x^2 - 4x - 5} \cdot \frac{x^2 - 25}{x^2 + 2x - 15}$

27. $\frac{x^2 + x - 12}{x^2 - x - 12} \cdot \frac{x^2 + 7x + 12}{x^2 - 7x + 12}$

28. $\frac{x^2 + 7x + 6}{x^2 - x - 6} \cdot \frac{x^2 + 5x + 6}{x^2 + 5x - 6}$

29. $\frac{1 - x^2}{1 + x^2} \cdot \frac{x^3 + x}{x^3 - x}$

30. $\frac{1 - x}{1 + x} \cdot \frac{x^2 + x}{x^2 - x}$

31. $\frac{4x^2 + 4x + 1}{x^2 + 4x + 4} \cdot \frac{x^2 - 4}{4x^2 - 1}$

32. $\frac{9x^2 - 1}{x^2 - 9} \cdot \frac{x^2 + 6x + 9}{9x^2 + 6x + 1}$

33. $\frac{2x^2 + x - 3}{2x^2 - x - 3} \cdot \frac{4x^2 - 9}{x^2 - 1}$

34. $\frac{2x^2 + x - 10}{2x^2 - x - 10} \cdot \frac{x^2 - 4}{2x^2 + 5x}$

35. $\frac{x^3 - 8}{25 - 4x^2} \cdot \frac{10 + 4x}{2x^2 - 9x + 10}$

36. $\frac{3x^2 + 7x + 2}{9 - x^2} \cdot \frac{6 - 2x}{x^3 + 8}$

37. $\frac{8x^3 + 27}{1 + x - 6x^2} \cdot \frac{4 - 10x - 6x^2}{4x^2 + 12x + 9}$

38. $\frac{9x^2 - 12x + 4}{3 - 8x - 3x^2} \cdot \frac{2 - 3x - 9x^2}{27x^3 - 8}$

In Problems 39–68, perform the indicated operation and simplify the result. Leave your answer in factored form.

39. $\frac{\frac{2x}{9}}{\frac{x}{3}}$

40. $\frac{\frac{3x^2}{5}}{\frac{9x}{10}}$

41. $\frac{\frac{3x^2}{x + 1}}{\frac{6x}{x^2 - 1}}$

42. $\frac{\frac{x^2 - 4}{3x^2}}{\frac{x - 2}{18x}}$

43. $\dfrac{\frac{12x}{x-5}}{\frac{2x^2}{3x-15}}$

44. $\dfrac{\frac{x+6}{6x^2}}{\frac{4x+24}{3x^3}}$

45. $\dfrac{\frac{4x-8}{3x+6}}{\frac{8}{x^2-4}}$

46. $\dfrac{\frac{4x-1}{10x}}{\frac{8x-2}{5x^3}}$

47. $\dfrac{\frac{6x}{x^2-4}}{\frac{3x-9}{2x+4}}$

48. $\dfrac{\frac{12x}{5x+20}}{\frac{4x^2}{x^2-16}}$

49. $\dfrac{\frac{8x}{x^2-1}}{\frac{10x}{x+1}}$

50. $\dfrac{\frac{x-2}{4x}}{\frac{x^2-4x+4}{12x}}$

51. $\dfrac{\frac{4-x}{4+x}}{\frac{4x}{x^2-16}}$

52. $\dfrac{\frac{3+x}{3-x}}{\frac{x^2-9}{9x^3}}$

53. $\dfrac{\frac{x^2+7x+12}{x^2-7x+12}}{\frac{x^2+x-12}{x^2-x-12}}$

54. $\dfrac{\frac{x^2+7x+6}{x^2+x-6}}{\frac{x^2+5x-6}{x^2+5x+6}}$

55. $\dfrac{\frac{1-x^2}{1+x^2}}{\frac{x-x^3}{x+x^3}}$

56. $\dfrac{\frac{1-x}{1+x}}{\frac{x-x^2}{x+x^2}}$

57. $\dfrac{\frac{2x^2-x-28}{3x^2-x-2}}{\frac{4x^2+16x+7}{3x^2+11x+6}}$

58. $\dfrac{\frac{9x^2+3x-2}{12x^2+5x-2}}{\frac{9x^2-6x+1}{8x^2-10x-3}}$

59. $\dfrac{\frac{8x^2-6x+1}{4x^2-1}}{\frac{12x^2+5x-2}{6x^2-x-2}}$

60. $\dfrac{\frac{3x^2+2x-1}{5x^2-9x-2}}{\frac{2x^2-x-3}{10x^2-13x-3}}$

61. $\dfrac{\frac{9x^2+6x+1}{x^2+6x+9}}{\frac{9x^2-1}{x^2-9}}$

62. $\dfrac{\frac{x^2-4}{4x^2-1}}{\frac{x^2+4x+4}{4x^2+4x+1}}$

63. $\dfrac{\frac{9-4x^2}{1-x^2}}{\frac{2x^2-x-3}{2x^2+x-3}}$

64. $\dfrac{\frac{4-x^2}{2x^2+5x}}{\frac{2x^2-3x-10}{2x^2+x-10}}$

65. $\dfrac{\frac{4x+10}{2x^2-9x+10}}{\frac{25-4x^2}{8-x^3}}$

66. $\dfrac{\frac{3x^2+7x+2}{9-x^2}}{\frac{8+x^3}{2x-6}}$

67. $\dfrac{\frac{9x^2-12x+4}{3-8x-3x^2}}{\frac{27x^3-8}{2-3x-9x^2}}$

68. $\dfrac{\frac{4-11x-6x^2}{4x^2+12x+9}}{\frac{1+x-6x^2}{27+8x^3}}$

In Problems 69–78, perform the indicated operations and simplify. Leave your answer in factored form.

69. $\dfrac{3x}{x+2}\cdot\dfrac{x^2-4}{12x^3}\cdot\dfrac{18x}{x-2}$

70. $\dfrac{x^2-9}{18x}\cdot\dfrac{3x^2}{x^2+5x+6}\cdot\dfrac{x^2+2x}{x+1}$

71. $\dfrac{5x^2-x}{3x+2}\cdot\dfrac{2x^2-x}{2x^2-x-1}\cdot\dfrac{10x^2+3x-1}{6x^2+x-2}$

72. $\dfrac{x^2-7x-8}{x^2+2x-15}\cdot\dfrac{x^2+8x+12}{x^2-6x-7}\cdot\dfrac{x^2+4x-5}{x^2+11x+18}$

73. $\dfrac{\frac{x+1}{x+2}\cdot\frac{x+3}{x-1}}{\frac{x^2-1}{x^2+2x}}$

74. $\dfrac{\frac{x-1}{x+1}\cdot\frac{x-3}{x+2}}{\frac{x^2-3x}{x^2-1}}$

75. $\dfrac{\frac{x^2+3x+2}{x^2-9}\cdot\frac{x^2+9}{x^2+6x+9}}{\frac{x^2+4}{x^2}\cdot\frac{3x^2+27}{2x^2-18}}$

76. $\dfrac{\frac{x^2-16}{x^2+7x+6}\cdot\frac{x^2+2}{x^2+8x+16}}{\frac{x^2+3}{x+4}\cdot\frac{3x^2+6}{x+6}}$

77. $\dfrac{\frac{x^4-x^8}{x^2+1}\cdot\frac{3x^2}{(x-2)^2}}{\frac{x^3+x^6}{x^2-1}\cdot\frac{12x}{x^4-1}}$

78. $\dfrac{\frac{x^3-x^6}{x^2-4}\cdot\frac{(x+1)^2}{12x}}{\frac{x^4+x^2}{(x-2)^2}\cdot\frac{x^2+x}{9x+18}}$

'Are You Prepared?' Answers

1. $\frac{8}{17}$ 2. $\frac{14}{9}$ 3. $\frac{5x+1}{x+3}$ 4. $\frac{x^2-2x+4}{(x+2)(x+2)}$ or $\frac{x^2-2x+4}{(x+2)^2}$ 5. $(2x+3)^2$ 6. $(4x+3)(x+3)$

3.4 Addition and Subtraction of Rational Expressions

PREPARING FOR THIS SECTION *Before getting started, review the following:*

- Factoring Polynomials (Sections 2.6, 2.7, 2.8)
- Adding and Subtracting Fractions with Equal Denominators (pp. 132–134)
- Adding and Subtracting Fractions with Unequal Denominators (pp. 134–136)
- Using the Least Common Multiple Method with Fractions (pp. 136–138)
- Reducing Rational Expressions (pp. 142–144)

Now work the 'Are You Prepared?' problems on page 162.

OBJECTIVES

1 Add and Subtract Rational Expressions with Equal Denominators
2 Add and Subtract Rational Expressions with Unequal Denominators
3 Use the Least Common Multiple Method with Rational Expressions

The rules for adding and subtracting rational expressions are the same as the rules for adding and subtracting fractions.

1 If the denominators of two rational expressions to be added (or subtracted) are equal, we add (or subtract) the numerators and keep the common denominator.

If $\frac{a}{b}$ and $\frac{c}{b}$, $b \neq 0$, are two rational expressions, then

$$\frac{a}{b}+\frac{c}{b}=\frac{a+c}{b} \qquad \frac{a}{b}-\frac{c}{b}=\frac{a-c}{b} \qquad \text{if } b \neq 0 \qquad \textbf{(1)}$$

As before, we will leave our answers in factored form.

EXAMPLE 1 **Adding Rational Expressions**

Perform the indicated operation and simplify the result. Leave your answer in factored form.

(a) $\frac{x}{x^2+1}+\frac{5}{x^2+1}$ (b) $\frac{2x^2-4}{2x+5}+\frac{x+3}{2x+5}$

Solution (a) $\frac{x}{x^2+1}+\frac{5}{x^2+1}=\frac{x+5}{x^2+1}$

(b) $\frac{2x^2-4}{2x+5}+\frac{x+3}{2x+5}=\frac{(2x^2-4)+(x+3)}{2x+5}$

$$= \frac{2x^2 + x - 1}{2x + 5}$$

$$= \frac{(2x - 1)(x + 1)}{2x + 5}$$

◀

EXAMPLE 2 **Adding Rational Expressions**

Perform the indicated operation and simplify the result. Leave your answer in factored form.

$$\frac{2x}{x - 3} + \frac{5}{3 - x}$$

Solution Notice that the denominators of the two rational expressions are different. However, the denominator of the second expression is just the additive inverse of the denominator of the first. That is,

$$3 - x = -x + 3 = -1 \cdot (x - 3) = -(x - 3)$$

METHOD 1:

$$\frac{2x}{x - 3} + \frac{5}{3 - x} = \frac{2x}{x - 3} + \frac{5}{-(x - 3)} \quad \text{Since } 3 - x = -(x - 3)$$

$$= \frac{2x}{x - 3} + \frac{-5}{x - 3} \quad \text{Since } \frac{a}{-b} = \frac{-a}{b}$$

$$= \frac{2x + (-5)}{x - 3}$$

$$= \frac{2x - 5}{x - 3}$$

METHOD 2:

$$\frac{2x}{x - 3} + \frac{5}{3 - x} = \frac{2x}{x - 3} + \frac{5}{-(x - 3)} \quad \text{Since } 3 - x = -(x - 3)$$

$$= \frac{2x}{x - 3} - \frac{5}{x - 3} \quad \text{Since } \frac{a}{-b} = -\frac{a}{b}$$

$$= \frac{2x - 5}{x - 3}$$

◀

EXAMPLE 3 **Subtracting Rational Expressions**

Perform the indicated operation and simplify the result. Leave your answer in factored form.

(a) $\frac{2x}{x^2 - 1} - \frac{2}{x^2 - 1}$ (b) $\frac{x}{x - 3} - \frac{3x + 2}{x - 3}$

Solution

(a) $$\frac{2x}{x^2 - 1} - \frac{2}{x^2 - 1} = \frac{2x - 2}{x^2 - 1} = \frac{2\cancel{(x - 1)}}{\cancel{(x - 1)}(x + 1)} = \frac{2}{x + 1}$$

(b) $$\frac{x}{x - 3} - \frac{3x + 2}{x - 3} = \frac{x - (3x + 2)}{x - 3} = \frac{x - 3x - 2}{x - 3}$$

$$= \frac{-2x - 2}{x - 3} = \frac{-2(x + 1)}{x - 3}$$

◀

Notice in Example 3(b) that we subtracted the quantity $(3x + 2)$ from x in the first step. When subtracting rational expressions, be careful to place parentheses around the numerator of the fraction being subtracted to ensure that the entire numerator is subtracted, so that the subtraction sign is distributed.

Practice Exercise 1 *In Problems 1–4, perform the indicated operation and simplify the result. Leave your answer in factored form.*

1. $\frac{3x}{x-3} + \frac{4x}{x-3}$ **2.** $\frac{x+2}{x^2+1} - \frac{3x-5}{x^2+1}$ **3.** $\frac{2x+3}{x-2} + \frac{3x-2}{2-x}$

4. $\frac{3x}{x^2-4} - \frac{6}{x^2-4}$

2 If the denominators of two rational expressions to be added or subtracted are not equal, we can use the general formulas for adding and subtracting quotients.

$$\frac{a}{b} + \frac{c}{d} = \frac{a \cdot d}{b \cdot d} + \frac{b \cdot c}{b \cdot d} = \frac{ad + bc}{bd} \quad \text{if } b \neq 0, d \neq 0 \quad \textbf{(2)}$$

$$\frac{a}{b} - \frac{c}{d} = \frac{a \cdot d}{b \cdot d} - \frac{b \cdot c}{b \cdot d} = \frac{ad - bc}{bd} \quad \text{if } b \neq 0, d \neq 0 \quad \textbf{(3)}$$

Notice that in each of the equations (2) and (3) all we are really doing to get a common denominator is multiplying $\frac{a}{b}$ by $\frac{d}{d} = 1$, the multiplicative identity, so we are not changing the value of $\frac{a}{b}$. Likewise, we are multiplying $\frac{c}{d}$ by $\frac{b}{b} = 1$, again the multiplicative identity.

EXAMPLE 4 **Adding and Subtracting Rational Expressions**

Perform the indicated operation and simplify the result. Leave your answer in factored form.

(a) $\frac{x-3}{x+4} + \frac{x}{x-2}$ (b) $\frac{x^2}{x^2-4} - \frac{1}{x}$

Solution (a)
$$\frac{x-3}{x+4} + \frac{x}{x-2} = \frac{x-3}{x+4} \cdot \frac{x-2}{x-2} + \frac{x+4}{x+4} \cdot \frac{x}{x-2} \quad \text{Use equation (2).}$$
$$= \frac{(x-3)(x-2) + (x+4)(x)}{(x+4)(x-2)}$$
$$= \frac{x^2 - 5x + 6 + x^2 + 4x}{(x+4)(x-2)}$$
$$= \frac{2x^2 - x + 6}{(x+4)(x-2)}$$

(b)
$$\frac{x^2}{x^2-4} - \frac{1}{x} = \frac{x^2}{x^2-4} \cdot \frac{x}{x} - \frac{x^2-4}{x^2-4} \cdot \frac{1}{x} \quad \text{Use equation (3).}$$
$$= \frac{x^2 \cdot x - (x^2-4) \cdot 1}{(x^2-4)x}$$

$$= \frac{x^3 - (x^2 - 4)}{x(x^2 - 4)}$$

$$= \frac{x^3 - x^2 + 4}{x(x - 2)(x + 2)}$$

◀

EXAMPLE 5 **Adding and Subtracting Rational Expressions**

Perform the indicated operation and simplify the result. Leave your answer in factored form if possible.

(a) $x + \frac{1}{x}$ (b) $\frac{x - 1}{x + 2} - \frac{x + 4}{x - 5}$

Solution (a) $x + \frac{1}{x} = \frac{x}{1} + \frac{1}{x}$ Write x as the ratio $\frac{x}{1}$.

$$= \frac{x}{1} \cdot \frac{x}{x} + \frac{1}{1} \cdot \frac{1}{x}$$ Use equation (2).

$$= \frac{x \cdot x + 1 \cdot 1}{1 \cdot x}$$

$$= \frac{x^2 + 1}{x}$$

(b) $\frac{x - 1}{x + 2} - \frac{x + 4}{x - 5} = \frac{x - 1}{x + 2} \cdot \frac{x - 5}{x - 5} - \frac{x + 2}{x + 2} \cdot \frac{x + 4}{x - 5}$ Use equation (3).

$$= \frac{(x - 1)(x - 5) - (x + 2)(x + 4)}{(x + 2)(x - 5)}$$

$$= \frac{(x^2 - 6x + 5) - (x^2 + 6x + 8)}{(x + 2)(x - 5)}$$ Multiply.

$$= \frac{x^2 - 6x + 5 - x^2 - 6x - 8}{(x + 2)(x - 5)}$$ Distribute negative sign.

$$= \frac{-12x - 3}{(x + 2)(x - 5)}$$ Combine like terms.

$$= \frac{-3(4x + 1)}{(x + 2)(x - 5)}$$ Factor. ◀

Practice Exercise 2 *In Problems 1–3, perform the indicated operation and simplify the result. Leave your answer in factored form.*

1. $\frac{x + 2}{x^2 - 1} + \frac{3x - 2}{x^2 + 1}$ **2.** $\frac{3}{x} - \frac{2x + 1}{x - 2}$ **3.** $1 + \frac{1}{x}$

◀

Least Common Multiple (LCM)

If the denominators of two rational expressions to be added (or subtracted) have common factors, we usually do not use the general rules given by equations (2) and (3). Just as with fractions, we apply the **least common multiple (LCM) method.** The LCM method uses the polynomial of least degree that has each denominator polynomial as a factor.

The LCM Method for Adding or Subtracting Rational Expressions

The least common multiple (LCM) method requires four steps:

STEP 1: Factor completely the polynomial in the denominator of each rational expression.

STEP 2: The LCM of the denominator is the product of each of these factors raised to a power equal to the greatest number of times that the factor occurs in any one of the polynomials.

STEP 3: Write each rational expression using the LCM as the common denominator.

STEP 4: Add or subtract the rational expressions using equation (1).

Let's work two examples that only require Steps 1 and 2.

EXAMPLE 6 Finding the Least Common Multiple

Find the least common multiple of the following pair of polynomials:

$$x(x-1)^2(x+1) \quad \text{and} \quad 4(x-1)(x+1)^3$$

Solution **STEP 1:** The polynomials are already factored completely as

$$x(x-1)^2(x+1) \quad \text{and} \quad 4(x-1)(x+1)^3$$

STEP 2: Start by writing the factors of the left-hand polynomial. (Or you could start with the one on the right.)

$$x(x-1)^2(x+1)$$

Now look at the right-hand polynomial. Its first factor, 4, does not appear in our list, so we insert it.

$$4x(x-1)^2(x+1)$$

The next factor, $x - 1$, is already in our list, so no change is necessary. The final factor is $(x+1)^3$. Since our list has $x + 1$ to the first power only, we replace $x + 1$ in the list by $(x+1)^3$. The LCM is

$$4x(x-1)^2(x+1)^3$$ ◀

Notice that the LCM is, in fact, the polynomial of least degree that contains $x(x-1)^2(x+1)$ and $4(x-1)(x+1)^3$ as factors.

EXAMPLE 7 Finding the Least Common Multiple

Find the least common multiple of the following pairs of polynomials:

$$2x^2 - 2x - 12 \quad \text{and} \quad x^3 - 3x^2$$

Solution **STEP 1:** Factor completely each polynomial:

$$2x^2 - 2x - 12 = 2(x^2 - x - 6) = 2(x-3)(x+2)$$
$$x^3 - 3x^2 = x^2(x-3)$$

Step 2: Start by writing the factors of one of the polynomials:

$$2(x - 3)(x + 2)$$

Now look at the other polynomial. Its first factor, x^2, does not appear in our list, so we insert it.

$$2x^2(x - 3)(x + 2)$$

The remaining factor, $x - 3$, is already in our list, so no change is necessary. The LCM is $2x^2(x - 3)(x + 2)$. ◀

Practice Exercise 3 *In Problems 1–3, find the least common multiple (LCM) of each pair of polynomials.*

1. $3x(x - 1)(x - 2)$ and $6x(x + 1)(x - 1)$

2. $6x^2 + 9x - 6$ and $9x^2 + 18x$

3. $x^3 - 1$ and $x^2 - 1$

◀

The next three examples illustrate how the LCM is used for adding and subtracting rational expressions. Once again, we will leave our answers in factored form.

EXAMPLE 8 Using the Least Common Multiple with Rational Expressions

Perform the indicated operation and simplify the result. Leave your answer in factored form.

$$\frac{x}{x^2 + 3x + 2} + \frac{2x - 3}{x^2 - 1}$$

Solution **Step 1:** Factor completely the polynomials in the denominators.

$$x^2 + 3x + 2 = (x + 2)(x + 1)$$
$$x^2 - 1 = (x - 1)(x + 1)$$

Step 2: The LCM is $(x + 2)(x + 1)(x - 1)$. Do you see why?

Step 3: Write each rational expression using the LCM as the denominator.

$$\frac{x}{x^2 + 3x + 2} = \frac{x}{(x + 2)(x + 1)} = \frac{x}{(x + 2)(x + 1)} \cdot \frac{x - 1}{x - 1} = \frac{x(x - 1)}{(x + 2)(x + 1)(x - 1)}$$

↑ Multiply numerator and denominator by $x - 1$ to get the LCM in the denominator.

$$\frac{2x - 3}{x^2 - 1} = \frac{2x - 3}{(x - 1)(x + 1)} = \frac{2x - 3}{(x - 1)(x + 1)} \cdot \frac{x + 2}{x + 2} = \frac{(2x - 3)(x + 2)}{(x - 1)(x + 1)(x + 2)}$$

↑ Multiply numerator and denominator by $x + 2$ to get the LCM in the denominator.

STEP 4: Now we can add by using equation (1).

$$\frac{x}{x^2+3x+2}+\frac{2x-3}{x^2-1}=\frac{x(x-1)}{(x+2)(x+1)(x-1)}+\frac{(2x-3)(x+2)}{(x+2)(x+1)(x-1)}$$
$$=\frac{x(x-1)+(2x-3)(x+2)}{(x+2)(x+1)(x-1)}$$
$$=\frac{(x^2-x)+(2x^2+x-6)}{(x+2)(x+1)(x-1)}$$
$$=\frac{3x^2-6}{(x+2)(x+1)(x-1)}$$
$$=\frac{3(x^2-2)}{(x+2)(x+1)(x-1)}$$

◀

If we had not used the LCM method to add the rational expressions in Example 8, but decided instead to use the general rule of equation (2), we would have obtained a more complicated expression, as follows:

$$\frac{x}{x^2+3x+2}+\frac{2x-3}{x^2-1}=\frac{x(x^2-1)+(x^2+3x+2)(2x-3)}{(x^2+3x+2)(x^2-1)}$$
$$=\frac{3x^3+3x^2-6x-6}{(x^2+3x+2)(x^2-1)}=\frac{3(x^3+x^2-2x-2)}{(x^2+3x+2)(x^2-1)}$$

Now we are faced with a more complicated problem of expressing this quotient in lowest terms. It is always best to first look for common factors in the denominators of expressions to be added or subtracted and use the LCM if any common factors are found.

EXAMPLE 9 Using the Least Common Multiple with Rational Expressions

Perform the indicated operation and simplify the result. Leave your answer in factored form if possible.

$$\frac{x^2}{6x+6}-\frac{x}{3x^2-3}$$

Solution **STEP 1:** Factor completely the polynomials in the denominators.

$$6x+6=6(x+1)=2\cdot 3\cdot(x+1)$$
$$3x^2-3=3(x^2-1)=3(x-1)(x+1)$$

STEP 2: The LCM is $2\cdot 3\cdot(x+1)(x-1)$.

STEP 3: Write each rational expression using the LCM as the denominator.

$$\frac{x^2}{6x+6}=\frac{x^2}{(2)(3)(x+1)}=\frac{x^2(x-1)}{(2)(3)(x+1)(x-1)}$$

↑
Multiply numerator and denominator by $x-1$ to get the LCM in the denominator.

$$\frac{x}{3x^2 - 3} = \frac{x}{3(x-1)(x+1)} = \frac{(2)x}{(2)(3)(x-1)(x+1)}$$

↑ Multiply numerator and denominator by 2 to get the LCM in the denominator.

STEP 4: Now the denominators of the two rational expressions are the same, so we can subtract using equation (1).

$$\frac{x^2}{6x+6} - \frac{x}{3x^2-3} = \frac{x^2(x-1)}{(2)(3)(x-1)(x+1)} - \frac{2x}{(2)(3)(x-1)(x+1)}$$
$$= \frac{x^2(x-1) - 2x}{(2)(3)(x-1)(x+1)}$$
$$= \frac{x^3 - x^2 - 2x}{(2)(3)(x-1)(x+1)} = \frac{x(x^2 - x - 2)}{(2)(3)(x-1)(x+1)}$$
$$= \frac{x\cancel{(x+1)}(x-2)}{(2)(3)(x-1)\cancel{(x+1)}} = \frac{x(x-2)}{6(x-1)}$$

◀

EXAMPLE 10 Using the Least Common Multiple to Subtract Rational Expressions

Perform the indicated operations and simplify the result. Leave your answer in factored form.

$$\frac{3}{x^2+x} - \frac{x+4}{x^2+2x+1}$$

Solution **STEP 1:** Factor completely the polynomials in the denominators.

$$x^2 + x = x(x+1)$$
$$x^2 + 2x + 1 = (x+1)^2$$

STEP 2: The LCM is $x(x+1)^2$.

STEP 3: Write each rational expression using the LCM as the denominator.

$$\frac{3}{x^2+x} = \frac{3}{x(x+1)} = \frac{3}{x(x+1)} \cdot \frac{x+1}{x+1} = \frac{3(x+1)}{x(x+1)^2}$$
$$\frac{x+4}{x^2+2x+1} = \frac{x+4}{(x+1)^2} = \frac{x+4}{(x+1)^2} \cdot \frac{x}{x} = \frac{x(x+4)}{x(x+1)^2}$$

STEP 4: Subtract, using equation (1).

$$\frac{3}{x^2+x} - \frac{x+4}{x^2+2x+1} = \frac{3(x+1)}{x(x+1)^2} - \frac{x(x+4)}{x(x+1)^2}$$
$$= \frac{3(x+1) - x(x+4)}{x(x+1)^2}$$
$$= \frac{3x + 3 - x^2 - 4x}{x(x+1)^2}$$
$$= \frac{-x^2 - x + 3}{x(x+1)^2}$$

◀

Practice Exercise 4 *In Problems 1 and 2, perform the indicated operation and simplify the result. Leave your answer in factored form.*

1. $\dfrac{x}{x^2 + 3x - 4} + \dfrac{2x + 5}{x^2 - 1}$
2. $\dfrac{x}{9x^2 + 30x + 25} - \dfrac{x}{3x + 5}$

Answers to Practice Exercises

Practice Exercise 1 1. $\dfrac{7x}{x - 3}$ 2. $\dfrac{-2x + 7}{x^2 + 1}$ 3. $\dfrac{-x + 5}{x - 2}$ 4. $\dfrac{3}{x + 2}$

Practice Exercise 2 1. $\dfrac{2(2x^3 - x + 2)}{(x - 1)(x + 1)(x^2 + 1)}$ 2. $\dfrac{-2(x^2 - x + 3)}{x(x - 2)}$ 3. $\dfrac{x + 1}{x}$

Practice Exercise 3 1. $6x(x - 1)(x - 2)(x + 1)$ 2. $9x(2x - 1)(x + 2)$
3. $(x - 1)(x^2 + x + 1)(x + 1)$

Practice Exercise 4 1. $\dfrac{3x^2 + 14x + 20}{(x + 4)(x - 1)(x + 1)}$ 2. $\dfrac{-x(3x + 4)}{(3x + 5)^2}$

3.4 Assess Your Understanding

'Are You Prepared?' *Answers are given at the end of these exercises. If you get a wrong answer, read the pages listed in parentheses.*

1. Simplify: $\dfrac{7}{9} - \dfrac{16}{9}$ (pp. 132–134)
2. Simplify: $\dfrac{3}{5} + \dfrac{1}{3} - \dfrac{5}{6}$ (pp. 136–138)
3. Simplify: $\dfrac{4}{7} + \dfrac{2}{3}$ (pp. 134–136)
4. Simplify: $\dfrac{5}{12} + \dfrac{4}{12}$ (pp. 136–138)
5. Reduce: $\dfrac{x^4 - x}{x^3 - x^2}$ (pp. 142–144)
6. Factor completely: $20x^3 + 34x^2 + 6x$ (Sections: 2.6, 2.7)

Concepts and Vocabulary

7. LCM is an abbreviation for ________ ________ ________.
8. The LCM of $18x(x - 1)$ and $12(x^2 - 1)$ is ________.
9. *True or False:* To add two rational expressions with the same denominator, keep the denominator and add the numerators.
10. The LCM of $(2x + 1)$ and $(x + 5)$ is ________.

Exercises

In Problems 11–36, perform the indicated operation(s) and simplify the result. Leave your answer in factored form.

11. $\dfrac{x^2}{2} + \dfrac{1}{2}$
12. $\dfrac{x}{5} + \dfrac{2}{5}$
13. $\dfrac{3}{x} + \dfrac{5}{x}$
14. $\dfrac{8}{x^2} + \dfrac{1}{x^2}$
15. $\dfrac{x}{x^2 - 4} + \dfrac{6}{x^2 - 4}$
16. $\dfrac{3x}{x^2 - 9} + \dfrac{2}{x^2 - 9}$
17. $\dfrac{x^2}{x^2 + 1} - \dfrac{1}{x^2 + 1}$
18. $\dfrac{x}{x + 2} - \dfrac{4}{x + 2}$
19. $\dfrac{3}{x} - \dfrac{5}{x}$
20. $\dfrac{8}{x^2} - \dfrac{14}{x^2}$
21. $\dfrac{x}{x + 2} - \dfrac{x + 1}{x + 2}$
22. $\dfrac{3x + 2}{x^2 + 1} - \dfrac{3x - 2}{x^2 + 1}$
23. $\dfrac{x^2}{x^2 + 4} - \dfrac{x^2 + 1}{x^2 + 4}$
24. $\dfrac{x}{2x + 3} - \dfrac{x^2 + x}{2x + 3}$
25. $\dfrac{3x + 1}{x - 2} + \dfrac{2x - 3}{2 - x}$

26. $\dfrac{4x+5}{x-4} - \dfrac{2x+4}{4-x}$

27. $\dfrac{2x-5}{x-7} - \dfrac{3x+4}{7-x}$

28. $\dfrac{3x-4}{x-6} + \dfrac{x-5}{6-x}$

29. $\dfrac{x-4}{x-1} - \dfrac{x+2}{1-x}$

30. $\dfrac{2x-1}{x-3} - \dfrac{x-8}{3-x}$

31. $\dfrac{x-7}{x-5} + \dfrac{2x-9}{5-x}$

32. $\dfrac{3x+5}{x-8} + \dfrac{x+21}{8-x}$

33. $\dfrac{3x+7}{x^2-4} - \dfrac{2x+1}{x^2-4} + \dfrac{x+3}{x^2-4}$

34. $\dfrac{2x+3}{x^2+4} - \dfrac{3x+4}{x^2+4} + \dfrac{2x}{4+x^2}$

35. $\dfrac{x^2}{x-3} - \dfrac{x^2+2x}{3-x} + \dfrac{x^2-2x+4}{3-x}$

36. $\dfrac{x^2+4}{x-4} - \dfrac{x^2+x-2}{x-4} + \dfrac{2x^2+x-4}{4-x}$

In Problems 37–60, perform the indicated operation(s) and simplify the result. Leave your answer in factored form.

37. $\dfrac{x}{2} + \dfrac{3}{x}$

38. $\dfrac{x^2}{3} + \dfrac{x}{x+1}$

39. $\dfrac{x-3}{4} + \dfrac{4}{x}$

40. $\dfrac{x+5}{x} + \dfrac{3}{4}$

41. $\dfrac{x-3}{x} - \dfrac{4}{3}$

42. $\dfrac{x}{x-2} - \dfrac{1}{2}$

43. $\dfrac{x}{3} - \dfrac{x+1}{x}$

44. $\dfrac{x}{4} - \dfrac{2x+1}{x}$

45. $\dfrac{3}{x+1} + \dfrac{4}{x-2}$

46. $\dfrac{6}{x-1} + \dfrac{1}{x}$

47. $\dfrac{4}{x-1} - \dfrac{1}{x+2}$

48. $\dfrac{2}{x+5} - \dfrac{3}{x-5}$

49. $\dfrac{x}{x+1} + \dfrac{2x-3}{x-1}$

50. $\dfrac{3x}{x-4} + \dfrac{2x}{x+3}$

51. $\dfrac{x-2}{x+2} - \dfrac{x+2}{x-2}$

52. $\dfrac{2x-1}{x-1} - \dfrac{2x+1}{x+1}$

53. $\dfrac{x}{x^2-4} + \dfrac{1}{x}$

54. $\dfrac{x-1}{x^3} + \dfrac{1}{x^2+1}$

55. $\dfrac{x^3}{(x-1)^2} - \dfrac{x^2+1}{x}$

56. $\dfrac{3x^2}{4} - \dfrac{x^3}{x^2-1}$

57. $\dfrac{x}{x+1} + \dfrac{x-2}{x-1} - \dfrac{x+1}{x-2}$

58. $\dfrac{3x+1}{x} + \dfrac{x}{x-1} - \dfrac{2x}{x+1}$

59. $\dfrac{1}{x} + \dfrac{1}{x+1} - \dfrac{1}{x-1}$

60. $\dfrac{1}{x} - \dfrac{1}{x-1} - \dfrac{1}{x-2}$

In Problems 61–68, find the LCM of the given polynomials.

61. $x^2-4;\ \ x^2-x-2$

62. $x^2-x-12;\ \ x^2-8x+16$

63. $x^3-x;\ \ x^2-x$

64. $3x^2-27;\ \ 2x^2-x-15$

65. $4x^3-4x^2+x;\ \ 2x^3-x^2;\ \ x^3$

66. $x-3;\ \ x^2+3x;\ \ x^3-9x$

67. $x^3-x;\ \ x^3-2x^2+x;\ \ x^3-1$

68. $x^2+4x+4;\ \ x^3+2x^2;\ \ (x+2)^3$

In Problems 69–90, perform the indicated operation(s) and simplify the result. Leave your answer in factored form.

69. $\dfrac{3}{x(x+1)} + \dfrac{4}{(x+1)(x+2)}$

70. $\dfrac{8}{(x+1)(x-1)} + \dfrac{4}{x(x-1)}$

71. $\dfrac{x+3}{(x+1)(x-2)} + \dfrac{2x-6}{(x+1)(x+2)}$

72. $\dfrac{x-1}{(x+1)(x+2)} + \dfrac{3x+4}{(x+2)(x-1)}$

73. $\dfrac{2x}{(3x+1)(x-2)} - \dfrac{4x}{(3x+1)(x+2)}$

74. $\dfrac{3}{(2x+3)(x-3)} - \dfrac{6}{(3x+1)(x-3)}$

75. $\dfrac{4x+1}{x^2+7x+12} + \dfrac{2x+3}{x^2+5x+4}$

76. $\dfrac{3x-2}{x^2-x-6} + \dfrac{4x-3}{x^2-9}$

77. $\dfrac{x^2-1}{3x^2-5x-2} + \dfrac{x^2-x}{2x^2-3x-2}$

78. $\dfrac{x^2+4x}{2x^2-x-1} + \dfrac{x^2-9x}{3x^2-2x-1}$

79. $\dfrac{x}{x^2-7x+6} - \dfrac{x}{x^2-2x-24}$

80. $\dfrac{x}{x-3} - \dfrac{x+1}{x^2+5x-24}$

81. $\dfrac{4}{x^2-4} - \dfrac{2}{x^2+x-6}$

82. $\dfrac{3}{x-1} - \dfrac{x-4}{x^2-2x+1}$

83. $\dfrac{3}{(x-1)^2(x+1)} + \dfrac{2}{(x-1)(x+1)^2}$

84. $\dfrac{2}{(x+2)^2(x-1)} - \dfrac{6}{(x+2)(x-1)^2}$

85. $\dfrac{x+4}{x^2-x-2} - \dfrac{2x+3}{x^2+2x-8}$

86. $\dfrac{2x-3}{x^2+8x+7} - \dfrac{x-2}{(x+1)^2}$

87. $\dfrac{1}{x} - \dfrac{2}{x^2+x} + \dfrac{3}{x^3-x^2}$

88. $\dfrac{x}{(x-1)^2} + \dfrac{2}{x} - \dfrac{x+1}{x^3-x^2}$

89. $\dfrac{1}{h}\left(\dfrac{1}{x+h} - \dfrac{1}{x}\right)$

90. $\dfrac{1}{h}\left[\dfrac{1}{(x+h)^2} - \dfrac{1}{x^2}\right]$

91. Write a few paragraphs that outline your strategy for adding two rational expressions.

'Are You Prepared?' Answers

1. -1 **2.** $\dfrac{1}{10}$ **3.** $\dfrac{26}{21}$ **4.** $\dfrac{3}{4}$ **5.** $\dfrac{x^2+x+1}{x}$ **6.** $2x(2x+3)(5x+1)$

3.5 Mixed Quotients

PREPARING FOR THIS SECTION *Before getting started review, the following:*

- Dividing Rational Expressions (pp. 149–151)
- Adding and Subtracting Rational Expressions (pp. 154–162)

Now work the 'Are You Prepared?' problems on page 169.

OBJECTIVES

1 Simplify Mixed Quotients

1 When sums and/or differences of rational expressions appear as the numerator and/or denominator of a quotient, the quotient is called a **mixed quotient.***

For example, the expressions

$$\frac{1+\dfrac{1}{x}}{1-\dfrac{1}{x}} \quad \text{and} \quad \frac{\dfrac{x^2}{x^2-4}-3}{\dfrac{x-3}{x+2}-1}$$

are mixed quotients. To **simplify** a mixed quotient means to write it as a rational expression reduced to lowest terms. This can be accomplished in either of two ways:

Simplifying a Mixed Quotient

Method 1: Treat the numerator and denominator of the mixed quotient separately, performing whatever operations are indicated and simplifying the results. Follow this by simplifying the resulting rational expression.

Method 2: Find the LCM of the denominators of all rational expressions that appear in the mixed quotient. Multiply the numerator and denominator of the mixed quotient by the LCM and simplify the result.

*Some texts use the term **complex fraction.**

We will use both methods in the next three examples. By carefully studying each method, you can discover situations in which one method may be easier to use than the other.

EXAMPLE 1 Simplifying a Mixed Quotient

Simplify: $\dfrac{\dfrac{1}{2} + \dfrac{3}{x}}{\dfrac{x+3}{4}}$

Solution **Method 1:** First, we perform the indicated operation in the numerator, and then we divide:

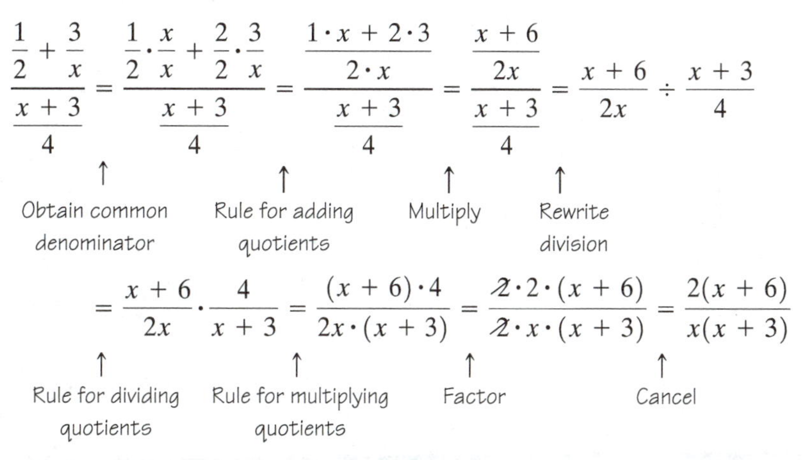

Method 2: The rational expressions that appear in the mixed quotient are

$$\frac{1}{2} \quad \frac{3}{x} \quad \frac{x+3}{4}$$

The LCM of their denominators is $4x$. We multiply the numerator and denominator of the mixed quotient by $4x$ and then simplify:

$$\frac{\dfrac{1}{2} + \dfrac{3}{x}}{\dfrac{x+3}{4}} = \frac{4x \cdot \left(\dfrac{1}{2} + \dfrac{3}{x}\right)}{4x \cdot \left(\dfrac{x+3}{4}\right)} = \frac{\dfrac{4x}{1} \cdot \left(\dfrac{1}{2} + \dfrac{3}{x}\right)}{\dfrac{4x}{1} \cdot \left(\dfrac{x+3}{4}\right)} = \frac{\dfrac{4x}{1} \cdot \dfrac{1}{2} + \dfrac{4x}{1} \cdot \dfrac{3}{x}}{\dfrac{4x}{1} \cdot \dfrac{x+3}{4}}$$

Multiply by $\dfrac{4x}{4x} = 1$ — Rewrite each $4x$ as $\dfrac{4x}{1}$ — Distributive Property in numerator

$$= \frac{\dfrac{4x \cdot 1}{1 \cdot 2} + \dfrac{4x \cdot 3}{1 \cdot x}}{\dfrac{4x \cdot (x+3)}{1 \cdot 4}} = \frac{\dfrac{\cancel{2} \cdot 2 \cdot x \cdot 1}{1 \cdot \cancel{2}} + \dfrac{4\cancel{x} \cdot 3}{1 \cdot \cancel{x}}}{\dfrac{\cancel{4}x \cdot (x+3)}{1 \cdot \cancel{4}}} = \frac{2x + 12}{x(x+3)} = \frac{2(x+6)}{x(x+3)}$$

Multiply — Factor — Cancel — Factor ◀

EXAMPLE 2 Simplifying a Mixed Quotient

Simplify: $\dfrac{4 + \dfrac{2}{x}}{4 - \dfrac{2}{x}}$

Solution **METHOD 1:** We perform the indicated operations in the numerator and the denominator, and then we divide:

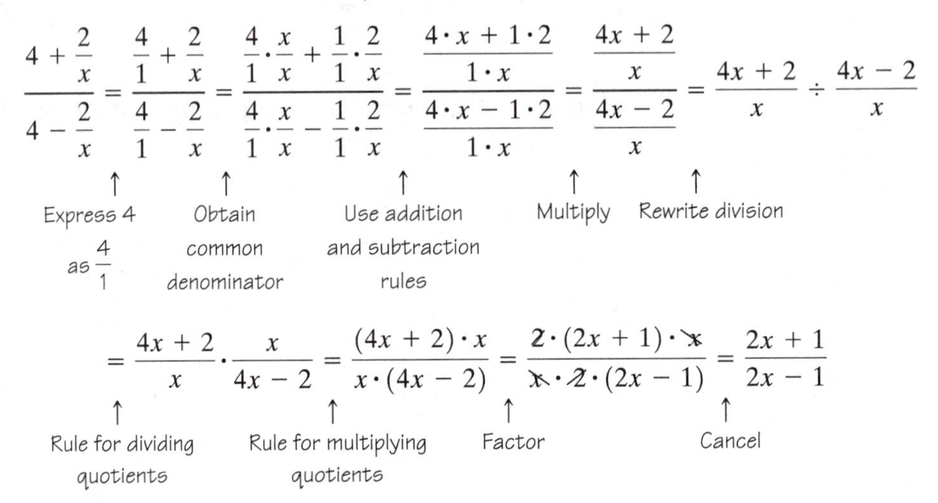

METHOD 2: The rational expressions that appear in the mixed quotient are

$$4 = \frac{4}{1} \quad \frac{2}{x}$$

The LCM of their denominators is x, so we multiply the numerator and denominator of the mixed quotient by x:

$$\frac{4 + \dfrac{2}{x}}{4 - \dfrac{2}{x}} = \frac{x \cdot \left(4 + \dfrac{2}{x}\right)}{x \cdot \left(4 - \dfrac{2}{x}\right)} = \frac{x \cdot 4 + x \cdot \dfrac{2}{x}}{x \cdot 4 - x \cdot \dfrac{2}{x}}$$

Distributive property

$$= \frac{x \cdot 4 + \dfrac{x}{1} \cdot \dfrac{2}{x}}{x \cdot 4 - \dfrac{x}{1} \cdot \dfrac{2}{x}} \qquad \text{Rewrite } x = \frac{x}{1}$$

$$= \frac{4x + \dfrac{2\cancel{x}}{\cancel{x}}}{4x - \dfrac{2\cancel{x}}{\cancel{x}}} = \frac{4x + 2}{4x - 2} = \frac{\cancel{2} \cdot (2x + 1)}{\cancel{2} \cdot (2x - 1)} = \frac{2x + 1}{2x - 1}$$

Cancel Factor Cancel ◀

EXAMPLE 3 Simplifying a Mixed Quotient

Simplify: $\dfrac{\dfrac{x^2}{x^2 - 4} - 3}{\dfrac{x - 3}{x + 2} - 1}$

Solution **Method 1:**

$$\frac{\dfrac{x^2}{x^2-4}-3}{\dfrac{x-3}{x+2}-1} = \frac{\dfrac{x^2}{x^2-4}-\dfrac{3(x^2-4)}{x^2-4}}{\dfrac{x-3}{x+2}-\dfrac{x+2}{x+2}} = \frac{\dfrac{x^2-3(x^2-4)}{x^2-4}}{\dfrac{(x-3)-(x+2)}{x+2}}$$

$$= \frac{\dfrac{x^2-3x^2+12}{x^2-4}}{\dfrac{x-3-x-2}{x+2}} = \frac{\dfrac{-2x^2+12}{x^2-4}}{\dfrac{-5}{x+2}}$$

$$= \frac{-2x^2+12}{x^2-4} \div \frac{-5}{x+2} = \frac{-2x^2+12}{x^2-4}\cdot\frac{x+2}{-5}$$

$$= \frac{-2(x^2-6)}{(x-2)(x+2)}\cdot\frac{x+2}{-5} = \frac{-2(x^2-6)\cancel{(x+2)}}{(x-2)\cancel{(x+2)}(-5)} = \frac{2(x^2-6)}{5(x-2)}$$

Method 2: The rational expressions that appear in the mixed quotient are

$$\frac{x^2}{x^2-4} \quad \frac{3}{1} \quad \frac{x-3}{x+2} \quad \frac{1}{1}$$

The LCM of the denominators x^2-4, $x+2$, and 1 is x^2-4. Multiply the numerator and denominator of the mixed quotient by x^2-4.

$$\frac{\dfrac{x^2}{x^2-4}-3}{\dfrac{x-3}{x+2}-1} = \frac{(x^2-4)\left(\dfrac{x^2}{x^2-4}-3\right)}{(x^2-4)\left(\dfrac{x-3}{x+2}-1\right)}$$

$$= \frac{(x^2-4)\left(\dfrac{x^2}{x^2-4}\right)-(x^2-4)\cdot 3}{(x^2-4)\left(\dfrac{x-3}{x+2}\right)-(x^2-4)\cdot 1}$$

↑ Distributive property

$$= \frac{\cancel{(x^2-4)}\left(\dfrac{x^2}{\cancel{x^2-4}}\right)-(x^2-4)\cdot 3}{(x-2)\cancel{(x+2)}\left(\dfrac{x-3}{\cancel{x+2}}\right)-(x^2-4)\cdot 1}$$

$$= \frac{x^2-3\cdot(x^2-4)}{(x-2)(x-3)-(x^2-4)}$$

$$= \frac{x^2-3x^2+12}{x^2-5x+6-x^2+4}$$

$$= \frac{-2x^2+12}{-5x+10}$$

$$= \frac{-2(x^2-6)}{-5(x-2)}$$

$$= \frac{2(x^2-6)}{5(x-2)}$$

◀

Practice Exercise 1 *Simplify each mixed quotient. Use both Method 1 and Method 2.*

1. $\dfrac{x + \dfrac{1}{x}}{\dfrac{x^2 - 1}{5x}}$ **2.** $\dfrac{\dfrac{x^2 - 9x}{x^2 - 1} - 1}{\dfrac{x}{x + 1} + \dfrac{x + 2}{x - 1}}$

EXAMPLE 4 **Simplifying a Mixed Quotient**

Simplify: $\dfrac{\dfrac{x^2 + 1}{x^2 - x - 12} - \dfrac{x + 2}{x - 4}}{x - \dfrac{1}{x^3}}$

Solution We will use Method 1.

$$\frac{\dfrac{x^2 + 1}{x^2 - x - 12} - \dfrac{x + 2}{x - 4}}{x - \dfrac{1}{x^3}} = \frac{\dfrac{x^2 + 1}{(x - 4)(x + 3)} - \dfrac{x + 2}{x - 4}}{\dfrac{x}{1} - \dfrac{1}{x^3}}$$

$$= \frac{\dfrac{x^2 + 1}{(x - 4)(x + 3)} - \dfrac{(x + 2)(x + 3)}{(x - 4)(x + 3)}}{\dfrac{x}{1} \cdot \dfrac{x^3}{x^3} - \dfrac{1}{x^3}}$$

$$= \frac{\dfrac{x^2 + 1}{(x - 4)(x + 3)} - \dfrac{x^2 + 5x + 6}{(x - 4)(x + 3)}}{\dfrac{x^4}{x^3} - \dfrac{1}{x^3}}$$

$$= \frac{\dfrac{(x^2 + 1) - (x^2 + 5x + 6)}{(x - 4)(x + 3)}}{\dfrac{x^4 - 1}{x^3}}$$

$$= \frac{\dfrac{x^2 + 1 - x^2 - 5x - 6}{(x - 4)(x + 3)}}{\dfrac{x^4 - 1}{x^3}}$$

$$= \frac{\dfrac{-5x - 5}{(x - 4)(x + 3)}}{\dfrac{x^4 - 1}{x^3}}$$

$$= \frac{-5x - 5}{(x - 4)(x + 3)} \div \frac{x^4 - 1}{x^3}$$

$$= \frac{-5x - 5}{(x - 4)(x + 3)} \cdot \frac{x^3}{x^4 - 1}$$

$$= \frac{-5(x + 1)}{(x - 4)(x + 3)} \cdot \frac{x^3}{(x^2 + 1)(x - 1)(x + 1)}$$

$$= \frac{-5(\cancel{x+1}) \cdot x^3}{(x-4)(x+3) \cdot (x^2+1)(x-1)(\cancel{x+1})}$$

$$= \frac{-5x^3}{(x-4)(x+3)(x^2+1)(x-1)}$$ ◀

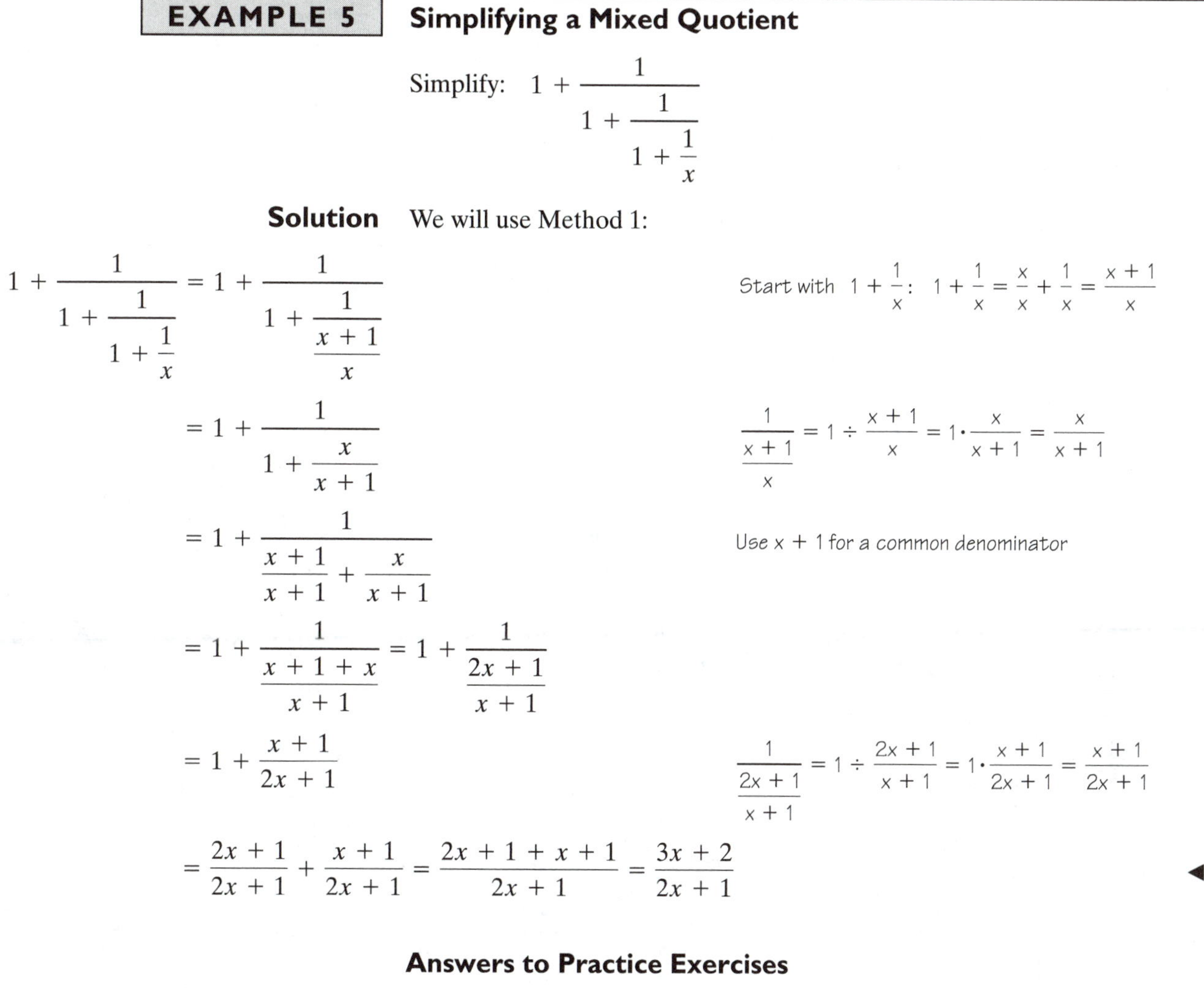

EXAMPLE 5 Simplifying a Mixed Quotient

Simplify: $1 + \cfrac{1}{1 + \cfrac{1}{1 + \cfrac{1}{x}}}$

Solution We will use Method 1:

$$1 + \cfrac{1}{1 + \cfrac{1}{1 + \cfrac{1}{x}}} = 1 + \cfrac{1}{1 + \cfrac{1}{\cfrac{x+1}{x}}}$$

Start with $1 + \frac{1}{x}$: $1 + \frac{1}{x} = \frac{x}{x} + \frac{1}{x} = \frac{x+1}{x}$

$$= 1 + \cfrac{1}{1 + \cfrac{x}{x+1}}$$

$\frac{1}{\frac{x+1}{x}} = 1 \div \frac{x+1}{x} = 1 \cdot \frac{x}{x+1} = \frac{x}{x+1}$

$$= 1 + \cfrac{1}{\cfrac{x+1}{x+1} + \cfrac{x}{x+1}}$$

Use $x + 1$ for a common denominator

$$= 1 + \cfrac{1}{\cfrac{x+1+x}{x+1}} = 1 + \cfrac{1}{\cfrac{2x+1}{x+1}}$$

$$= 1 + \frac{x+1}{2x+1}$$

$\frac{1}{\frac{2x+1}{x+1}} = 1 \div \frac{2x+1}{x+1} = 1 \cdot \frac{x+1}{2x+1} = \frac{x+1}{2x+1}$

$$= \frac{2x+1}{2x+1} + \frac{x+1}{2x+1} = \frac{2x+1+x+1}{2x+1} = \frac{3x+2}{2x+1}$$ ◀

Answers to Practice Exercises

Practice Exercise 1 **1.** $\frac{5(x^2+1)}{(x-1)(x+1)}$ **2.** $\frac{-9x+1}{2(x^2+x+1)}$

3.5 Assess Your Understanding

'Are You Prepared?' *Answers are given at the end of these exercises. If you get a wrong answer, read the pages listed in parentheses.*

1. Divide: $\cfrac{\cfrac{3x-6}{2x+1}}{\cfrac{3x-3}{2x-1}}$ (pp. 149–151)

2. Subtract: $\frac{x}{x+3} - \frac{x}{x-3}$ (pp. 154–162)

Concepts and Vocabulary

3. *True or False:* When simplifying mixed quotients, Method 1 is always the easiest method to use.

4. To simplify $\dfrac{\dfrac{x-3}{4}}{\dfrac{2x}{x-1}-\dfrac{6}{x-1}}$ with Method 2, we can multiply the numerator and denominator of the mixed quotient by ________.

Exercises

In Problems 5–36, perform the indicated operations and simplify the result. Leave your answer in factored form. Use whichever method you prefer, Method 1 or Method 2, for simplifying the mixed quotients.

5. $\dfrac{\dfrac{x}{x+1}+\dfrac{4}{x+1}}{\dfrac{x+4}{2}}$

6. $\dfrac{\dfrac{x}{x-3}+\dfrac{2}{x-3}}{\dfrac{x+2}{3}}$

7. $\dfrac{\dfrac{x-1}{3}}{\dfrac{x}{x+4}-\dfrac{1}{x+4}}$

8. $\dfrac{\dfrac{x-3}{4}}{\dfrac{2x}{x-1}-\dfrac{6}{x-1}}$

9. $\dfrac{\dfrac{x}{x+1}+\dfrac{5}{x+1}}{\dfrac{x^2}{x-1}-\dfrac{25}{x-1}}$

10. $\dfrac{\dfrac{x}{x-2}-\dfrac{4}{x-2}}{\dfrac{x^2}{x+2}-\dfrac{16}{x+2}}$

11. $\dfrac{\dfrac{1}{3}+\dfrac{2}{x}}{\dfrac{x+1}{4}}$

12. $\dfrac{\dfrac{2}{3}+\dfrac{4}{x}}{\dfrac{x-3}{2}}$

13. $\dfrac{\dfrac{4}{x}+\dfrac{3}{x+1}}{\dfrac{4}{x}}$

14. $\dfrac{\dfrac{5}{x+2}+\dfrac{4}{x+3}}{\dfrac{6}{x+2}}$

15. $\dfrac{\dfrac{x}{x+2}-\dfrac{3}{x+2}}{\dfrac{x}{x^2-4}-\dfrac{1}{x^2-4}}$

16. $\dfrac{\dfrac{x}{x-2}+\dfrac{4}{x-2}}{\dfrac{x}{x^2-4}+\dfrac{1}{x^2-4}}$

17. $\dfrac{4+\dfrac{3}{x}}{1-\dfrac{2}{x}}$

18. $\dfrac{3-\dfrac{2}{x}}{2+\dfrac{1}{x}}$

19. $\dfrac{x-\dfrac{1}{x}}{x+\dfrac{1}{x}}$

20. $\dfrac{1-\dfrac{x}{x+1}}{2-\dfrac{x-1}{x}}$

21. $\dfrac{3-\dfrac{x^2}{x+1}}{1+\dfrac{x}{x^2-1}}$

22. $\dfrac{3x-\dfrac{3}{x^2}}{\dfrac{1}{(x-1)^2}-1}$

23. $\dfrac{\dfrac{x+4}{x-2}-\dfrac{x-3}{x+1}}{x+1}$

24. $\dfrac{\dfrac{x-2}{x+1}-\dfrac{x}{x-2}}{x+3}$

25. $\dfrac{\dfrac{x-2}{x+2}+\dfrac{x-1}{x+1}}{\dfrac{x}{x+1}-\dfrac{2x-3}{x}}$

26. $\dfrac{\dfrac{2x+5}{x}-\dfrac{x}{x-3}}{\dfrac{x^2}{x-3}-\dfrac{(x+1)^2}{x+3}}$

27. $\dfrac{\dfrac{x}{x-5}-\dfrac{4}{5-x}}{\dfrac{3x}{x+1}+\dfrac{x}{x-1}-\dfrac{2}{x^2-1}}$

28. $\dfrac{\dfrac{x}{x+2}-\dfrac{x}{x-2}+\dfrac{4}{x^2-4}}{\dfrac{2x}{x-1}-\dfrac{4}{1-x}}$

29. $\dfrac{1+\dfrac{1}{x}+\dfrac{1}{x^2}}{1-\dfrac{1}{x}+\dfrac{1}{x^2}}$

30. $\dfrac{x+\dfrac{1}{x}+\dfrac{1}{x^2}}{x-\dfrac{1}{x}+\dfrac{1}{x^2}}$

31. $1-\dfrac{1}{1-\dfrac{1}{x}}$

32. $1-\dfrac{1}{1-\dfrac{1}{1-x}}$

33. $\dfrac{\dfrac{x+h-2}{x+h+2}-\dfrac{x-2}{x+2}}{h}$

34. $\dfrac{\dfrac{x+h+1}{x+h-1} - \dfrac{x+1}{x-1}}{h}$

35. $\dfrac{x}{x + \dfrac{1}{x + \dfrac{1}{x+1}}}$

36. $\dfrac{x}{x - \dfrac{1}{x - \dfrac{1}{x-1}}}$

37. An electrical circuit contains two resistors connected in parallel, as shown in the figure. If the resistance of each one is R_1 and R_2 ohms, their combined resistance R is given by the formula

$$R = \frac{1}{\dfrac{1}{R_1} + \dfrac{1}{R_2}}$$

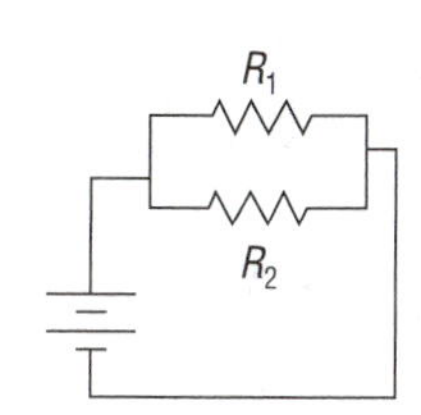

(a) Express R as a rational expression; that is, simplify the right-hand side of this formula.
(b) Evaluate the rational expression if $R_1 = 6$ ohms and $R_2 = 10$ ohms.

38. The combined resistance R in an electrical circuit containing three resistors connected in parallel is given by the formula

$$R = \frac{1}{\dfrac{1}{R_1} + \dfrac{1}{R_2} + \dfrac{1}{R_3}}$$

where R_1, R_2, and R_3 are the resistances of the three resistors.

(a) Express R as a rational expression.

(b) Find R if $R_1 = 4$ ohms, $R_2 = 6$ ohms, and $R_3 = 12$ ohms.

39. The expressions below are called **continued fractions:**

$$1 + \frac{1}{x}, \quad 1 + \frac{1}{1 + \dfrac{1}{x}}, \quad 1 + \frac{1}{1 + \dfrac{1}{1 + \dfrac{1}{x}}}, \quad 1 + \frac{1}{1 + \dfrac{1}{1 + \dfrac{1}{1 + \dfrac{1}{x}}}}, \ldots$$

Each one simplifies to an expression of the form

$$\frac{ax + b}{bx + c}$$

Trace the successive values of a, b, and c as you "continue" the fraction. Can you discover the pattern these values follow?

40. Explain what is wrong with the following calculation.

$$\frac{\dfrac{2}{x} + \dfrac{x}{2}}{x} = \frac{1}{2} + \frac{1}{2} = 1$$

'Are You Prepared?' Answers

1. $\dfrac{(x-2)(2x-1)}{(2x+1)(x-1)}$ **2.** $\dfrac{-6x}{(x-3)(x+3)}$

Chapter Review

Things to Know

Rational number, fraction	Quotient of two integers
Rational expression	Quotient of two polynomials
Reduced to lowest terms	Numerator and denominator have no common factors, except 1 and -1
Cancellation property	$\dfrac{ac}{bc} = \dfrac{a}{b}$ if $b \neq 0, c \neq 0$

Formulas

Addition	$\frac{a}{b}+\frac{c}{b}=\frac{a+c}{b}$ $\quad \frac{a}{b}+\frac{c}{d}=\frac{ad+bc}{bd}$ if $b \neq 0, d \neq 0$
Subtraction	$\frac{a}{b}-\frac{c}{b}=\frac{a-c}{b}$ $\quad \frac{a}{b}-\frac{c}{d}=\frac{ad-bc}{bd}$ if $b \neq 0, d \neq 0$
Multiplication	$\frac{a}{b}\cdot\frac{c}{d}=\frac{a\cdot c}{b\cdot d}$ if $b \neq 0, d \neq 0$
Division	$\frac{\frac{a}{b}}{\frac{c}{d}}=\frac{a}{b}\div\frac{c}{d}=\frac{a}{b}\cdot\frac{d}{c}=\frac{ad}{bc}$ if $b \neq 0, c \neq 0, d \neq 0$

Objectives

Section		You Should Be Able To . . .	Review Exercises
3.1	1	Show equivalency of fractions (p. 126)	1–6
	2	Reduce fractions to lowest terms (p. 128)	7–10
	3	Multiply fractions (p. 129)	31, 32
	4	Divide fractions (p. 130)	41, 42
	5	Add and subtract fractions with equal denominators (p. 132)	33–36
	6	Add and subtract fractions with unequal denominators (p. 134)	37–40
	7	Use the least common multiple method with fractions (p. 136)	38–40
3.2	1	Reduce rational expressions (p. 142)	11–20
	2	Evaluate rational expressions (p. 144)	21–26
3.3	1	Multiply rational expressions (p. 148)	43–48
	2	Divide rational expressions (p. 149)	49–54
3.4	1	Add and subtract rational expressions with equal denominators (p. 154)	55–60
	2	Add and subtract rational expressions with unequal denominators (p. 156)	61–68
	3	Use the least common multiple method with rational expressions (p. 157)	63–68
3.5	1	Simplify mixed quotients (p. 164)	69–80

Review Exercises

In Problems 1–6, show that the fractions are equivalent by:
(a) *Showing that each one has the same decimal form*
(b) *Demonstrating that property (1), page 127, is satisfied*
(c) *Using the Cancellation Property (2), page 129*

1. $\frac{14}{5}$ and $\frac{84}{30}$

2. $\frac{9}{10}$ and $\frac{54}{60}$

3. $\frac{-5}{18}$ and $\frac{-30}{108}$

4. $\frac{24}{-9}$ and $\frac{16}{-6}$

5. $\frac{-30}{-54}$ and $\frac{5}{9}$

6. $\frac{-72}{-32}$ and $\frac{9}{4}$

In Problems 7–20, reduce each expression to lowest terms.

7. $\frac{36}{28}$
8. $\frac{-42}{36}$
9. $\frac{-24}{-18}$
10. $\frac{21}{-72}$
11. $\frac{3x^2}{9x}$
12. $\frac{x^2 + x}{x^3 - x}$
13. $\frac{4x^2 + 4x + 1}{4x^2 - 1}$
14. $\frac{x^2 + 5x + 6}{x^2 - 9}$
15. $\frac{2x^2 + 11x + 14}{x^2 - 4}$
16. $\frac{x^2 - 5x - 14}{4 - x^2}$
17. $\frac{x^3 + x^2}{x^3 + 1}$
18. $\frac{x - x^3}{x^4 - 1}$
19. $\frac{(4x - 12x^2)(x - 2)^2}{(x^2 - 4)(1 - 9x^2)}$
20. $\frac{(25 - 9x^2)(1 - x)^3}{(x^2 - 1)(3x - 5)^2}$

In Problems 21–26, evaluate each rational expression for the given value of the variable.

21. $\frac{x^2 + x - 1}{(x - 2)^2}$ for $x = 3$
22. $\frac{x^2 - x + 1}{(x - 3)^2}$ for $x = 2$
23. $\frac{(x + 1)^3}{x^2 + x + 1}$ for $x = 1$
24. $\frac{x^3 + x^2 + 1}{3x + 5}$ for $x = 0$
25. $\frac{x^3 - 3x^2 + 4}{(x + 1)^2}$ for $x = -2$
26. $\frac{-2x^2 + 1}{(x + 1)(x - 1)}$ for $x = -2$

In Problems 27–30, determine which of the value(s) given below, if any, must be excluded from the domain of the variable in each rational expression.

(a) $x = 2$ (b) $x = 1$ (c) $x = 0$ (d) $x = -1$

27. $\frac{x^2 - 4}{x}$
28. $\frac{9x - 18}{x^2 - 1}$
29. $\frac{x(x - 1)}{(x + 2)(x + 3)}$
30. $\frac{x^2}{x^2 + 1}$

In Problems 31–80, perform the indicated operation(s) and simplify the result.

31. $\frac{9}{8} \cdot \frac{2}{27}$
32. $\frac{18}{25} \cdot \frac{45}{14}$
33. $\frac{3}{4} + \frac{8}{4}$
34. $\frac{9}{8} + \frac{13}{8}$
35. $\frac{8}{3} - \frac{4}{3}$
36. $\frac{9}{2} - \frac{5}{2}$
37. $\frac{3}{4} - \frac{2}{3}$
38. $\frac{5}{18} - \frac{5}{12}$
39. $\frac{7}{24} + \frac{1}{9} - \frac{5}{6}$
40. $\frac{8}{15} - \frac{2}{9} + \frac{5}{6}$
41. $\frac{\frac{1}{2} + \frac{1}{3}}{\frac{3}{8} - \frac{1}{4}}$
42. $\frac{\frac{2}{3} + \frac{7}{8}}{\frac{5}{12} + \frac{1}{18}}$
43. $\frac{x + 3}{8x} \cdot \frac{6x^2}{3x^2 + 9x}$
44. $\frac{x^2 - x}{4x^3} \cdot \frac{6x^2}{x^2 - 1}$
45. $\frac{x^2 + 7x + 6}{x^2 + 5x - 14} \cdot \frac{(x + 2)^2}{1 - x^2}$
46. $\frac{x^2 + x - 6}{x^2 - 11x - 12} \cdot \frac{(x + 1)^2}{4 - x^2}$
47. $\frac{6x^2 + 7x - 3}{3x^2 + 11x - 4} \cdot \frac{x^2 - 16}{9 - 4x^2}$
48. $\frac{2x^2 + 11x + 12}{4x^2 + 4x + 1} \cdot \frac{3 - 5x - 2x^2}{4x^2 - 9}$
49. $\frac{\frac{4x + 20}{9x^2}}{\frac{x^2 - 25}{3x}}$
50. $\frac{\frac{9x^2}{x^2 - 4}}{\frac{3x + 9}{2x - 4}}$
51. $\frac{\frac{4x^2 + 4x + 1}{4 - x^2}}{\frac{4x^2 - 1}{x^2 + 4x + 4}}$

52. $\dfrac{\dfrac{9-x^2}{x^2+6x+8}}{\dfrac{x^2+6x+9}{16-x^2}}$

53. $\dfrac{\dfrac{x^2-25}{x^2-x-6}\cdot\dfrac{9-x^2}{x^2+10x+25}}{\dfrac{5-x}{3+x}\cdot\dfrac{x^2+x}{x^2+4x+4}}$

54. $\dfrac{\dfrac{9x^2-16}{(x-4)^2}\cdot\dfrac{x^2-2x-8}{4-3x}}{\dfrac{x^3}{4-x}\cdot\dfrac{3x+4}{x+1}}$

55. $\dfrac{3x+1}{x-2}+\dfrac{2x-1}{x-2}$

56. $\dfrac{x^2-1}{x+3}+\dfrac{x^2+1}{x+3}$

57. $\dfrac{x^2-1}{x^2+4}-\dfrac{1-x+x^2}{x^2+4}$

58. $\dfrac{x^2-2x}{4x^2-9}-\dfrac{2-x+x^2}{4x^2-9}$

59. $\dfrac{5x-3}{4-x}+\dfrac{1-2x}{x-4}$

60. $\dfrac{x^2+1}{x^2-4}+\dfrac{1+x^2}{4-x^2}$

61. $\dfrac{x+1}{x}+\dfrac{x-1}{x+1}$

62. $\dfrac{x^2}{x+4}+\dfrac{x-1}{x}$

63. $\dfrac{x}{(x-1)(x+1)}-\dfrac{4x}{(x-1)(x+2)}$

64. $\dfrac{x+1}{x(x-2)}-\dfrac{x+2}{x(x+1)}$

65. $\dfrac{3x+5}{x^2+x-6}+\dfrac{2x-1}{x^2-9}$

66. $\dfrac{4x+1}{x^2-x-12}+\dfrac{1-3x}{x^2-16}$

67. $\dfrac{x^2+4}{3x^2-5x-2}-\dfrac{x^2+9}{2x^2-3x-2}$

68. $\dfrac{2x^2+4x}{2x^2-x-1}-\dfrac{3x^2}{3x^2-2x-1}$

69. $\dfrac{\dfrac{x}{3}+\dfrac{4}{x}}{\dfrac{x+4}{x}}$

70. $\dfrac{\dfrac{x}{x+1}+\dfrac{x}{x+1}}{\dfrac{x+4}{x}}$

71. $\dfrac{1-\dfrac{3}{x}}{1-\dfrac{2}{x}}$

72. $\dfrac{3+\dfrac{2}{x}}{3-\dfrac{2}{x}}$

73. $\dfrac{\dfrac{x^2}{3x-4}-\dfrac{1-x^2}{4-3x}}{\dfrac{x^2}{9x^2-16}}$

74. $\dfrac{\dfrac{x}{4x-3}-\dfrac{2x-3}{3-4x}}{\dfrac{x+2}{16x^2-9}}$

75. $\dfrac{\dfrac{x^2}{x-2}-\dfrac{x^2}{x+2}}{\dfrac{x}{x+2}+\dfrac{x}{x-2}}$

76. $\dfrac{\dfrac{2x}{x^2-4}-\dfrac{2}{x+2}}{\dfrac{x^3}{x^2+4x+4}-x}$

77. $\dfrac{3-\dfrac{x^2}{x^2+1}}{4-\dfrac{x^2}{x^2+1}}$

78. $\dfrac{1-\dfrac{x}{x^2-1}}{2+\dfrac{x}{x^2-1}}$

79. $\dfrac{\dfrac{1}{x}-\dfrac{1}{2}}{x+1+\dfrac{1}{x}}$

80. $\dfrac{x+1-\dfrac{1}{x}}{\dfrac{1}{x}+\dfrac{1}{3}}$

4 Exponents, Radicals; Complex Numbers

OUTLINE

In this chapter, we continue to study exponents, beginning with negative integer exponents. After discussing radicals, we then extend the definition of exponents to include rational numbers. We conclude the chapter with a brief introduction to complex numbers.

4.1 Negative Integer Exponents; Scientific Notation

PREPARING FOR THIS SECTION *Before getting started, review the following:*

- Laws of Exponents (pp. 62–68)

Now work the 'Are You Prepared?' problems on page 186.

OBJECTIVES

1. Work with Negative Exponents
2. Review the Laws of Exponents
3. Use Scientific Notation

In Section 2.1 we studied Laws of Exponents as they related to nonnegative integer exponents. In this section we extend the Laws of Exponents to include exponents that are negative integers.

We begin by repeating the example given at the end of Section 2.1.

EXAMPLE 1 **Defining a Negative Exponent**

Simplify $\frac{4^2}{4^5}$

Solution A: $\frac{4^2}{4^5} = 4^{2-5} = 4^{-3}$

Solution B: $\frac{4^2}{4^5} = \frac{1}{4^{5-2}} = \frac{1}{4^3}$ ◀

By comparing Solutions A and B and using the Transitive Property of Equality, we see that $4^{-3} = \frac{1}{4^3}$. This leads us to formulate the following definition:

If $a \neq 0$ and if n is a positive integer, then

$$a^{-n} = \frac{1}{a^n} \qquad \text{if } a \neq 0 \qquad \textbf{(1)}$$

1 **EXAMPLE 2** **Simplifying Expressions Containing Negative Exponents**

Simplify each expression:

(a) 8^{-1} (b) 2^{-3} (c) $(-4)^{-2}$ (d) -4^{-2}

Solution (a) $8^{-1} = \frac{1}{8}$ (b) $2^{-3} = \frac{1}{2^3} = \frac{1}{8}$

(c) $(-4)^{-2} = \frac{1}{(-4)^2} = \frac{1}{16}$ (d) $-4^{-2} = -\frac{1}{4^2} = -\frac{1}{16}$ ◀

EXAMPLE 3 Simplifying Expressions Containing Negative Exponents

Simplify each expression. Express answers so that all exponents are positive. Assume that no denominator is 0.

(a) $4x^2y^{-3}$ (b) $x^{-5}y^8$

Solution

(a) $4x^2y^{-3} = 4x^2 \cdot y^{-3} = 4x^2 \cdot \frac{1}{y^3} = \frac{4x^2}{1} \cdot \frac{1}{y^3} = \frac{4x^2 \cdot 1}{1 \cdot y^3} = \frac{4x^2}{y^3}$

(b) $x^{-5}y^8 = x^{-5} \cdot y^8 = \frac{1}{x^5} \cdot \frac{y^8}{1} = \frac{1 \cdot y^8}{x^5 \cdot 1} = \frac{y^8}{x^5}$ ◀

Notice in Example 3 that the factors raised to a negative power in the original problem are raised to a positive power in the denominator of the final answer. This shortcut is used in the next example.

EXAMPLE 4 Simplifying Expressions Containing Negative Exponents

Simplify each expression. Express answers so that all exponents are positive. Assume that no denominator is 0.

(a) $8x^4y^{-5}z^{-1}$ (b) $3x^{-5}y^{-2}$ (c) $-3x^{-2}y^{-4}$ (d) $x^{-1}y^{-3}$

Solution

(a) $8x^4y^{-5}z^{-1} = \frac{8x^4}{y^5z}$ (b) $3x^{-5}y^{-2} = \frac{3}{x^5y^2}$

(c) $-3x^{-2}y^{-4} = \frac{-3}{x^2y^4}$ (d) $x^{-1}y^{-3} = \frac{1}{xy^3}$ ◀

WARNING: Look again at Example 4(c). We are simplifying expressions so that all exponents are positive. A negative coefficient, like -3 in this case, is acceptable in the final answer. Many students get caught in a trap and once they see a negative number, *any negative number*, they start trying to make it a positive number, as in writing $-3x^{-5}$ as $\frac{1}{3x^5}$. This is not correct! Instead, $-3x^{-5} = \frac{-3}{x^5}$. ■

Practice Exercise 1 *In Problems 1–9, simplify each expression. Express answers so that all exponents are positive. Assume that no denominator is 0.*

1. 3^{-2} **2.** -3^{-2} **3.** $(-3)^{-2}$ **4.** -3^2 **5.** $(-3)^2$ **6.** x^{-3}
7. $2x^{-5}y$ **8.** $-2x^{-5}y$ **9.** $7x^{-2}y^{-1}$

When a negative exponent appears in the denominator, as in $\frac{1}{4^{-2}}$, we use properties of division and proceed as follows:

$$\frac{1}{4^{-2}} = \frac{1}{\frac{1}{4^2}} = 1 \div \frac{1}{4^2} = \frac{1}{1} \cdot \frac{4^2}{1} = \frac{4^2}{1} = 4^2$$

Notice that

$$\frac{1}{4^{-2}} = 4^2$$

This leads us to the following result:

If $a \neq 0$ and if n is a positive integer, then

$$\frac{1}{a^{-n}} = a^n \qquad \text{if } a \neq 0 \tag{2}$$

EXAMPLE 5 Simplifying Expressions Containing Negative Exponents

Simplify each expression.

(a) $\dfrac{1}{2^{-3}}$ (b) $\dfrac{1}{\left(\dfrac{3}{4}\right)^{-2}}$ (c) $\dfrac{4^{-2}}{3^{-4}}$

Solution (a) $\dfrac{1}{2^{-3}} = 2^3 = 8$

(b) $\dfrac{1}{\left(\dfrac{3}{4}\right)^{-2}} = \left(\dfrac{3}{4}\right)^2 = \dfrac{9}{16}$

(c) $\dfrac{4^{-2}}{3^{-4}} = \dfrac{\dfrac{1}{4^2}}{\dfrac{1}{3^4}} = \dfrac{1}{4^2} \div \dfrac{1}{3^4} = \dfrac{1}{4^2} \cdot \dfrac{3^4}{1} = \dfrac{1 \cdot 3^4}{4^2 \cdot 1} = \dfrac{3^4}{4^2} = \dfrac{81}{16}$ ◀

Notice in Example 5(c) that we had expressions in both the numerator and denominator raised to negative powers. The net result in simplifying 5(c) was that the expression raised to the negative power in the numerator ended up raised to a positive power in the denominator and the expression raised to the negative power in the denominator ended up raised to a positive power in the numerator. An alternative solution to 5(c) is $\dfrac{4^{-2}}{3^{-4}} = \dfrac{3^4}{4^2} = \dfrac{81}{16}$. We will use this shortcut in the next example.

EXAMPLE 6 Simplifying Expressions Containing Negative Exponents

Simplify each expression. Express answers so that all exponents are positive. Assume that no denominator is 0.

(a) $\dfrac{2x^{-3}}{y^{-4}}$ (b) $\dfrac{-2x^{-3}}{y^{-4}}$ (c) $\dfrac{-5x^{-2}y^2}{z^{-3}}$ (d) $\dfrac{5^{-1}}{x^2}$ (e) $\dfrac{4xy^{-3}}{5z^{-2}w^0}$

Solution (a) $\dfrac{2x^{-3}}{y^{-4}} = \dfrac{2y^4}{x^3}$

(b) $\dfrac{-2x^{-3}}{y^{-4}} = \dfrac{-2y^4}{x^3}$

(c) $\dfrac{-5x^{-2}y^2}{z^{-3}} = \dfrac{-5y^2z^3}{x^2}$

(d) $\dfrac{5^{-1}}{x^2} = \dfrac{1}{5x^2}$

(e) $\dfrac{4xy^{-3}}{5z^{-2}w^0} = \dfrac{4xz^2}{5y^3}$ ◀

Practice Exercise 2 *In Problems 1–9, simplify each expression. Express answers so that all exponents are positive. Assume that no denominator is 0.*

1. $\frac{1}{3^{-2}}$ **2.** $\frac{1}{4^{-3}}$ **3.** $\frac{5^{-2}}{2^{-5}}$ **4.** $\frac{(-2)^{-2}}{(-3)^{-3}}$ **5.** $\frac{x^{-3}}{3^2}$

6. $\frac{4^{-2}}{x^{-3}}$ **7.** $\frac{4^{-2}}{x^3}$ **8.** $\frac{8x^{-5}y^{-1}}{z^{-6}}$ **9.** $\frac{-3x^{-1}z^2}{y^{-5}}$

Laws of Exponents

2 It can be shown that the Laws of Exponents developed in Section 2.1 are true whether the exponents are positive integers, negative integers, or zero. Following is a brief review of the Laws of Exponents, allowing for integer exponents to be positive, negative, or zero. Examples and practice exercises are also provided.

To multiply two expressions having the same base, we retain the base and add the exponents. That is,

$$a^m a^n = a^{m+n} \quad (3)$$

If m, n, or $m + n$ is 0 or negative in equation (3), then a cannot be 0.

EXAMPLE 7 **Using Equation (3)**

Simplify each expression. Express answers so that all exponents are positive. Assume that no denominator is 0.

(a) $2 \cdot 2^4$ (b) $(-3)^2(-3)^{-3}$ (c) $4^{-3} \cdot 4^5$

(d) $(-2)^{-1}(-2)^{-3}$ (e) $x^{-4} \cdot x^6$ (f) $x^{-2} \cdot x^{-1}$

Solution (a) $2 \cdot 2^4 = 2^{1+4} = 2^5 = 32$

(b) $(-3)^2(-3)^{-3} = (-3)^{2+(-3)} = (-3)^{-1} = \frac{1}{(-3)^1} = \frac{1}{-3} = -\frac{1}{3}$

(c) $4^{-3} \cdot 4^5 = 4^{-3+5} = 4^2 = 16$

(d) $(-2)^{-1}(-2)^{-3} = (-2)^{-1+(-3)} = (-2)^{-4} = \frac{1}{(-2)^4} = \frac{1}{16}$

(e) $x^{-4} \cdot x^6 = x^{-4+6} = x^2$

(f) $x^{-2} \cdot x^{-1} = x^{-2+(-1)} = x^{-3} = \frac{1}{x^3}$

Practice Exercise 3 *In Problems 1–5, simplify each expression. Express answers so that all exponents are positive. Assume that no denominator is 0.*

1. $5^{-4} \cdot 5^2$ **2.** $x^{-9} \cdot x^5$ **3.** $(-2)^{-5} \cdot (-2)^4$ **4.** $y^{-7} \cdot y^{10}$ **5.** $3^{-2} \cdot 2^{-2}$

If an expression containing a power is raised to a power, we retain the base and multiply the powers. That is,

$$(a^m)^n = a^{mn} \quad (4)$$

Again, if m or n is 0 or negative, then a cannot be 0.

EXAMPLE 8 **Using Equation (4)**

Simplify each expression. Express answers so that all exponents are positive. Assume that no denominator is 0.

(a) $[(-2)^3]^2$ (b) $(3^{-4})^0$ (c) $(2^{-3})^2$ (d) $(5^{-1})^{-2}$
(e) $(x^{-3})^2$ (f) $(y^{-2})^{-5}$ (g) $x^3 \cdot (x^{-4})^2$

Solution (a) $[(-2)^3]^2 = (-2)^{(3)\cdot(2)} = (-2)^6 = 64$

(b) $(3^{-4})^0 = 3^{(-4)(0)} = 3^0 = 1$

(c) $(2^{-3})^2 = 2^{(-3)(2)} = 2^{-6} = \dfrac{1}{2^6} = \dfrac{1}{64}$

(d) $(5^{-1})^{-2} = 5^{(-1)(-2)} = 5^2 = 25$

(e) $(x^{-3})^2 = x^{(-3)(2)} = x^{-6} = \dfrac{1}{x^6}$

(f) $(y^{-2})^{-5} = y^{(-2)(-5)} = y^{10}$

(g) $x^3 \cdot (x^{-4})^2 = x^3 \cdot x^{(-4)(2)} = x^3 \cdot x^{-8} = x^{3+(-8)} = x^{-5} = \dfrac{1}{x^5}$ ◀

Practice Exercise 4 *In Problems 1–4, simplify each expression. Express answers so that all exponents are positive. Assume that no denominator is 0.*

1. $(3^{-2})^{-1}$ **2.** $(5^{-3})^2$ **3.** $(x^2)^{-6}$ **4.** $x \cdot (x^{-1})^{-3}$ ◀

If a product is raised to a power, the result equals the product of each factor raised to that power. That is,

$$(a \cdot b)^n = a^n \cdot b^n \qquad \textbf{(5)}$$

If n is 0 or negative, neither a nor b can be 0.

EXAMPLE 9 **Using Equation (5)**

Simplify each expression. Express answers so that all exponents are positive. Assume that no denominator is 0.

(a) $(2x)^3$ (b) $(-2x)^0$ (c) $(ax)^{-2}$ (d) $(3x^{-2})^{-4}$ (e) $(2xy^2)^{-3}(x^5y)$

Solution (a) $(2x)^3 = 2^3 \cdot x^3 = 8x^3$

(b) $(-2x)^0 = (-2)^0 \cdot x^0 = 1 \cdot 1 = 1$

(c) $(ax)^{-2} = a^{-2} \cdot x^{-2} = \dfrac{1}{a^2} \cdot \dfrac{1}{x^2} = \dfrac{1}{a^2x^2}$

or alternatively, $(ax)^{-2} = \dfrac{1}{(ax)^2} = \dfrac{1}{a^2x^2}$

(d) $(3x^{-2})^{-4} = 3^{-4} \cdot (x^{-2})^{-4} = 3^{-4} \cdot x^8 = \dfrac{x^8}{3^4} = \dfrac{x^8}{81}$

(e) $(2xy^2)^{-3}(x^5y) = 2^{-3}x^{-3}(y^2)^{-3}x^5y = \dfrac{x^{-3}y^{-6}x^5y}{2^3} = \dfrac{x^2y^{-5}}{8} = \dfrac{x^2}{8y^5}$ ◀

Practice Exercise 5 *In Problems 1–4, simplify each expression. Express answers so that all exponents are positive. Assume that no denominator is 0.*

1. $(3x)^{-4}$ **2.** $(-2x)^3 \cdot x^{-4}$ **3.** $(4ax)^{-3}$ **4.** $(-5x^2)^{-1} \cdot x^{-4}$

To divide two expressions having the same base, we retain the base and subtract the exponents. That is,

$$\frac{a^m}{a^n} = a^{m-n} \quad \text{if } a \neq 0 \tag{6}$$

Similarly,

$$\frac{a^m}{a^n} = \frac{1}{a^{n-m}} \quad \text{if } a \neq 0 \tag{7}$$

In using equations (6) and (7) to simplify $\frac{a^m}{a^n}$, we usually choose the form that results in a positive exponent for a.

EXAMPLE 10 **Using Equations (6) and (7)**

Simplify each expression. Express answers so that all exponents are positive. Assume that no denominator is 0.

(a) $\frac{2^6}{2^4}$ (b) $\frac{3^2}{3^5}$ (c) $\frac{3^{-2}}{3^{-5}}$ (d) $\frac{x^4}{x^9}$

(e) $\frac{x^{-5}}{x^3}$ (f) $\frac{x^5y^{-2}}{x^3y}$ (g) $\frac{-3x^{-2}y^8}{4x^{-1}y^2}$

Solution (a) $\frac{2^6}{2^4} = 2^{6-4} = 2^2 = 4$

(b) $\frac{3^2}{3^5} = \frac{1}{3^{5-2}} = \frac{1}{3^3} = \frac{1}{27}$

(c) $\frac{3^{-2}}{3^{-5}} = 3^{-2-(-5)} = 3^{-2+5} = 3^3 = 27$

(d) $\frac{x^4}{x^9} = \frac{1}{x^{9-4}} = \frac{1}{x^5}$

(e) $\frac{x^{-5}}{x^3} = x^{-5-3} = x^{-8} = \frac{1}{x^8}$

(f) $\frac{x^5y^{-2}}{x^3y} = x^{5-3}y^{-2-1} = x^2y^{-3} = \frac{x^2}{y^3}$

(g) $\frac{-3x^{-2}y^8}{4x^{-1}y^2} = \frac{-3}{4}x^{-2-(-1)}y^{8-2} = \frac{-3}{4}x^{-2+1}y^6 = \frac{-3}{4}x^{-1}y^6 = \frac{-3y^6}{4x}$

Alternatively, we could simplify (g) as follows:

$$\frac{-3x^{-2}y^8}{4x^{-1}y^2} = \frac{-3x^1y^8}{4x^2y^2} = \frac{-3y^{8-2}}{4x^{2-1}} = \frac{-3y^6}{4x}$$

Example 10(g) was simplified with two different approaches to show that sometimes you have a choice on which Law of Exponents to begin with when simplifying expressions.

Practice Exercise 6 *In Problems 1–7, simplify each expression. Express answers so that all exponents are positive. Assume that no denominator is 0.*

1. $\frac{2}{2^{-3}}$ **2.** $\frac{2^{-2}}{2^3}$ **3.** $\frac{(-3)^4}{(-3)^2}$ **4.** $\frac{x^{-7}}{x^4}$ **5.** $\frac{x^{-6}}{x^{-8}}$

6. $\frac{xy^{-2}}{x^{-5}y^{-4}}$ **7.** $\frac{-2x^{-1}y^{-3}}{5x^{-3}y}$

A second law for quotients is stated next.

If a and b are real numbers and if n is an integer, then

$$\left(\frac{a}{b}\right)^n = \frac{a^n}{b^n} \quad \text{if } b \neq 0 \qquad \textbf{(8)}$$

If n is zero or negative, then $a \neq 0$.

EXAMPLE 11 **Using Equation (8)**

Simplify each expression. Express answers so that all exponents are positive. Assume that no denominator is 0.

(a) $\left(\frac{2}{3}\right)^4$ (b) $\left(\frac{2x}{5}\right)^2$ (c) $\left(\frac{5}{2}\right)^{-2}$ (d) $\left(\frac{2x^{-1}}{y}\right)^3$

Solution (a) $\left(\frac{2}{3}\right)^4 = \frac{2^4}{3^4} = \frac{16}{81}$

(b) $\left(\frac{2x}{5}\right)^2 = \frac{(2x)^2}{5^2} = \frac{4x^2}{25}$

(c) $\left(\frac{5}{2}\right)^{-2} = \frac{5^{-2}}{2^{-2}} = \frac{2^2}{5^2} = \frac{4}{25}$

(d) $\left(\frac{2x^{-1}}{y}\right)^3 = \frac{(2x^{-1})^3}{y^3} = \frac{2^3(x^{-1})^3}{y^3} = \frac{8x^{-3}}{y^3} = \frac{8}{x^3y^3}$

You may have observed a shortcut for problems like Example 11(c), namely,

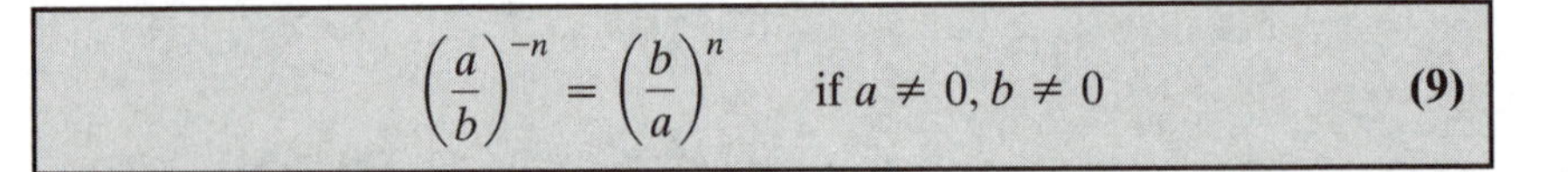

$$\left(\frac{a}{b}\right)^{-n} = \left(\frac{b}{a}\right)^n \quad \text{if } a \neq 0, b \neq 0 \qquad \textbf{(9)}$$

EXAMPLE 12 **Using Equation (9)**

Simplify each expression. Express answers so that all exponents are positive. Assume that no denominator is 0.

(a) $\left(\frac{2}{3}\right)^{-2}$ (b) $\left(\frac{3}{2}\right)^{-3}$ (c) $\left(\frac{x^2}{y^5}\right)^{-3}$ (d) $\left(\frac{2x^{-1}}{y^4}\right)^{-5}$

Solution (a) $\left(\frac{2}{3}\right)^{-2} = \left(\frac{3}{2}\right)^2 = \frac{9}{4}$

(b) $\left(\frac{3}{2}\right)^{-3} = \left(\frac{2}{3}\right)^{3} = \frac{8}{27}$

(c) $\left(\frac{x^2}{y^5}\right)^{-3} = \left(\frac{y^5}{x^2}\right)^{3} = \frac{y^{15}}{x^6}$

(d) $\left(\frac{2x^{-1}}{y^4}\right)^{-5} = \left(\frac{y^4}{2x^{-1}}\right)^{5} = \frac{y^{20}}{2^5x^{-5}} = \frac{x^5y^{20}}{32}$ ◀

Practice Exercise 7 *In Problems 1–4, simplify each expression. Express answers so that all exponents are positive. Assume that no denominator is 0.*

1. $\left(\frac{4x}{5}\right)^{2}$ **2.** $\left(\frac{5}{3}\right)^{-2}$ **3.** $\left(\frac{2x}{3}\right)^{-3}$ **4.** $\left(\frac{x^{-5}}{y^2}\right)^{-3}$ ◀

EXAMPLE 13 Simplifying an Expression Containing Exponents

Write the expression below so that all exponents are positive.

$$\frac{xy}{x^{-1} - y^{-1}} \qquad x \neq 0,\ y \neq 0$$

Solution

$$\frac{xy}{x^{-1} - y^{-1}} = \frac{xy}{\frac{1}{x} - \frac{1}{y}} = \frac{xy}{\frac{1}{x}\cdot\frac{y}{y} - \frac{x}{x}\cdot\frac{1}{y}} = \frac{xy}{\frac{y - x}{xy}}$$

$$= \frac{xy}{1} \div \frac{y - x}{xy} = \frac{xy}{1}\cdot\frac{xy}{y - x} = \frac{xy\cdot xy}{1\cdot(y - x)} = \frac{x^2y^2}{y - x}$$ ◀

Practice Exercise 8 *In Problems 1–3, simplify each expression. Express answers so that all exponents are positive. Assume that no denominator is 0.*

1. $\left(\frac{2^4x^{-2}y^3}{2^5x^{-3}y^4}\right)^{-1}$ **2.** $x^{-2} + y^{-2}$ **3.** $\frac{xy^{-1} + yx^{-1}}{(xy)^{-1}}$ ◀

Scientific Notation

3 Measurements of physical quantities can range from very small to very large. For example, the mass of a proton is approximately 0.00000000000000000000000000167 kilogram and the mass of the Earth is about 5,980,000,000,000,000,000,000,000 kilograms. These numbers obviously are tedious to write down and difficult to read, so we use exponents to rewrite each one.

> When a number has been written as the product of a number a, where $1 \leq a < 10$, times a power of 10, it is said to be written in **scientific notation**.

In scientific notation:

Mass of a proton $= 1.67 \times 10^{-27}$ kilogram

Mass of the Earth $= 5.98 \times 10^{24}$ kilograms

Converting a Decimal to Scientific Notation

To change a positive number into scientific notation, do the following:

STEP 1: Count the number N of places the decimal point must be moved in order to arrive at a number a, where $1 \le a < 10$, that is, a number between 1 and 10.

STEP 2: If the original number is greater than or equal to 1, the scientific notation is $a \times 10^N$, resulting in a positive, or possibly 0, exponent for 10. If the original number is between 0 and 1, the scientific notation is $a \times 10^{-N}$, resulting in a negative exponent for 10.

EXAMPLE 14 Writing a Number in Scientific Notation

Write each number in scientific notation.

(a) 9582 (b) 1.245 (c) 0.285 (d) 0.000561

Solution (a) The decimal point in 9582 follows the 2. We count

9 5 8 2 .
3 2 1

stopping after three moves because 9.582 is a number between 1 and 10. Since 9582 is greater than 1, we write

$$9582 = 9.582 \times 10^3$$

(b) The decimal point in 1.245 is between the 1 and the 2. Since the number is already between 1 and 10, the scientific notation for it is $1.245 \times 10^0 = 1.245$.

(c) The decimal point in 0.285 is between the 0 and the 2. We count

0 . 2 8 5
1

stopping after one move because 2.85 is a number between 1 and 10. Since 0.285 is between 0 and 1, we write

$$0.285 = 2.85 \times 10^{-1}$$

(d) The decimal point in 0.000561 is moved as follows:

0 . 0 0 0 5 6 1
1 2 3 4

Then,

$$0.000561 = 5.61 \times 10^{-4}$$

◄

Practice Exercise 9 *Write each number in scientific notation.*

1. 0.00146 **2.** 48,000,000

EXAMPLE 15 Changing From Scientific Notation to a Decimal

Write each number as a decimal.

(a) 2.1×10^4 (b) 3.26×10^{-5} (c) 1×10^{-2}

Solution (a) $2.1 \times 10^4 = 2\,.\,1\,0\,0\,0 \times 10^4 = 21{,}000$

1 2 3 4

(b) $3.26 \times 10^{-5} = 0\,0\,0\,0\,0\,3\,.\,26 \times 10^{-5} = 0.0000326$

5 4 3 2 1

(c) $1 \times 10^{-2} = 0\,0\,1\,. \times 10^{-2} = 0.01$

2 1

◀

Practice Exercise 10 *Write each number as a decimal.*

1. 3.29×10^4 **2.** 8.7×10^{-7}

◀

EXAMPLE 16 Using Scientific Notation

(a) The diameter of the smallest living cell is only about 0.00001 centimeter (cm). Express this number in scientific notation.

Source: *Powers of Ten*, Philip and Phylis Morrison

(b) The surface area of Earth is about 1.97×10^8 square miles. Express the surface area as a whole number.

Source: *1998 Information Please Almanac*

Solution (a) 0.00001 cm $= 1 \times 10^{-5}$ cm because the decimal point is moved five places and the number is less than 1.

(b) 1.97×10^8 square miles $= 197{,}000{,}000$ square miles. ◀

Notice how your calculator handles scientific notation. Enter into your calculator $(38{,}900{,}000) \cdot (62{,}000)$ and see how the answer is displayed. On a TI-83 you get the following display:

```
38900000*62000
        2.4118 E12
```

to be interpreted as 2.4118×10^{12}. On the TI-30Xa the result is 2.4118^{12}, again to be interpreted as 2.4118×10^{12}. On the TI-30X IIS the display shows $2.4118_{\text{X10}}{}^{12}$.

Summary

We close this section by summarizing the laws of exponents. In the list that follows, a and b are real numbers and m and n are integers. Also, we assume that no denominator is 0 and that all expressions are defined.

Laws of Exponents

$$a^0 = 1 \qquad a^{-n} = \frac{1}{a^n} \qquad \frac{1}{a^{-n}} = a^n$$

$$a^m a^n = a^{m+n} \qquad (a^m)^n = a^{mn} \qquad (ab)^n = a^n b^n$$

$$\frac{a^m}{a^n} = a^{m-n} = \frac{1}{a^{n-m}} \qquad \left(\frac{a}{b}\right)^n = \frac{a^n}{b^n} \qquad \left(\frac{a}{b}\right)^{-n} = \left(\frac{b}{a}\right)^n$$

Answers to Practice Exercises

Practice Exercise 1 **1.** $\frac{1}{9}$ **2.** $-\frac{1}{9}$ **3.** $\frac{1}{9}$ **4.** -9 **5.** 9 **6.** $\frac{1}{x^3}$ **7.** $\frac{2y}{x^5}$ **8.** $\frac{-2y}{x^5}$ **9.** $\frac{7}{x^2y}$

Practice Exercise 2 **1.** 9 **2.** 64 **3.** $\frac{32}{25}$ **4.** $\frac{-27}{4}$ **5.** $\frac{1}{9x^3}$ **6.** $\frac{x^3}{16}$ **7.** $\frac{1}{16x^3}$ **8.** $\frac{8z^6}{x^5y}$ **9.** $\frac{-3y^5z^2}{x}$

Practice Exercise 3 **1.** $\frac{1}{25}$ **2.** $\frac{1}{x^4}$ **3.** $\frac{1}{-2} = -\frac{1}{2}$ **4.** y^3 **5.** $\frac{1}{36}$

Practice Exercise 4 **1.** 9 **2.** $\frac{1}{5^6}$ **3.** $\frac{1}{x^{12}}$ **4.** x^4

Practice Exercise 5 **1.** $\frac{1}{81x^4}$ **2.** $\frac{-8}{x}$ **3.** $\frac{1}{64a^3x^3}$ **4.** $\frac{1}{-5x^6}$

Practice Exercise 6 **1.** 16 **2.** $\frac{1}{32}$ **3.** 9 **4.** $\frac{1}{x^{11}}$ **5.** x^2 **6.** x^6y^2 **7.** $\frac{-2x^2}{5y^4}$

Practice Exercise 7 **1.** $\frac{16x^2}{25}$ **2.** $\frac{9}{25}$ **3.** $\frac{27}{8x^3}$ **4.** $x^{15}y^6$

Practice Exercise 8 **1.** $\frac{2y}{x}$ **2.** $\frac{y^2 + x^2}{x^2y^2}$ **3.** $x^2 + y^2$

Practice Exercise 9 **1.** 1.46×10^{-3} **2.** 4.8×10^7

Practice Exercise 10 **1.** 32,900 **2.** 0.00000087

4.1 Assess Your Understanding

'Are You Prepared?' *Answers are given at the end of these exercises. If you get a wrong answer, read the pages listed in parentheses.*

1. Simplify: $\frac{24x^6y^2}{15y^9}$ (pp. 62–68)

2. Simplify: $(2x^4y)^3$ (pp. 62–68)

3. Simplify: $\frac{(5x)(x^7)}{x^2}$ (pp. 62–68)

Concepts and Vocabulary

4. x^{-3} simplifies to ________ .

5. *True or False:* If $a \neq 0$ and $b \neq 0$, then $\left(\frac{a}{b}\right)^{-n} = \left(\frac{b}{a}\right)^n$

6. *True or False:* In scientific notation, $38{,}000 = 3.8 \times 10^{-4}$

Exercises

In Problems 7–26, simplify each expression.

7. 3^0 **8.** 3^2 **9.** 4^{-2} **10.** $(-3)^2$ **11.** $\left(\frac{2}{3}\right)^2$

12. $\left(\frac{-4}{5}\right)^3$ **13.** $3^0 \cdot 2^{-3}$ **14.** $(-2)^{-3} \cdot 2^0$ **15.** $2^{-3} + \left(\frac{1}{2}\right)^3$ **16.** $3^{-2} + \frac{1}{3}$

17. $3^{-6} \cdot 3^4$ **18.** $4^{-2} \cdot 4^3$ **19.** $\dfrac{8^2}{2^3}$ **20.** $\dfrac{4^3}{2^2}$ **21.** $\left(\dfrac{2}{3}\right)^{-2}$

22. $\left(\dfrac{3}{2}\right)^{-3}$ **23.** $\dfrac{2^3 \cdot 3^2}{2 \cdot 3^{-2}}$ **24.** $\dfrac{3^{-2} \cdot 5^3}{3 \cdot 5}$ **25.** $\left(\dfrac{9}{2}\right)^{-2}$ **26.** $\left(\dfrac{6}{5}\right)^{-3}$

In Problems 27–64, simplify each expression so that all exponents are positive. Whenever an exponent is negative or 0, we assume the base does not equal 0.

27. x^0y^2 **28.** $x^{-1}y$ **29.** $x^{-2}y$ **30.** x^4y^0 **31.** $(8x^3)^{-2}$

32. $(-8x^3)^{-2}$ **33.** $-4x^{-1}$ **34.** $(-4x)^{-1}$ **35.** $5x^0$ **36.** $(5x)^0$

37. $\dfrac{x^{-2}y^3}{xy^4}$ **38.** $\dfrac{x^{-2}y}{xy^2}$ **39.** $x^{-1}y^{-1}$ **40.** $\dfrac{x^{-2}y^{-3}}{x}$ **41.** $\dfrac{x^{-1}}{y^{-1}}$

42. $\left(\dfrac{2x}{3}\right)^{-1}$ **43.** $\left(\dfrac{4x}{5y}\right)^{-2}$ **44.** $(xy^2)^{-2}$ **45.** $x^{-1} + y^{-2}$ **46.** $x^{-1} + y^{-1}$

47. $\dfrac{x^{-1}y^{-2}z}{x^2yz^3}$ **48.** $\dfrac{3x^{-2}yz^2}{x^4y^{-3}z}$ **49.** $\dfrac{(-2)^3x^4(yz)^2}{3^2xy^3z^4}$ **50.** $\dfrac{4x^{-2}(yz)^{-1}}{(-5)^2x^4y^2z^{-2}}$ **51.** $\dfrac{x^{-2}}{x^{-2} + y^{-2}}$

52. $\dfrac{x^{-1} + y^{-1}}{x^{-1} - y^{-1}}$ **53.** $\dfrac{\left(\dfrac{x}{y}\right)^{-2} \cdot \left(\dfrac{y}{x}\right)^{4}}{x^2y^3}$ **54.** $\dfrac{\left(\dfrac{y}{x}\right)^{2}}{x^{-2}y}$ **55.** $\left(\dfrac{3x^{-1}}{4y^{-1}}\right)^{-2}$ **56.** $\left(\dfrac{5x^{-2}}{6y^{-2}}\right)^{-3}$

57. $\dfrac{(xy^{-1})^{-2}}{xy}$ **58.** $\dfrac{(3xy^{-1})^2}{(2x^{-1}y)^3}$ **59.** $\dfrac{\left(\dfrac{x^2}{y}\right)^{3}}{\left(\dfrac{x}{y^2}\right)^{2}}$ **60.** $\dfrac{\left(\dfrac{2x}{3y^2}\right)^{-2}}{\dfrac{1}{y^3}}$ **61.** $\left(\dfrac{x}{y^2}\right)^{-2} \cdot (y^2)^{-1}$

62. $\dfrac{(x^2)^{-3}y^3}{(x^3y)^{-2}}$ **63.** $(x^2y^3)^{-1}(xy)^5$ **64.** $\dfrac{(3x)^{-2}}{9x^2}$

In Problems 65–72, evaluate each expression. Round your answers to three decimal places.

65. $(8.2)^5$ **66.** $(3.7)^4$ **67.** $(6.1)^{-3}$ **68.** $(2.2)^{-5}$

69. $(-2.8)^6$ **70.** $-(2.8)^6$ **71.** $-(9.25)^{-2}$ **72.** $(-9.25)^{-2}$

In Problems 73–80, write each number in scientific notation.

73. 454.2 **74.** 32.14 **75.** 0.013 **76.** 0.00421

77. 32,155 **78.** 21,210 **79.** 0.000423 **80.** 0.0514

In Problems 81–88, write each number as a decimal.

81. 2.15×10^4 **82.** 6.7×10^3 **83.** 1.215×10^{-3} **84.** 9.88×10^{-4}

85. 1.1×10^8 **86.** 4.112×10^2 **87.** 8.1×10^{-2} **88.** 6.453×10^{-1}

---

89. One light-year is defined by astronomers to be the distance a beam of light will travel in 1 year (365 days). If the speed of light is 186,000 miles per second, how many miles are in a light-year? Express your answer in scientific notation.

90. How long does it take a beam of light to reach the Earth from the Sun, when the Sun is 93,000,000 miles from Earth? Express your answer in seconds, using scientific notation, and round your answer to two decimal places.

91. Write a paragraph to justify the definition given in the text that $a^{-n} = \dfrac{1}{a^n}$ where $a \neq 0$.

'Are You Prepared?' Answers

1. $\dfrac{8x^6}{5y^7}$ **2.** $8x^{12}y^3$ **3.** $5x^6$

4.2 Square Roots

PREPARING FOR THIS SECTION *Before getting started, review the following:*

- Distributive Property (p. 28)
- Combine Like Terms (pp. 71–72)
- Multiplying Binominals (pp. 85–87)
- Difference of Two Squares (p. 87–88)

Now work the 'Are You Prepared?' problems on page 197.

OBJECTIVES

1 Evaluate Square Roots
2 Simplify Square Roots
3 Multiply Square Roots
4 Add and Subtract Square Roots
5 Multiply Expressions Containing Square Roots
6 Rationalize Denominators
7 Simplify Quotients of Square Roots

A real number is squared when it is raised to the power 2. The inverse of squaring is finding a **square root.** For example, since $6^2 = 36$ and $(-6)^2 = 36$, the numbers 6 and -6 are square roots of 36.

The symbol $\sqrt{\ }$, called a **radical sign,** is used to denote the **principal,** or nonnegative, square root. So, $\sqrt{36} = 6$.

In general, if a is a nonnegative real number, the nonnegative number b such that $b^2 = a$ is the **principal square root** of a and is denoted by $b = \sqrt{a}$.

The following comments are noteworthy:

1. Negative numbers do not have square roots (in the real number system), because the square of any real number is *nonnegative.* For example, $\sqrt{-4}$ is not a real number, because there is no real number whose square is -4.
2. The principal square root of 0 is 0, since $0^2 = 0$. That is, $\sqrt{0} = 0$.
3. The principal square root of a positive number is positive.
4. If $c \geq 0$, then $\left(\sqrt{c}\right)^2 = c$. For example, $\left(\sqrt{2}\right)^2 = 2$ and $\left(\sqrt{3}\right)^2 = 3$.

1 **EXAMPLE 1** **Evaluating Square Roots**

Simplify the following.

(a) $\sqrt{64}$ (b) $\sqrt{\dfrac{1}{16}}$ (c) $\left(\sqrt{9}\right)^2$ (d) $\sqrt{9^2}$ (e) $\left(\sqrt{1.4}\right)^2$

Solution (a) $\sqrt{64} = 8$ (b) $\sqrt{\dfrac{1}{16}} = \dfrac{1}{4}$ (c) $\left(\sqrt{9}\right)^2 = 3^2 = 9$

(d) $\sqrt{9^2} = \sqrt{81} = 9$ (e) $\left(\sqrt{1.4}\right)^2 = 1.4$ ◀

Examples 1(a) and 1(b) are examples of **square roots of perfect squares.** That is, 64 is a **perfect square,** since $64 = 8^2$, and $\frac{1}{16}$ is a **perfect square,** since $\frac{1}{16} = \left(\frac{1}{4}\right)^2$.

Do you see the difference between 1(c) and 1(d)? Due to the grouping symbols in 1(c), $\sqrt{9}$ is simplified first to 3 and then that result is squared to get 9. We could have also worked 1(c) by applying comment 4 on the previous page. For 1(d), 9 is squared to get 81 and then we evaluate the square root of 81 to get 9.

Practice Exercise 1 *Find the value of each expression.*

1. $\sqrt{9}$ **2.** $\sqrt{\frac{1}{4}}$ **3.** $(\sqrt{5.1})^2$ **4.** $\sqrt{4^2}$ **5.** $\sqrt{(-4)^2}$

In working Problems 4 and 5, you should have found that

$$\sqrt{4^2} = \sqrt{16} = 4 \qquad \text{and} \qquad \sqrt{(-4)^2} = \sqrt{16} = 4$$

In general, we have

$$\sqrt{a^2} = |a| \qquad \textbf{(1)}$$

Notice the need for the absolute value in equation (1). Since $a^2 \geq 0$, the principal square root of a^2 is defined whether $a > 0$ or $a < 0$. However, since the principal square root is nonnegative, we need the absolute value to ensure the nonnegative result.

EXAMPLE 2 **Using Equation (1)**

(a) $\sqrt{(2.3)^2} = |2.3| = 2.3$

(b) $\sqrt{(-2.3)^2} = |-2.3| = 2.3$

(c) $\sqrt{x^2} = |x|$

Properties of Square Roots

We begin with the following observation:

$$\sqrt{4 \cdot 25} = \sqrt{100} = 10 \qquad \text{and} \qquad \sqrt{4}\sqrt{25} = 2 \cdot 5 = 10$$

This suggests the following property of square roots:

Product Property of Square Roots

If a and b are each nonnegative real numbers, then

$$\sqrt{ab} = \sqrt{a}\sqrt{b} \qquad \textbf{(2)}$$

2 When used in connection with square roots, the direction to "simplify" means to remove from the square root any perfect squares that occur as factors. We can use equation (2) to simplify a square root that contains a perfect square as a factor, as illustrated by the following examples.

EXAMPLE 3 **Simplifying Square Roots**

(a) $\sqrt{32} = \sqrt{16 \cdot 2} = \sqrt{16}\sqrt{2} = 4\sqrt{2}$

16 is a perfect square. (2)

(b) $\sqrt{135} = \sqrt{9 \cdot 15} = \sqrt{9}\sqrt{15} = 3\sqrt{15}$ ◀

Factor out the perfect square. (2)

Notice in Example 3(a) that if we had written $\sqrt{32} = \sqrt{4 \cdot 8} = \sqrt{4} \cdot \sqrt{8} = 2\sqrt{8}$, we would have more to do since $8 = 4 \cdot 2$ and 4 is a perfect square. That is, $\sqrt{32} = 2\sqrt{8} = 2\sqrt{4 \cdot 2} = 2\sqrt{4} \cdot \sqrt{2} = 2 \cdot 2 \cdot \sqrt{2} = 4\sqrt{2}$. It is more efficient to factor out the greatest perfect square possible at the beginning.

EXAMPLE 4 **Simplifying Square Roots**

Simplify:

(a) $-3\sqrt{72}$ (b) $\sqrt{75x^2}$

Solution (a) $-3\sqrt{72} = -3\sqrt{36 \cdot 2} = -3\sqrt{36}\sqrt{2} = -3 \cdot 6\sqrt{2} = -18\sqrt{2}$

(b) $\sqrt{75x^2} = \sqrt{(25x^2)(3)} = \sqrt{25x^2}\sqrt{3} = \sqrt{25}\sqrt{x^2}\sqrt{3} = 5|x|\sqrt{3}$ ◀

(1)

In Example 4(b) we encountered $\sqrt{x^2}$. Since the variable could be any real number, we put absolute value signs around the final x to ensure a nonnegative result.

Practice Exercise 2 *Simplify:*

1. $\sqrt{18}$ **2.** $-5\sqrt{48}$ **3.** $\sqrt{16x^2}$ **4.** $\sqrt{12x^2}$ ◀

Products of Square Roots

3 To find the product of two square roots, we use equation (2).

EXAMPLE 5 **Multiplying Square Roots**

Simplify:

(a) $\sqrt{12}\sqrt{3}$ (b) $\sqrt{8}\sqrt{6}$ (c) $\sqrt{2x}\sqrt{6x}$

Solution (a) $\sqrt{12}\sqrt{3} = \sqrt{12 \cdot 3} = \sqrt{36} = 6$

(b) $\sqrt{8}\sqrt{6} = \sqrt{8 \cdot 6} = \sqrt{48} = \sqrt{16 \cdot 3} = \sqrt{16}\sqrt{3} = 4\sqrt{3}$

(c) $\sqrt{2x}\sqrt{6x} = \sqrt{2x \cdot 6x} = \sqrt{12x^2} = \sqrt{3 \cdot 4x^2} = \sqrt{3} \cdot 2x$ ◀

In Example 5(c) we did not need absolute value signs around x in the final answer. This is because the domain of x includes the same restrictions at the end of the problem that it had at the beginning of the problem. In the original problem, x must be nonnegative since the square root of a negative number is not defined (as a real number).

Based on this discussion, we state the following special case of equation (2).

$$\sqrt{a}\sqrt{a} = a \qquad \text{if } a \geq 0 \qquad \textbf{(3)}$$

EXAMPLE 6 **Multiplying Expressions Containing Square Roots**

Simplify: $-4\sqrt{30} \cdot 5\sqrt{30}$

Solution $-4\sqrt{30} \cdot 5\sqrt{30} = -4 \cdot 5\sqrt{30}\sqrt{30} = -20 \cdot 30 = -600$ ◀

↑ Equation (3)

Practice Exercise 3 *In Problems 1–4, simplify each expression.*

1. $\sqrt{18}\sqrt{2}$ **2.** $\sqrt{27}\sqrt{6}$ **3.** $-2\sqrt{10} \cdot 4\sqrt{10}$ **4.** $\sqrt{3x}\sqrt{4x}$ ◀

Sums and Differences of Square Roots

4 In Chapter 2 we saw how to simplify algebraic expressions by combining like terms. The same idea is used to add or subtract square roots. Two square roots can be combined, provided each has the same radicand. Such square roots are called **like square roots.**

EXAMPLE 7 **Combining Like Square Roots**

$$\begin{aligned} 2\sqrt{3} + 6\sqrt{3} + 4\sqrt{5} &= \left(2\sqrt{3} + 6\sqrt{3}\right) + 4\sqrt{5} \\ &= (2 + 6)\sqrt{3} + 4\sqrt{5} = 8\sqrt{3} + 4\sqrt{5} \end{aligned}$$

The expression $8\sqrt{3} + 4\sqrt{5}$ is in its simplest form, since $8\sqrt{3}$ and $4\sqrt{5}$ are not "like square roots" and so cannot be combined. ◀

In the next example, it appears that the square roots are not "like square roots," but upon simplifying them, we discover that they are "like square roots" and can be combined.

EXAMPLE 8 **Combining Like Roots**

$$\begin{aligned} 3\sqrt{8} + 5\sqrt{50} - 4\sqrt{32} &= 3\sqrt{4 \cdot 2} + 5\sqrt{25 \cdot 2} - 4\sqrt{16 \cdot 2} \\ &= 3\sqrt{4}\sqrt{2} + 5\sqrt{25}\sqrt{2} - 4\sqrt{16}\sqrt{2} \\ &= 3 \cdot 2\sqrt{2} + 5 \cdot 5\sqrt{2} - 4 \cdot 4\sqrt{2} \\ &= 6\sqrt{2} + 25\sqrt{2} - 16\sqrt{2} \\ &= (6 + 25 - 16)\sqrt{2} = 15\sqrt{2} \end{aligned}$$ ◀

Practice Exercise 4 *In Problems 1–3, simplify each expression.*

1. $2\sqrt{3} + 4\sqrt{5} - 8\sqrt{3}$ **2.** $4\sqrt{18} + 5\sqrt{2}$

3. $3\sqrt{5} - 4\sqrt{20} + \sqrt{45}$ ◀

More on Products of Square Roots

5 **EXAMPLE 9** **Multiplying Expressions Containing Square Roots**

Find the product and simplify the result:

$$3\sqrt{2}(2\sqrt{8} + 2\sqrt{3})$$

Solution We use the Distributive Property and then simplify with the Commutative and Associative Properties.

$$\begin{aligned} 3\sqrt{2}(2\sqrt{8} + 2\sqrt{3}) &= 3\sqrt{2}\cdot 2\sqrt{8} + 3\sqrt{2}\cdot 2\sqrt{3} \\ &= 3\cdot 2\sqrt{2}\sqrt{8} + 3\cdot 2\sqrt{2}\sqrt{3} \\ &= 6\sqrt{16} + 6\sqrt{6} \\ &= 6\cdot 4 + 6\sqrt{6} \\ &= 24 + 6\sqrt{6} \end{aligned}$$

◀

EXAMPLE 10 **Multiplying Binomials Containing Square Roots**

Find the product and simplify the result.

(a) $(3 - \sqrt{2})(5 - \sqrt{2})$ (b) $(4 + \sqrt{3})(2 - \sqrt{5})$

(c) $(4 - \sqrt{5})(4 + \sqrt{5})$ (d) $(\sqrt{6} - \sqrt{5})^2$

Solution We will use the FOIL method to carry out the multiplication and then simplify the result.

(a)
$$\begin{aligned} (3 - \sqrt{2})(5 - \sqrt{2}) &= \overset{F}{3\cdot 5} - \overset{O}{3\cdot\sqrt{2}} - \overset{I}{5\cdot\sqrt{2}} + \overset{L}{\sqrt{2}\cdot\sqrt{2}} \\ &= 15 - 3\sqrt{2} - 5\sqrt{2} + (\sqrt{2})^2 \\ &= 15 - 8\sqrt{2} + 2 \\ &= 17 - 8\sqrt{2} \end{aligned}$$

(b)
$$\begin{aligned} (4 + \sqrt{3})(2 - \sqrt{5}) &= 4\cdot 2 - 4\sqrt{5} + 2\sqrt{3} - \sqrt{3}\cdot\sqrt{5} \\ &= 8 - 4\sqrt{5} + 2\sqrt{3} - \sqrt{3\cdot 5} \\ &= 8 - 4\sqrt{5} + 2\sqrt{3} - \sqrt{15} \end{aligned}$$

(c)
$$\begin{aligned} (4 - \sqrt{5})(4 + \sqrt{5}) &= 4\cdot 4 + 4\sqrt{5} - 4\sqrt{5} - \sqrt{5}\cdot\sqrt{5} \\ &= 16 - (\sqrt{5})^2 = 16 - 5 = 11 \end{aligned}$$

(d)
$$\begin{aligned} (\sqrt{6} - \sqrt{5})^2 &= (\sqrt{6} - \sqrt{5})(\sqrt{6} - \sqrt{5}) \\ &= \sqrt{6}\cdot\sqrt{6} - \sqrt{6}\cdot\sqrt{5} - \sqrt{5}\cdot\sqrt{6} + \sqrt{5}\cdot\sqrt{5} \\ &= (\sqrt{6})^2 - \sqrt{30} - \sqrt{30} + (\sqrt{5})^2 \\ &= 6 - 1\cdot\sqrt{30} - 1\cdot\sqrt{30} + 5 \\ &= 11 - 2\sqrt{30} \end{aligned}$$

◀

The Special Products from Chapter 2, listed below, can also be helpful for finding the products of certain sums and differences of square roots. In fact, these Special Products could have been used in Examples 10(a), 10(c), and 10(d).

Difference of Two Squares

$$(x - a)(x + a) = x^2 - a^2 \quad \textbf{(4)}$$

Squares of Binomials, or Perfect Squares

$$(x + a)^2 = x^2 + 2ax + a^2$$
$$(x - a)^2 = x^2 - 2ax + a^2 \qquad (5)$$

Second-Degree Trinomials

$$(x + a)(x + b) = x^2 + (a + b)x + ab \qquad (6)$$

EXAMPLE 11 Multiplying Binomials Containing Square Roots

Find the product and simplify the result.

(a) $(\sqrt{5} - 3)(\sqrt{5} + 3)$ (b) $(\sqrt{3} + \sqrt{7})^2$ (c) $(\sqrt{5} - 3)(\sqrt{5} + 2)$

Solution (a) $(\sqrt{5} - 3)(\sqrt{5} + 3) = (\sqrt{5})^2 - 3^2 = 5 - 9 = -4$

↑ Equation (4)

(b) $(\sqrt{3} + \sqrt{7})^2 = (\sqrt{3} + \sqrt{7})(\sqrt{3} + \sqrt{7})$

$= (\sqrt{3})^2 + 2 \cdot \sqrt{3} \cdot \sqrt{7} + (\sqrt{7})^2$

↑ Equation (5)

$= 3 + 2\sqrt{3 \cdot 7} + 7 = 10 + 2\sqrt{21}$

(c) $(\sqrt{5} - 3)(\sqrt{5} + 2) = (\sqrt{5})^2 + 2\sqrt{5} - 3\sqrt{5} - 3 \cdot 2$

↑ Equation (6)

$= 5 - 1 \cdot \sqrt{5} - 6 = 5 - 6 - \sqrt{5} = -1 - \sqrt{5}$

◀

Practice Exercise 5 *Find each product and simplify the result.*

1. $2\sqrt{3}(4\sqrt{18} - 2\sqrt{12})$

2. $3\sqrt{5}(4\sqrt{50} - \sqrt{20})$

3. $(\sqrt{2} + 4)(\sqrt{2} - 4)$

4. $(\sqrt{7} - 5)^2$

5. $(\sqrt{3} + 2\sqrt{2})(\sqrt{3} - 3\sqrt{2})$

◀

Examples like 10(c) and 11(a), which fit the special product formula for the Difference of Two Squares, are special cases in multiplying expressions that contain square roots. Notice that in examples like 10(c) and 11(a), the square roots reduced out of the problem.

EXAMPLE 12 Multiplying Binomials Containing Square Roots

Find the product and simplify the result.

(a) $(5 - \sqrt{8})(5 + \sqrt{8})$

(b) $(\sqrt{5} - \sqrt{3})(\sqrt{5} + \sqrt{3})$

(c) $(\sqrt{6} + 2\sqrt{3})(\sqrt{6} - 2\sqrt{3})$

Solution (a) $(5 - \sqrt{8})(5 + \sqrt{8}) = 5^2 - (\sqrt{8})^2 = 25 - 8 = 17$

(b) $(\sqrt{5} - \sqrt{3})(\sqrt{5} + \sqrt{3}) = (\sqrt{5})^2 - (\sqrt{3})^2 = 5 - 3 = 2$

(c) $(\sqrt{6} + 2\sqrt{3})(\sqrt{6} - 2\sqrt{3}) = (\sqrt{6})^2 - (2\sqrt{3})^2 = 6 - 4 \cdot 3$
$= 6 - 12 = -6$ ◀

The solution obtained in Example 12 leads to the following results.

If x is any real number with $a \geq 0$ and $b \geq 0$, then

$$(x - \sqrt{a})(x + \sqrt{a}) = x^2 - (\sqrt{a})^2 = x^2 - a$$
$$(\sqrt{a} - x)(\sqrt{a} + x) = (\sqrt{a})^2 - x^2 = a - x^2 \qquad (7)$$
$$(\sqrt{a} - \sqrt{b})(\sqrt{a} + \sqrt{b}) = (\sqrt{a})^2 - (\sqrt{b})^2 = a - b$$

We use the facts expressed in (7) in the following discussion.

Rationalizing

6 When square roots occur in quotients, it is customary to rewrite the quotient so that the denominator contains no square roots. This process is referred to as **rationalizing the denominator.**

The idea is to multiply by an appropriate expression so that the new denominator contains no square roots. For example:

If the Denominator Contains the Factor	Multiply by	To Obtain a Denominator Free of Radicals
$\sqrt{3}$	$\sqrt{3}$	$\sqrt{3} \cdot \sqrt{3} = (\sqrt{3})^2 = 3$
$\sqrt{3} + 1$	$\sqrt{3} - 1$	$(\sqrt{3} + 1)(\sqrt{3} - 1) = (\sqrt{3})^2 - 1^2 = 3 - 1 = 2$
$\sqrt{2} - 3$	$\sqrt{2} + 3$	$(\sqrt{2} - 3)(\sqrt{2} + 3) = (\sqrt{2})^2 - 3^2 = 2 - 9 = -7$
$\sqrt{5} - \sqrt{3}$	$\sqrt{5} + \sqrt{3}$	$(\sqrt{5} - \sqrt{3})(\sqrt{5} + \sqrt{3}) = (\sqrt{5})^2 - (\sqrt{3})^2 = 5 - 3 = 2$

In rationalizing the denominator of a quotient, be sure to multiply both the numerator and denominator by the appropriate expression so that a quotient equivalent to the original is obtained. In other words, be sure that the net result from the multiplication is simply multiplication by 1, the multiplicative identity, so that the value of the original expression is unchanged.

EXAMPLE 13 Rationalizing Denominators

Rationalize the denominator of each expression.

(a) $\dfrac{1}{\sqrt{3}}$ (b) $\dfrac{5}{4\sqrt{2}}$ (c) $\dfrac{5\sqrt{2}}{3\sqrt{5}}$

Solution (a) The denominator contains the factor $\sqrt{3}$, so we multiply the original expression by 1, written as $\frac{\sqrt{3}}{\sqrt{3}}$, as follows:

$$\frac{1}{\sqrt{3}} = \frac{1}{\sqrt{3}}\cdot 1 = \frac{1}{\sqrt{3}}\cdot\frac{\sqrt{3}}{\sqrt{3}} = \frac{1\cdot\sqrt{3}}{\sqrt{3}\cdot\sqrt{3}} = \frac{\sqrt{3}}{3}$$

The new denominator, 3, contains no square roots, so the original quotient has been rationalized.

(b) The denominator contains the factor $\sqrt{2}$, so we multiply the original expression by 1, written as $\frac{\sqrt{2}}{\sqrt{2}}$, as follows:

$$\frac{5}{4\sqrt{2}} = \frac{5}{4\sqrt{2}}\cdot 1 = \frac{5}{4\sqrt{2}}\cdot\frac{\sqrt{2}}{\sqrt{2}} = \frac{5\cdot\sqrt{2}}{4\sqrt{2}\cdot\sqrt{2}} = \frac{5\sqrt{2}}{4\cdot 2} = \frac{5\sqrt{2}}{8}$$

(c) The denominator contains the factor $\sqrt{5}$, so we multiply the original expression by 1, written as $\frac{\sqrt{5}}{\sqrt{5}}$, as follows:

$$\frac{5\sqrt{2}}{3\sqrt{5}} = \frac{5\sqrt{2}}{3\sqrt{5}}\cdot 1 = \frac{5\sqrt{2}}{3\sqrt{5}}\cdot\frac{\sqrt{5}}{\sqrt{5}} = \frac{5\sqrt{2}\cdot\sqrt{5}}{3\sqrt{5}\cdot\sqrt{5}} = \frac{\cancel{5}\sqrt{2}\cdot\sqrt{5}}{3\cdot\cancel{5}} = \frac{\sqrt{10}}{3}$$ ◀

Practice Exercise 6 *Rationalize the denominator of each expression.*

1. $\frac{3}{\sqrt{5}}$ **2.** $\frac{4}{\sqrt{6}}$ **3.** $\frac{4}{3\sqrt{2}}$ **4.** $\frac{2\sqrt{7}}{3\sqrt{6}}$

EXAMPLE 14 **Rationalizing Denominators**

Rationalize the denominator of each expression.

(a) $\frac{5}{\sqrt{3}+2}$ (b) $\frac{8}{1+\sqrt{6}}$ (c) $\frac{\sqrt{2}}{\sqrt{3}-\sqrt{2}}$

Solution (a) The denominator contains the factor $\sqrt{3}+2$, so we multiply the original expression by 1 written as $\frac{\sqrt{3}-2}{\sqrt{3}-2}$ to obtain:

$$\begin{aligned}\frac{5}{\sqrt{3}+2} &= \frac{5}{\sqrt{3}+2}\cdot\frac{\sqrt{3}-2}{\sqrt{3}-2}\\ &= \frac{5(\sqrt{3}-2)}{(\sqrt{3}+2)(\sqrt{3}-2)} = \frac{5\sqrt{3}-10}{(\sqrt{3})^2-4}\\ &= \frac{5\sqrt{3}-10}{3-4} = \frac{5\sqrt{3}-10}{-1}\\ &= \frac{-1(5\sqrt{3}-10)}{1}\\ &= -5\sqrt{3}+10\end{aligned}$$

We can also express this answer in the form $10 - 5\sqrt{3}$.

(b) The denominator contains the factor $1 + \sqrt{6}$, so we multiply the numerator and denominator by $1 - \sqrt{6}$ to obtain

$$\begin{aligned}\frac{8}{1+\sqrt{6}} &= \frac{8}{1+\sqrt{6}}\cdot\frac{1-\sqrt{6}}{1-\sqrt{6}}\\ &= \frac{8(1-\sqrt{6})}{(1+\sqrt{6})(1-\sqrt{6})} = \frac{8-8\sqrt{6}}{1^2-(\sqrt{6})^2}\\ &= \frac{8-8\sqrt{6}}{1-6}\\ &= \frac{8-8\sqrt{6}}{-5} = \frac{-8+8\sqrt{6}}{5}\end{aligned}$$

(c) The denominator contains the factor $\sqrt{3} - \sqrt{2}$, so we multiply the numerator and denominator by $\sqrt{3} + \sqrt{2}$ to obtain

$$\begin{aligned}\frac{\sqrt{2}}{\sqrt{3}-\sqrt{2}} &= \frac{\sqrt{2}}{\sqrt{3}-\sqrt{2}}\cdot\frac{\sqrt{3}+\sqrt{2}}{\sqrt{3}+\sqrt{2}}\\ &= \frac{\sqrt{2}(\sqrt{3}+\sqrt{2})}{(\sqrt{3}-\sqrt{2})(\sqrt{3}+\sqrt{2})} = \frac{\sqrt{2}\cdot\sqrt{3}+\sqrt{2}\cdot\sqrt{2}}{(\sqrt{3})^2-(\sqrt{2})^2}\\ &= \frac{\sqrt{6}+\sqrt{4}}{3-2}\\ &= \frac{\sqrt{6}+2}{1} = \sqrt{6}+2\end{aligned}$$

◀

Practice Exercise 7 *Rationalize the denominator of each expression.*

1. $\frac{6}{\sqrt{2}-1}$ **2.** $\frac{6}{\sqrt{6}+5}$ **3.** $\frac{\sqrt{3}}{\sqrt{5}+\sqrt{3}}$

Quotients of Square Roots

A property of square roots analogous to equation (2) is suggested by the examples below:

$$\sqrt{\frac{36}{9}} = \sqrt{4} = 2 \quad \text{and} \quad \frac{\sqrt{36}}{\sqrt{9}} = \frac{6}{3} = 2$$

Quotient Property of Square Roots

If a is a nonnegative real number and b is a positive real number, then

$$\sqrt{\frac{a}{b}} = \frac{\sqrt{a}}{\sqrt{b}} \qquad \textbf{(8)}$$

7 **EXAMPLE 15** **Simplifying Quotients Involving Square Roots**

(a) $\sqrt{\dfrac{81}{25}} = \dfrac{\sqrt{81}}{\sqrt{25}} = \dfrac{9}{5}$

(b) $\dfrac{\sqrt{24}}{\sqrt{3}} = \sqrt{\dfrac{24}{3}} = \sqrt{8} = \sqrt{4 \cdot 2} = \sqrt{4}\sqrt{2} = 2\sqrt{2}$

(c) $\dfrac{\sqrt{x^5 y}}{\sqrt{x^3 y^3}} = \sqrt{\dfrac{x^5 y}{x^3 y^3}} = \sqrt{\dfrac{x^2}{y^2}} = \sqrt{\left(\dfrac{x}{y}\right)^2} = \left|\dfrac{x}{y}\right|$ ◀

Notice in Example 14(c), to get from $\sqrt{\dfrac{x^2}{y^2}}$ to $\sqrt{\left(\dfrac{x}{y}\right)^2}$ we used a property of exponents: $\dfrac{a^n}{b^n} = \left(\dfrac{a}{b}\right)^n$

Practice Exercise 8 *Simplify:*

1. $\sqrt{\dfrac{16}{9}}$ **2.** $\sqrt{\dfrac{32}{25}}$ **3.** $\dfrac{\sqrt{48}}{\sqrt{6}}$ **4.** $\dfrac{\sqrt{x^3 y^5}}{\sqrt{xy}}$ ◀

Answers to Practice Exercises

Practice Exercise 1 **1.** 3 **2.** $\dfrac{1}{2}$ **3.** 5.1 **4.** 4 **5.** 4

Practice Exercise 2 **1.** $3\sqrt{2}$ **2.** $-20\sqrt{3}$ **3.** $4|x|$ **4.** $2x\sqrt{3}$

Practice Exercise 3 **1.** 6 **2.** $9\sqrt{2}$ **3.** -80 **4.** $2\sqrt{3}x$

Practice Exercise 4 **1.** $-6\sqrt{3} + 4\sqrt{5}$ **2.** $17\sqrt{2}$ **3.** $-2\sqrt{5}$

Practice Exercise 5 **1.** $24\sqrt{6} - 24$ **2.** $60\sqrt{10} - 30$ **3.** -14 **4.** $32 - 10\sqrt{7}$ **5.** $-9 - \sqrt{6}$

Practice Exercise 6 **1.** $\dfrac{3\sqrt{5}}{5}$ **2.** $\dfrac{2\sqrt{6}}{3}$ **3.** $\dfrac{2\sqrt{2}}{3}$ **4.** $\dfrac{\sqrt{42}}{9}$

Practice Exercise 7 **1.** $6(\sqrt{2} + 1) = 6\sqrt{2} + 6$

2. $\dfrac{6(\sqrt{6} - 5)}{-19} = \dfrac{6\sqrt{6} - 30}{-19} = \dfrac{30 - 6\sqrt{6}}{19}$ **3.** $\dfrac{\sqrt{15} - 3}{2}$

Practice Exercise 8 **1.** $\dfrac{4}{3}$ **2.** $\dfrac{4\sqrt{2}}{5}$ **3.** $2\sqrt{2}$ **4.** $|x| \cdot y^2$

4.2 Assess Your Understanding

'Are You Prepared?' *Answers are given at the end of these exercises. If you get a wrong answer, read the pages listed in parentheses.*

1. Combine like terms: $6x^2 + 9x^2 - 2x$ (pp. 71–72)

2. Multiply: $4x(5x - 8)$ (p. 28)

3. Multiply: $(x + 4)(x - 6)$ (pp. 85–87)

4. Multiply: $(2x - 3)(2x + 3)$ (pp. 87–88)

Concepts and Vocabulary

5. *True or False:* $\sqrt{4} = \pm 2$
6. *True or False:* $\sqrt{a^2} = |a|$
7. To rationalize the denominator of $\dfrac{3}{2 - \sqrt{3}}$ you would multiply the numerator and denominator by ________.
8. *True or False:* The square root of a negative number is defined as zero.
9. *True or False:* In its simplest form $\sqrt{128}$ is $4\sqrt{8}$.
10. *True or False:* $6\sqrt{2} + 3\sqrt{5} = 9\sqrt{7}$
11. Multiply: $\sqrt{2}(3 + \sqrt{6})$
12. Simplify: $\dfrac{\sqrt{18}}{\sqrt{2}}$

Exercises

In Problems 13–26, find the value of each expression.

13. $\sqrt{4}$ **14.** $\sqrt{16}$ **15.** $\sqrt{\dfrac{1}{9}}$ **16.** $\sqrt{\dfrac{1}{25}}$ **17.** $\sqrt{25} - \sqrt{9}$

18. $\sqrt{9} + \sqrt{16}$ **19.** $\sqrt{25 - 9}$ **20.** $\sqrt{9 + 16}$ **21.** $\left(\sqrt{49}\right)^2$ **22.** $\sqrt{49^2}$

23. $\sqrt{(-3)^2}$ **24.** $\sqrt{(-10)^2}$ **25.** $\sqrt{(-2.4)^2}$ **26.** $\sqrt{(-6.2)^2}$

In Problems 27–52, simplify each expression. Assume that no denominator equals 0.

27. $\sqrt{12}$ **28.** $\sqrt{8}$ **29.** $-4\sqrt{27}$ **30.** $-3\sqrt{24}$ **31.** $\sqrt{50x^2}$

32. $\sqrt{32x^2y}$ **33.** $4\sqrt{18x^4y^3}$ **34.** $3\sqrt{20x^3y^2}$ **35.** $\sqrt{\dfrac{9}{16}}$ **36.** $\sqrt{\dfrac{25}{4}}$

37. $\sqrt{\dfrac{4}{9}}$ **38.** $\sqrt{\dfrac{25}{36}}$ **39.** $\sqrt{\dfrac{8}{9}}$ **40.** $\sqrt{\dfrac{12}{25}}$ **41.** $\sqrt{\dfrac{x^2}{9}}$

42. $\sqrt{\dfrac{16x^2}{y^2}}$ **43.** $\dfrac{\sqrt{32}}{\sqrt{2}}$ **44.** $\dfrac{\sqrt{28}}{\sqrt{7}}$ **45.** $\dfrac{\sqrt{18x^3y^5}}{\sqrt{50xy}}$ **46.** $\dfrac{\sqrt{27x^5y}}{\sqrt{12xy^3}}$

47. $\sqrt{6}\sqrt{12}$ **48.** $\sqrt{3}\sqrt{27}$ **49.** $\sqrt{5x}\sqrt{45x}$ **50.** $\sqrt{3x}\sqrt{12x}$ **51.** $2\sqrt{5}\cdot 4\sqrt{5}$

52. $-3\sqrt{7}\cdot 2\sqrt{7}$

In Problems 53–84, perform the indicated operation(s) and simplify.

53. $5\sqrt{3} - 9\sqrt{3} + 4\sqrt{2}$ **54.** $6\sqrt{5} + 4\sqrt{10} - \sqrt{5}$ **55.** $3\sqrt{2} + 4\sqrt{8}$

56. $9\sqrt{6} - 2\sqrt{24}$ **57.** $2\sqrt{12} - 3\sqrt{6} + 5\sqrt{27}$ **58.** $2\sqrt{50} - 10\sqrt{18} + \sqrt{98}$

59. $8\sqrt{3} + \sqrt{75} - \sqrt{100}$ **60.** $3\sqrt{12} - 2\sqrt{3} + 5\sqrt{2}$ **61.** $2\sqrt{3}\left(\sqrt{3} - \sqrt{2}\right)$

62. $4\sqrt{5}\left(\sqrt{2} + \sqrt{5}\right)$ **63.** $5\sqrt{2}\left(\sqrt{2} + \sqrt{6}\right)$ **64.** $3\sqrt{7}\left(2\sqrt{14} - \sqrt{21}\right)$

65. $\left(\sqrt{3} - \sqrt{2}\right)\left(\sqrt{3} + \sqrt{2}\right)$ **66.** $\left(\sqrt{7} - 2\right)\left(\sqrt{7} + 2\right)$ **67.** $\left(\sqrt{5} + \sqrt{3}\right)\left(\sqrt{5} - \sqrt{3}\right)$

68. $\left(3 + \sqrt{2}\right)\left(3 - \sqrt{2}\right)$ **69.** $\left(\sqrt{8} - 2\right)\left(\sqrt{8} + 2\right)$ **70.** $\left(1 + \sqrt{2}\right)\left(1 - \sqrt{2}\right)$

71. $\left(\sqrt{5} + 1\right)^2$ **72.** $\left(1 - \sqrt{2}\right)^2$ **73.** $\left(\sqrt{3} - \sqrt{2}\right)^2$

74. $\left(\sqrt{5} - \sqrt{3}\right)^2$ **75.** $\left(2 - \sqrt{3}\right)\left(1 + \sqrt{3}\right)$ **76.** $\left(5 - \sqrt{2}\right)\left(4 + \sqrt{2}\right)$

77. $\left(1 + \sqrt{6}\right)\left(3 + \sqrt{6}\right)$ **78.** $\left(2 - \sqrt{3}\right)\left(4 - \sqrt{3}\right)$ **79.** $\left(3 + 2\sqrt{5}\right)\left(4 + \sqrt{5}\right)$

80. $\left(2 + 4\sqrt{3}\right)\left(3 - 2\sqrt{3}\right)$ **81.** $\left(5 - \sqrt{2}\right)\left(5 - \sqrt{3}\right)$ **82.** $\left(3 + \sqrt{6}\right)\left(1 - \sqrt{2}\right)$

83. $\left(1 + \sqrt{5}\right)\left(2 - \sqrt{3}\right)$ **84.** $\left(6 - \sqrt{2}\right)\left(4 - \sqrt{3}\right)$

In Problems 85–100, rationalize the denominator of each expression.

85. $\dfrac{1}{\sqrt{2}}$ **86.** $\dfrac{2}{\sqrt{5}}$ **87.** $\dfrac{3}{2\sqrt{3}}$ **88.** $\dfrac{-2}{4\sqrt{5}}$

89. $\dfrac{\sqrt{3}}{\sqrt{5}}$ **90.** $\dfrac{2\sqrt{7}}{\sqrt{2}}$ **91.** $\dfrac{-8\sqrt{3}}{3\sqrt{2}}$ **92.** $\dfrac{5\sqrt{11}}{2\sqrt{5}}$

93. $\dfrac{1}{\sqrt{2}-1}$ 94. $\dfrac{2}{\sqrt{3}-1}$ 95. $\dfrac{\sqrt{2}}{1+\sqrt{5}}$ 96. $\dfrac{-3}{1-\sqrt{7}}$

97. $\dfrac{3}{\sqrt{3}-\sqrt{2}}$ 98. $\dfrac{2}{\sqrt{5}+\sqrt{3}}$ 99. $\dfrac{\sqrt{3}-\sqrt{2}}{2\sqrt{5}-\sqrt{7}}$ 100. $\dfrac{\sqrt{2}-2\sqrt{3}}{\sqrt{5}+3\sqrt{3}}$

101. Explain the difference between the facts that $\sqrt{4} = 2$ and that $x^2 = 4$ if $x = -2$ and if $x = 2$.

102. Write a few paragraphs that outline a procedure for rationalizing the denominator of an expression containing a square root. Be specific about the various possibilities that can occur.

'Are You Prepared?' Answers

1. $15x^2 - 2x$ **2.** $20x^2 - 32x$ **3.** $x^2 - 2x - 24$ **4.** $4x^2 - 9$

4.3 Radicals

PREPARING FOR THIS SECTION *Before getting started, review the following:*

- Square Roots (Section 4.2, pp. 188–197)

Now work the 'Are You Prepared?' problems on page 205.

OBJECTIVES

1 Work with *n*th Roots
2 Simplify *n*th Roots
3 Approximate *n*th Roots with a Calculator
4 Use Properties of Radicals
5 Combine Like Radicals

1 We have discussed the meaning of raising a number to a positive integer power. The inverse process is called **taking a root.** In Section 4.2, we specifically addressed the issue of taking a square root or second root. In this section, we address the issue as it relates not only to square roots, but to 3rd roots, 4th roots, 5th roots, 6th roots, and so on. To refer to an undetermined root, we use the phrase ***n*th root.**

Suppose $n \geq 2$ is an integer and a is a real number. An ***n*th root of *a*** is a number which, when raised to the power n, equals a.

Some examples will give you the idea.

EXAMPLE 1 **Finding Roots**

(a) A 3rd root of 8 is 2, since $2^3 = 8$.

(b) A 3rd root of -64 is -4, since $(-4)^3 = -64$.

(c) A 2nd root of 16 is 4, since $4^2 = 16$.

(d) Another 2nd root of 16 is -4, since $(-4)^2 = 16$.

(e) A 4th root of $\dfrac{1}{16}$ is $-\dfrac{1}{2}$, since $\left(-\dfrac{1}{2}\right)^4 = \dfrac{1}{16}$.

(f) Another 4th root of $\dfrac{1}{16}$ is $\dfrac{1}{2}$, since $\left(\dfrac{1}{2}\right)^4 = \dfrac{1}{16}$.

(g) An nth root of a is x if $x^n = a$. ◀

Let's examine the possibilities. A negative number raised to an odd power is negative, as in $(-2)^3 = -8$. Also, -2 is the only number whose cube is -8. A positive number raised to an odd power is positive, as in $4^3 = 64$. Also, 4 is the only number whose cube is 64. When n is odd, there is only one nth root of the number a; it is positive if $a > 0$, and it is negative if $a < 0$. Look back at Examples 1(a) and 1(b).

Because any real number raised to an even power is nonnegative, as in $(-2)^4 = 16$ and $3^2 = 9$, there is no real number x for which $x^n = a$ if n is even and a is negative. In other words, we cannot raise a real number to an even power and get a negative result. Consequently, even roots of negative numbers do not exist in the real number system.

A negative number raised to an even power is positive, as in $(-2)^4 = 16$. A positive number raised to an even power is also positive, as in $2^4 = 16$. As a result, 16 has two 4th roots. In general, when n is even and a is positive, there are two nth roots of a, one positive and the other negative. Look back at pairs of Examples 1(c) and 1(d), and 1(e) and 1(f).

Finally, if $a = 0$, the nth root of 0 is 0.

Based on this discussion, we state the following definition:

Principal *n*th Root of *a*

The **principal *n*th root of a number *a*,** symbolized by $\sqrt[n]{a}$, where $n \geq 2$ is an integer, is defined as follows:

(a) If a is positive and n is even, then $\sqrt[n]{a}$ is the *positive* nth root of a.

(b) If n is odd, then $\sqrt[n]{a}$ is the nth root of a.

(c) $\sqrt[n]{0} = 0$.

(d) If a is negative and n is even, then $\sqrt[n]{a}$ does not exist in the real number system.

The symbol $\sqrt[n]{a}$ for the principal nth root of a is sometimes called a **radical;** the integer n is called the **index,** and a is called the **radicand.** If the index of a radical is 2, we call $\sqrt[2]{a}$ the **square root** of a and omit the index 2 by simply writing $\sqrt{a}$. If the index is 3, we call $\sqrt[3]{a}$ the **cube root** of a.

To summarize:

$$\sqrt[n]{a} = b \quad \text{means} \quad b^n = a$$

where $a \geq 0$ and $b \geq 0$ if n is even
and a, b are any real numbers if n is odd

Notice that if a is negative and n is even, then $\sqrt[n]{a}$ is not defined in the real number system. When it is defined, the principal nth root of a number is **unique** (meaning there is one, and only one, result).

EXAMPLE 2 **Identifying the Radicand and the Index**

Identify the radicand and index in the following expressions.

(a) $\sqrt[5]{98}$ (b) $\sqrt[3]{-40}$ (c) $\sqrt{29}$ (d) $\sqrt[3]{5x - 4}$

Solution (a) The radicand is 98 and the index is 5.

(b) The radicand is -40 and the index is 3.

(c) The radicand is 29 and the index is 2.

(d) The radicand is $5x - 4$ and the index is 3. ◀

Practice Exercise 1 *In Problems 1–4, identify the radicand and index each in the expression.*

1. $\sqrt[7]{43}$ **2.** $\sqrt{82}$ **3.** $\sqrt[5]{-9}$ **4.** $\sqrt[4]{x-8}$

2 **EXAMPLE 3** **Evaluating Principal n^{th} Roots**

Find the value of each expression.

(a) $\sqrt[3]{8}$ (b) $\sqrt{64}$ (c) $\sqrt[3]{-64}$ (d) $\sqrt[4]{\frac{1}{16}}$

Solution (a) $\sqrt[3]{8} = 2$, since 2, when cubed, equals 8.

(b) $\sqrt{64} = 8$, since 8 is the positive number, when squared, that equals 64.

(c) $\sqrt[3]{-64} = -4$, since -4, when cubed, equals -64.

(d) $\sqrt[4]{\frac{1}{16}} = \frac{1}{2}$, since $\frac{1}{2}$ is the positive number, when raised to the 4th power, that equals $\frac{1}{16}$. ◀

These are examples of **perfect roots,** since each one simplifies to a rational number.

Practice Exercise 2 *Find the value of each expression.*

1. $\sqrt[3]{27}$ **2.** $\sqrt[4]{16}$ **3.** $\sqrt{\frac{1}{4}}$ **4.** $\sqrt[3]{-8}$

EXAMPLE 4 **Evaluating Principal n^{th} Roots**

(a) $\sqrt[3]{4^3} = \sqrt[3]{64} = 4$ (b) $\sqrt[3]{(-4)^3} = \sqrt[3]{-64} = -4$

(c) $\sqrt[4]{2^4} = \sqrt[4]{16} = 2$ (d) $\sqrt[4]{(-2)^4} = \sqrt[4]{16} = 2$ ◀

Notice the pattern in Example 4. In particular, notice how 4(a) and 4(b) are alike and how 4(c) and 4(d) are alike. This pattern suggests the following result.

If $n \geq 2$ is a positive integer and a is a real number, we have

$$\sqrt[n]{a^n} = a \quad \text{if } n \text{ is odd} \qquad \textbf{(1a)}$$

$$\sqrt[n]{a^n} = |a| \quad \text{if } n \text{ is even} \qquad \textbf{(1b)}$$

Notice the need for the absolute value in equation (1b). If n is even, then a^n is positive whether $a > 0$ or $a < 0$. And, if n is even, the principal nth root must be nonnegative. Hence the reason for using the absolute value—it gives a nonnegative result.

EXAMPLE 5 **Simplifying Principal n^{th} Roots**

(a) $\sqrt[3]{4^3} = 4$ (b) $\sqrt[5]{(-3)^5} = -3$ (c) $\sqrt[4]{3^4} = |3| = 3$

(d) $\sqrt[4]{(-3)^4} = |-3| = 3$ (e) $\sqrt{x^2} = |x|$ (f) $\sqrt[3]{x^3} = x$ ◀

Practice Exercise 3 *In Problems 1–4, find the exact value of each of the following expressions:*

1. $\sqrt[4]{(1.3)^4}$ **2.** $\sqrt[4]{(-1.3)^4}$ **3.** $\sqrt[3]{(1.3)^3}$ **4.** $\sqrt[3]{(-1.3)^3}$

3 Radicals provide a way of representing many irrational real numbers. For example, there is no rational number whose square is 2. Using decimals, we can only approximate the positive number whose square is 2. Using radicals, we can say that $\sqrt{2}$ is *the* positive number whose square is 2. Of course, to obtain a decimal approximation for a radical, we can use a calculator.

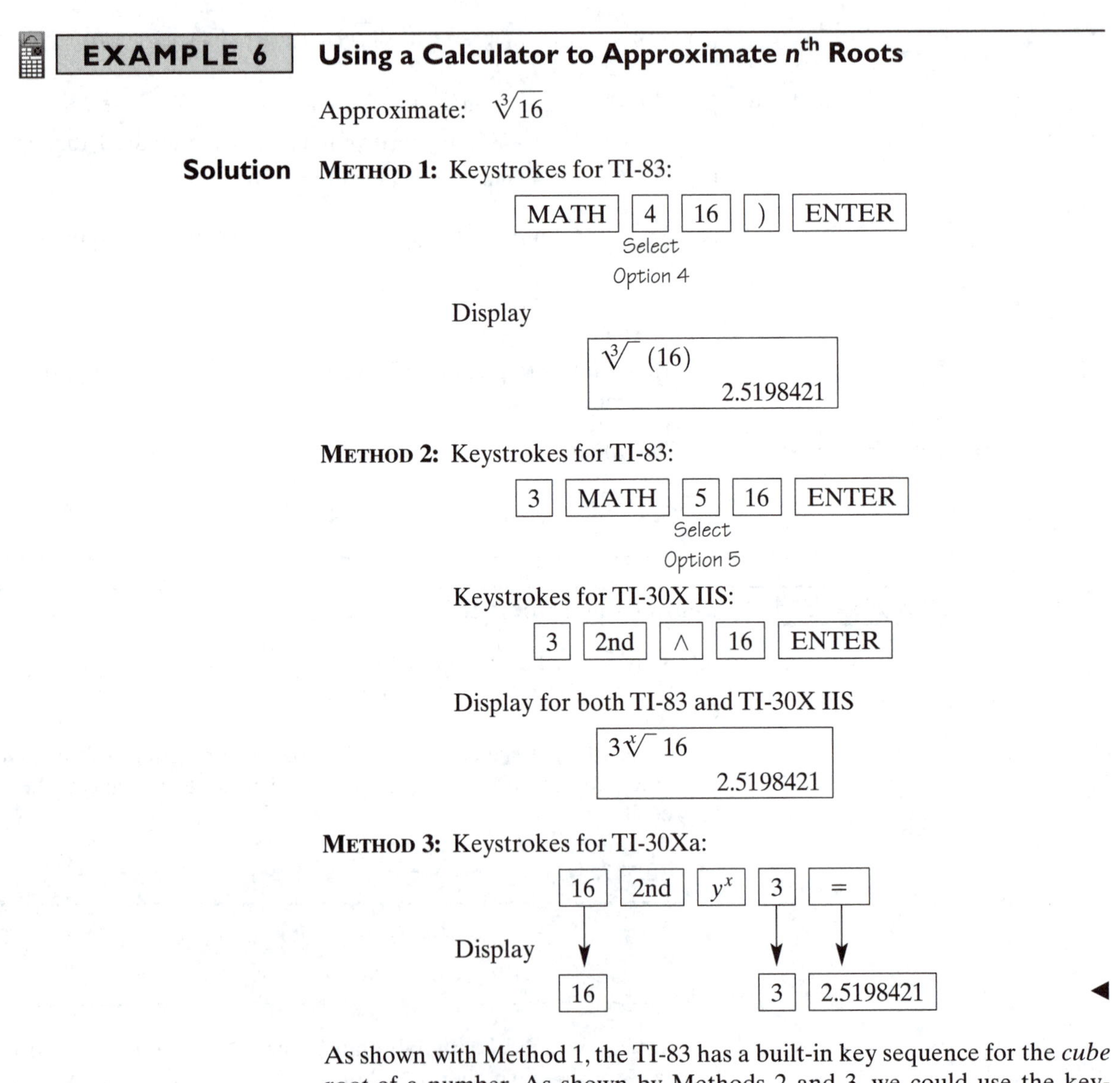

EXAMPLE 6 Using a Calculator to Approximate n^{th} Roots

Approximate: $\sqrt[3]{16}$

Solution **METHOD 1:** Keystrokes for TI-83:

[MATH] [4] [16] [)] [ENTER]

Select Option 4

Display

$\sqrt[3]{\ }(16)$
2.5198421

METHOD 2: Keystrokes for TI-83:

[3] [MATH] [5] [16] [ENTER]

Select Option 5

Keystrokes for TI-30X IIS:

[3] [2nd] [∧] [16] [ENTER]

Display for both TI-83 and TI-30X IIS

$3\sqrt[x]{\ }$ 16
2.5198421

METHOD 3: Keystrokes for TI-30Xa:

[16] [2nd] [y^x] [3] [=]

Display [16] [3] [2.5198421]

As shown with Method 1, the TI-83 has a built-in key sequence for the *cube root* of a number. As shown by Methods 2 and 3, we could use the keystrokes that allow us to key in any index we desire; again, in this case, the index is 3.

Practice Exercise 4 *In Problems 1–4, approximate each radical. Round your answer to two decimal places.*

1. $\sqrt{3}$ **2.** $\sqrt[3]{5}$ **3.** $\sqrt[5]{-2.6}$ **4.** $\sqrt[4]{21}$

Properties of Radicals

4 Let $n \geq 2$ and $m \geq 2$ denote positive integers, and let a and b represent real numbers. Assuming all radicals are defined, we have the following properties:

Properties of Radicals

$$\sqrt[n]{ab} = \sqrt[n]{a}\sqrt[n]{b} \qquad \textbf{(2a)}$$

$$\sqrt[n]{\frac{a}{b}} = \frac{\sqrt[n]{a}}{\sqrt[n]{b}} \qquad \textbf{(2b)}$$

$$\sqrt[n]{a^m} = \left(\sqrt[n]{a}\right)^m \qquad \textbf{(2c)}$$

We used properties (2a) and (2b) extensively in Section 4.2 as they applied to square roots. We also hinted at property (2c) with examples like $\sqrt{9^2} = \sqrt{81} = 9$ and $\left(\sqrt{9}\right)^2 = 3^2 = 9$.

As in Section 4.2, when used in reference to radicals, the direction to "simplify" will mean to remove from the radicals any perfect roots that occur as factors. Let's look at some examples of how the preceding rules are applied to simplify radicals.

EXAMPLE 7 **Simplifying Radicals**

Simplify each of the following:

(a) $\sqrt[4]{32}$ (b) $\sqrt[3]{16}$

Solution (a) $\sqrt[4]{32} = \sqrt[4]{16 \cdot 2} = \sqrt[4]{16}\sqrt[4]{2} = \sqrt[4]{2^4}\,\sqrt[4]{2} = 2\sqrt[4]{2}$

(first =: 16 is a perfect 4th power. second =: (2a))

(b) $\sqrt[3]{16} = \sqrt[3]{8 \cdot 2} = \sqrt[3]{8}\sqrt[3]{2} = \sqrt[3]{2^3}\,\sqrt[3]{2} = 2\sqrt[3]{2}$

(first =: Factor out perfect cube. second =: (2a)) ◀

EXAMPLE 8 **Simplifying Radicals**

Simplify each of the following:

(a) $\sqrt[3]{8x^4}$ (b) $\sqrt[3]{-16x^4y^7}$

Solution (a) $\sqrt[3]{8x^4} = \sqrt[3]{8 \cdot x^3 \cdot x} = \sqrt[3]{8} \cdot \sqrt[3]{x^3} \cdot \sqrt[3]{x} = 2x\sqrt[3]{x}$

(first =: Factor out perfect cubes. second =: (2a))

(b) $\sqrt[3]{-16x^4y^7} = \sqrt[3]{-8 \cdot 2 \cdot x^3 \cdot x \cdot y^3 \cdot y^3 \cdot y}$ Factor perfect cubes.

$= \sqrt[3]{-8} \cdot \sqrt[3]{2} \cdot \sqrt[3]{x^3} \cdot \sqrt[3]{x} \cdot \sqrt[3]{y^3} \cdot \sqrt[3]{y^3} \cdot \sqrt[3]{y}$ Use (2a).

$= -2xy^2\sqrt[3]{2xy}$ Simplify. ◀

With a little practice you can shorten the solution to Example 7(b) as follows:

$$\sqrt[3]{-16x^4y^7} = \sqrt[3]{-8\cdot 2\cdot x^3\cdot x\cdot y^6\cdot y} = -2xy^2\sqrt[3]{2xy}$$

EXAMPLE 9 Simplifying Radicals

Simplify each of the following. Assume all variables are positive.

(a) $\sqrt[3]{\frac{8x^5}{27}}$ (b) $\sqrt[4]{\frac{32x^9y}{16x^2y^5}}$

Solution (a) $\sqrt[3]{\frac{8x^5}{27}} = \frac{\sqrt[3]{8x^5}}{\sqrt[3]{27}} = \frac{\sqrt[3]{8\cdot x^3\cdot x^2}}{\sqrt[3]{27}} = \frac{\sqrt[3]{2^3\cdot x^3\cdot x^2}}{\sqrt[3]{3^3}} = \frac{2x\sqrt[3]{x^2}}{3}$

↑ (2b) ↑ Factor out perfect cubes.

(b) $\sqrt[4]{\frac{32x^9y}{16x^2y^5}} = \sqrt[4]{\frac{2x^7}{y^4}} = \frac{\sqrt[4]{2x^7}}{\sqrt[4]{y^4}} = \frac{\sqrt[4]{2\cdot x^4\cdot x^3}}{y} = \frac{x\sqrt[4]{2x^3}}{y}$

↑ Simplify inside the radical ↑ (2b) ↑ y is positive ↑ x is positive

◀

Notice in Example 8(b), since we assumed all variables are positive, we do not need to put absolute value signs around the variables that simplified out of the even radicals.

Practice Exercise 5 *Simplify each radical. Assume all variables are positive.*

1. $\sqrt[3]{24}$ **2.** $\sqrt[3]{-8x^3}$ **3.** $\sqrt[4]{162x^5y^{11}}$ **4.** $\sqrt[3]{\frac{27x^8}{64y^6}}$ **5.** $\sqrt{\frac{8x^2y}{45x}}$ ◀

5 Two or more radicals can be combined with addition and subtraction, provided they have the same index and the same radicand. Such radicals are called **like radicals.**

EXAMPLE 10 Combining Like Radicals

Simplify each expression.

(a) $4\sqrt{27} - 8\sqrt{12} + \sqrt{3}$ (b) $\sqrt[3]{8x^4} + \sqrt[3]{-x} + 4\sqrt[3]{27x}$

Solution (a)
$$\begin{aligned}4\sqrt{27} - 8\sqrt{12} + \sqrt{3} &= 4\sqrt{9\cdot 3} - 8\sqrt{4\cdot 3} + \sqrt{3}\\ &= 4\cdot 3\cdot\sqrt{3} - 8\cdot 2\cdot\sqrt{3} + \sqrt{3}\\ &= 12\sqrt{3} - 16\sqrt{3} + 1\sqrt{3} = -3\sqrt{3}\end{aligned}$$

(b)
$$\begin{aligned}\sqrt[3]{8x^4} + \sqrt[3]{-x} + 4\sqrt[3]{27x} &= \sqrt[3]{8\cdot x^3\cdot x} + \sqrt[3]{-1\cdot x} + 4\sqrt[3]{27\cdot x}\\ &= 2x\sqrt[3]{x} + (-1)\sqrt[3]{x} + 4\cdot 3\sqrt[3]{x}\\ &= 2x\sqrt[3]{x} + (-1)\sqrt[3]{x} + 12\sqrt[3]{x}\\ &= 2x\sqrt[3]{x} + 11\sqrt[3]{x} = (2x + 11)\sqrt[3]{x}\end{aligned}$$
◀

Practice Exercise 6 *Simplify each expression:*

1. $3\sqrt{18} - 5\sqrt{8} + 6\sqrt{50}$ **2.** $5\sqrt[3]{x^3} - 8\sqrt[3]{-x^3} + \sqrt[3]{8x^6}$ ◀

Historical Comment

The radical sign, $\sqrt{}$, was first used in print by Coss in 1525. It is thought to be the manuscript form of the letter *r* (for the Latin word *radix* = *root*), although this is not quite conclusively proved. It took a long time for $\sqrt{}$ to become the standard symbol for a square root and much longer to standardize $\sqrt[3]{}$, $\sqrt[4]{}$, $\sqrt[5]{}$ and so on. The indices of the root were placed in every conceivable position, with

$$\overset{3}{\sqrt{8}}, \quad \sqrt{}\ ③\,8, \quad \text{and} \quad \underset{3}{\sqrt{8}}$$

all being variants for $\sqrt[3]{8}$. The notation $\sqrt{\sqrt{16}}$ was popular for $\sqrt[4]{16}$. By the 1700's, the index had settled where we now put it.

The bar on top of the present radical symbol, as shown below,

$$\sqrt{a^2 + 2ab + b^2}$$

is the last survivor of the *vinculum*, a bar placed atop an expression to indicate what we would now indicate with parentheses. For example,

$$a\overline{b + c} = a(b + c)$$

Answers to Practice Exercises

Practice Exercise 1 **1.** The radicand is 43; the index is 7. **2.** The radicand is 82; the index is 2. **3.** The radicand is -9; the index is 5. **4.** The radicand is $x - 8$; the index is 4.

Practice Exercise 2 **1.** 3 **2.** 2 **3.** $\frac{1}{2}$ **4.** -2

Practice Exercise 3 **1.** 1.3 **2.** 1.3 **3.** 1.3 **4.** -1.3

Practice Exercise 4 **1.** 1.73 **2.** 1.71 **3.** -1.21 **4.** 2.14

Practice Exercise 5 **1.** $2\sqrt[3]{3}$ **2.** $-2x$ **3.** $3xy^2\sqrt[4]{2xy^3}$ **4.** $\frac{3x^2\sqrt[3]{x^2}}{4y^2}$ **5.** $\frac{2\sqrt{2xy}}{3\sqrt{5}} = \frac{2\sqrt{10xy}}{15}$

Practice Exercise 6 **1.** $29\sqrt{2}$ **2.** $13x + 2x^2$

4.3 Assess Your Understanding

'Are You Prepared?' *Answers are given at the end of these exercises. If you get a wrong answer, read the pages listed in parentheses.*

1. Simplify. Assume all variables are positive. $\sqrt{25x^6y^3}$ (pp. 188–197)
2. Simplify. Assume all variables are positive. $\sqrt{\frac{72x^4y}{50xy^5}}$ (pp. 188–197)
3. Simplify: $8\sqrt{3} + \sqrt{75} - 2\sqrt{27}$ (pp. 188–197)

Concepts and Vocabulary

4. In the expression $\sqrt[4]{25}$, the number 4 is called the ________ and the number 25 is called the ________.
5. We call $\sqrt[3]{a}$ the ________ ________ of *a*.
6. *True or False:* $\sqrt[3]{-8}$ is undefined.
7. *True or False:* $\sqrt[5]{x}$ and $\sqrt[4]{x}$ are like radicals.
8. *True or False:* $\sqrt[5]{\frac{27x^3}{10}} = \frac{\sqrt[5]{27x^3}}{\sqrt[5]{10}}$

Exercises

In Problems 9–12, identify the radicand and index.

9. $\sqrt{348}$
10. $\sqrt[6]{400}$
11. $\sqrt[3]{2x - 9}$
12. $\sqrt[4]{4x + 12}$

In Problems 13–28, evaluate each perfect root.

13. $\sqrt[3]{-27}$ **14.** $\sqrt[4]{16}$ **15.** $\sqrt[4]{1}$ **16.** $\sqrt[3]{27}$

17. $\sqrt[3]{125}$ **18.** $\sqrt[3]{-8}$ **19.** $\sqrt[5]{-1}$ **20.** $\sqrt[5]{-32}$

21. $\sqrt[4]{0}$ **22.** $\sqrt[3]{\dfrac{8}{27}}$ **23.** $\sqrt[4]{81x^4}$ **24.** $\sqrt[3]{27x^6}$

25. $\sqrt[5]{32x^{10}}$ **26.** $\sqrt{25x^8}$ **27.** $\sqrt[3]{8(1+x)^3}$ **28.** $\sqrt{12(x+4)^2}$

In Problems 29–52, simplify each expression. Assume all variables are positive when they appear.

29. $\sqrt[3]{81}$ **30.** $\sqrt[4]{32}$ **31.** $\sqrt[3]{16x^4}$ **32.** $\sqrt{27x^3}$

33. $\sqrt[3]{7^3}$ **34.** $\sqrt[4]{6^4}$ **35.** $\sqrt{\dfrac{32x^3}{9x}}$ **36.** $\sqrt[3]{\dfrac{x}{8x^4}}$

37. $\sqrt[4]{x^{12}y^8}$ **38.** $\sqrt[5]{x^{10}y^5}$ **39.** $\sqrt[4]{\dfrac{x^9y^7}{xy^3}}$ **40.** $\sqrt{\dfrac{3xy^2}{81x^4y^2}}$

41. $\sqrt{36x}$ **42.** $\sqrt{9x^5}$ **43.** $\sqrt{3x^2}\sqrt{12x}$ **44.** $\sqrt{5x}\sqrt{20x^3}$

45. $\dfrac{\sqrt{3xy^3}\sqrt{2x^2y}}{\sqrt{6x^3y^4}}$ **46.** $\dfrac{\sqrt[3]{x^2y}\sqrt[3]{125x^3}}{\sqrt[3]{8x^3y^4}}$ **47.** $\sqrt{\dfrac{4}{9x^2y^4}}$ **48.** $\sqrt{\dfrac{9}{16x^4y^6}}$

49. $(\sqrt{5}\sqrt[3]{9})^2$ **50.** $(\sqrt[3]{3}\sqrt{10})^4$ **51.** $(\sqrt[3]{4})^3 + (\sqrt[4]{5})^4$ **52.** $(\sqrt[3]{2})^3 + (\sqrt[5]{3})^5$

In Problems 53–62, simplify each expression.

53. $3\sqrt[4]{2} + 2\sqrt[4]{2} - \sqrt[4]{2}$ **54.** $6\sqrt[3]{5} - \sqrt[3]{5} - 4\sqrt[3]{5}$ **55.** $3\sqrt{2} - \sqrt{18} + 2\sqrt{8}$

56. $5\sqrt{3} + 2\sqrt{12} - 3\sqrt{27}$ **57.** $\sqrt[3]{16} + 5\sqrt[3]{2} - 2\sqrt[3]{54}$ **58.** $9\sqrt[3]{24} - \sqrt[3]{81}$

59. $\sqrt{8x^3} - 3\sqrt{50x} + \sqrt{2x^5},\quad x \geq 0$ **60.** $\sqrt{x^2y} - 3x\sqrt{9y} + 4\sqrt{25y},\quad x \geq 0, y \geq 0$

61. $\sqrt[3]{16x^4y} - 3x\sqrt[3]{2xy} + 5\sqrt[3]{-2xy^4}$ **62.** $8xy - \sqrt{25x^2y^2} + \sqrt[3]{8x^3y^3},\quad x \geq 0, y \geq 0$

In Problems 63–74, perform the indicated operation and simplify the result.

63. $(3\sqrt[3]{6})(2\sqrt[3]{9})$ **64.** $(5\sqrt[4]{8})(3\sqrt[4]{2})$ **65.** $(\sqrt[4]{9} + 3)(\sqrt[4]{9} - 3)$

66. $(\sqrt[4]{4} - 2)(\sqrt[4]{4} + 2)$ **67.** $(3\sqrt{7} + 3)(2\sqrt{7} + 2)$ **68.** $(2\sqrt{6} + 3)^2$

69. $(\sqrt{x} - 1)^2,\quad x \geq 0$ **70.** $(\sqrt{x} + \sqrt{5})^2,\quad x \geq 0$ **71.** $(\sqrt[3]{x} - 1)^3$

72. $(\sqrt[3]{x} + \sqrt[3]{2})^3$ **73.** $(2\sqrt{x} - 3\sqrt{y})(2\sqrt{x} + 5\sqrt{y}),\quad x \geq 0, y \geq 0$

74. $(4\sqrt{x} - \sqrt{y})(\sqrt{x} + 3\sqrt{y}),\quad x \geq 0, y \geq 0$

In Problems 75–82, approximate each expression. Round your answer to two decimal places.

75. $\sqrt{2}$ **76.** $\sqrt{7}$ **77.** $\sqrt[3]{4}$

78. $\sqrt[3]{-5}$ **79.** $\dfrac{2 + \sqrt{3}}{3 - \sqrt{5}}$ **80.** $\dfrac{\sqrt{5} - 2}{\sqrt{2} + 4}$

81. $\dfrac{3\sqrt[3]{5} - \sqrt{2}}{\sqrt{3}}$ **82.** $\dfrac{2\sqrt{3} - \sqrt[3]{4}}{\sqrt{2}}$

83. How would you explain to someone the meaning of the symbolism $\sqrt[q]{x^p}$, where p and $q \geq 2$ are integers?

'Are You Prepared?' Answers

1. $5x^3y\sqrt{y}$ **2.** $\dfrac{6x\sqrt{x}}{5y^2}$ **3.** $7\sqrt{3}$

4.4 Rational Exponents

PREPARING FOR THIS SECTION *Before getting started, review the following:*

- Laws of Exponents (Section 2.1, pp. 62–68 and Section 4.1, pp. 176–183)
- nth Roots (Section 4.3, pp. 199–204)

Now work the 'Are You Prepared?' problems on page 210.

OBJECTIVES

1 Simplify Expressions Containing Rational Exponents
2 Simplify Radicals Using Rational Exponents
3 Use Laws of Exponents with Rational Exponents

1 Our purpose in this section is to give a definition for "a raised to the power $\frac{m}{n}$," where a is a real number and $\frac{m}{n}$ is a rational number. However, we want the definition we give to ensure that the Laws of Exponents stated for integer exponents in Sections 2.1 and 4.1 remain true for rational exponents. For example, if the law of exponents $(a^r)^s = a^{rs}$ is to hold for rational numbers r and s, then it must be true that

$$(a^{1/n})^n = a^{(1/n)n} = a^1 = a$$

That is, $a^{1/n}$ is a number that, when raised to the power n, is a. But this was the definition we gave of the principal nth root of a in Section 4.3. As a result, we have

$$a^{1/2} = \sqrt{a} \qquad a^{1/3} = \sqrt[3]{a} \qquad a^{1/4} = \sqrt[4]{a}$$

and we can state the following definition:

If a is a real number and $n \geq 2$ is an integer, then

$$a^{1/n} = \sqrt[n]{a} \qquad \textbf{(1)}$$

provided $\sqrt[n]{a}$ exists.

If n is even and $a < 0$, then $\sqrt[n]{a}$ and $a^{1/n}$ are not defined.

EXAMPLE 1 Simplifying Fractional Exponents

Simplify each expression:

(a) $4^{1/2}$ (b) $(-27)^{1/3}$ (c) $(-32)^{1/5}$

Solution (a) $4^{1/2} = \sqrt{4} = 2$

(b) $(-27)^{1/3} = \sqrt[3]{-27} = -3$

(c) $(-32)^{1/5} = \sqrt[5]{-32} = -2$ ◀

EXAMPLE 2 **Simplifying Fractional Exponents**

Simplify each expression:

(a) $8^{1/2}$ (b) $16^{1/3}$ (c) $54^{1/3}$

Solution (a) $8^{1/2} = \sqrt{8} = \sqrt{4 \cdot 2} = \sqrt{4} \cdot \sqrt{2} = 2\sqrt{2}$

(b) $16^{1/3} = \sqrt[3]{16} = \sqrt[3]{8 \cdot 2} = \sqrt[3]{8} \cdot \sqrt[3]{2} = 2\sqrt[3]{2}$

(c) $54^{1/3} = \sqrt[3]{54} = \sqrt[3]{27 \cdot 2} = \sqrt[3]{27} \cdot \sqrt[3]{2} = 3\sqrt[3]{2}$ ◀

Practice Exercise 1 *Simplify each expression:*

1. $16^{1/4}$ **2.** $(-8)^{1/3}$ **3.** $18^{1/2}$ **4.** $(-24)^{1/3}$

We now seek a definition for $a^{m/n}$, where m and n are integers containing no common factors (except 1 and -1) and $n \geq 2$. Again, we want the definition to obey the Laws of Exponents stated earlier. For example,

$$a^{m/n} = a^{m(1/n)} = (a^m)^{1/n} \qquad \text{and} \qquad a^{m/n} = a^{(1/n)m} = (a^{1/n})^m$$

But by our previous comments, $(a^m)^{1/n} = \sqrt[n]{(a^m)} = \sqrt[n]{a^m}$ and $(a^{1/n})^m = \left(\sqrt[n]{a}\right)^m$. This leads us to formulate the following definition:

If a is a real number and if m and n are integers containing no common factors with $n \geq 2$, then

$$a^{m/n} = \sqrt[n]{a^m} = \left(\sqrt[n]{a}\right)^m \qquad \textbf{(2)}$$

provided $\sqrt[n]{a}$ exists.

We have two comments about equation (2):

1. The exponent m/n must be in lowest terms and $n \geq 2$.
2. In simplifying the rational expression $a^{m/n}$, either $\sqrt[n]{a^m}$ or $\left(\sqrt[n]{a}\right)^m$ may be used, the choice depending on which one is easier to simplify. Generally, taking the root first, as in $\left(\sqrt[n]{a}\right)^m$, is preferred.

EXAMPLE 3 **Simplifying Rational Exponents**

Simplify each expression:

(a) $4^{3/2}$ (b) $(-8)^{4/3}$ (c) $4^{6/4}$

Solution (a) $4^{3/2} = \sqrt[2]{4^3} = \left(\sqrt{4}\right)^3 = 2^3 = 8$

(b) $(-8)^{4/3} = \sqrt[3]{(-8)^4} = \left(\sqrt[3]{-8}\right)^4 = (-2)^4 = 16$

(c) $4^{6/4} = 4^{3/2} = (\sqrt{4})^3 = 2^3 = 8$ ◀

EXAMPLE 4 **Simplifying Rational Exponents**

Simplify each expression:

(a) $(32)^{-2/5}$ (b) $(-32)^{-3/5}$ (c) $(-27)^{-2/3}$ (d) $3^{-3/2}$

Solution (a) $(32)^{-2/5} = \dfrac{1}{32^{2/5}} = \dfrac{1}{\sqrt[5]{32^2}} = \dfrac{1}{(\sqrt[5]{32})^2} = \dfrac{1}{2^2} = \dfrac{1}{4}$

(b) $(-32)^{-3/5} = \dfrac{1}{(-32)^{3/5}} = \dfrac{1}{(\sqrt[5]{-32})^3} = \dfrac{1}{(-2)^3} = \dfrac{1}{-8} = -\dfrac{1}{8}$

(c) $(-27)^{-2/3} = \dfrac{1}{(-27)^{2/3}} = \dfrac{1}{(\sqrt[3]{-27})^2} = \dfrac{1}{(-3)^2} = \dfrac{1}{9}$

(d) $3^{-3/2} = \dfrac{1}{3^{3/2}} = \dfrac{1}{\sqrt{3^3}} = \dfrac{1}{\sqrt{3^2 \cdot 3}} = \dfrac{1}{3\sqrt{3}} = \dfrac{1}{3\sqrt{3}} \cdot \dfrac{\sqrt{3}}{\sqrt{3}} = \dfrac{\sqrt{3}}{9}$ ◀

Based on the definition of $a^{m/n}$, no meaning is given to $a^{m/n}$ if a is a negative real number and n is an even integer.

Practice Exercise 2 *Simplify each expression:*

1. $9^{3/2}$ **2.** $(-8)^{2/3}$ **3.** $16^{-3/4}$ **4.** $2^{-5/2}$

2 Rational exponents sometimes can be used to simplify radicals.

EXAMPLE 5 Simplify each expression:

(a) $(\sqrt[4]{7})^2$ (b) $\sqrt[9]{x^3}$ (c) $\sqrt[3]{4^9}$ (d) $\sqrt[3]{4}\sqrt{2}$ (e) $\sqrt[4]{y^3} \cdot \sqrt{y}$

Solution (a) $(\sqrt[4]{7})^2 = 7^{2/4} = 7^{1/2} = \sqrt{7}$

(b) $\sqrt[9]{x^3} = x^{3/9} = x^{1/3} = \sqrt[3]{x}$

(c) $\sqrt[3]{4^9} = 4^{9/3} = 4^3 = 64$

(d) $\sqrt[3]{4}\sqrt{2} = \sqrt[3]{2^2} \cdot \sqrt{2} = 2^{2/3} \cdot 2^{1/2} = 2^{\frac{2}{3}+\frac{1}{2}} = 2^{\frac{4}{6}+\frac{3}{6}} = 2^{7/6}$
$= \sqrt[6]{2^7} = \sqrt[6]{2^6 \cdot 2^1} = 2\sqrt[6]{2}$

(e) $\sqrt[4]{y^3} \cdot \sqrt{y} = y^{3/4} \cdot y^{1/2} = y^{\frac{3}{4}+\frac{1}{2}} = y^{\frac{3}{4}+\frac{2}{4}} = y^{\frac{5}{4}}$
$= \sqrt[4]{y^5} = \sqrt[4]{y^4 \cdot y} = y\sqrt[4]{y}$ ◀

Practice Exercise 3 *Simplify each expression:*

1. $(\sqrt[4]{6})^8$ **2.** $\sqrt[3]{2}\sqrt{2}$ **3.** $\sqrt[3]{9}\sqrt{3}$ **4.** $\sqrt[5]{x^2} \cdot \sqrt{x}$ **5.** $\sqrt[9]{y^6}$

3 The definitions in equations (1) and (2) were stated so that the Laws of Exponents would remain true for rational exponents. For convenience, we list again the Laws of Exponents.

Laws of Exponents

If a and b are real numbers and r and s are rational numbers, then

$$a^r a^s = a^{r+s} \qquad (a^r)^s = a^{rs} \qquad (ab)^r = a^r \cdot b^r$$

$$a^{-r} = \frac{1}{a^r} \qquad \left(\frac{a}{b}\right)^r = \frac{a^r}{b^r} \qquad \frac{a^r}{a^s} = a^{r-s} = \frac{1}{a^{s-r}} \qquad \left(\frac{a}{b}\right)^{-r} = \left(\frac{b}{a}\right)^r$$

where it is assumed that all expressions used are defined.

The remaining example in this section illustrates the use of the Laws of Exponents to simplify certain expressions containing rational exponents. In this example, we assume that the variables are positive.

3 **EXAMPLE 6** **Simplifying Expressions Containing Rational Exponents**

Simplify each expression. Express your answer so that only positive exponents occur. Assume that the variables are positive.

(a) $(x^{2/3}y)(x^{-2}y)^{1/2}$ (b) $\left(\dfrac{2x^{1/3}}{y^{2/3}}\right)^{-3}$ (c) $\left(\dfrac{9x^2y^{1/3}}{x^{1/3}y}\right)^{1/2}$

Solution (a)

$$(x^{2/3}y)(x^{-2}y)^{1/2} = (x^{2/3}y)[(x^{-2})^{1/2}y^{1/2}]$$
$$= x^{2/3}yx^{-1}y^{1/2} = (x^{2/3}\cdot x^{-1})(y\cdot y^{1/2})$$
$$= x^{2/3+(-1)}\,y^{1+1\neq 2} = x^{-1/3}y^{3/2}$$
$$= \frac{y^{3/2}}{x^{1/3}}$$

(b)

$$\left(\frac{2x^{1/3}}{y^{2/3}}\right)^{-3} = \left(\frac{y^{2/3}}{2x^{1/3}}\right)^{3} = \frac{(y^{2/3})^3}{(2x^{1/3})^3} = \frac{y^2}{2^3(x^{1/3})^3} = \frac{y^2}{8x}$$

(c)

$$\left(\frac{9x^2y^{1/3}}{x^{1/3}y}\right)^{1/2} = \left(\frac{9x^{2-(1/3)}}{y^{1-(1/3)}}\right)^{1/2} = \left(\frac{9x^{5/3}}{y^{2/3}}\right)^{1/2} = \frac{9^{1/2}(x^{5/3})^{1/2}}{(y^{2/3})^{1/2}} = \frac{3x^{5/6}}{y^{1/3}}$$

or alternatively,

(c)

$$\left(\frac{9x^2y^{1/3}}{x^{1/3}y}\right)^{1/2} = \frac{9^{1/2}\cdot(x^2)^{1/2}\cdot(y^{1/3})^{1/2}}{(x^{1/3})^{1/2}\cdot y^{1/2}} = \frac{3\cdot x\cdot y^{1/6}}{x^{1/6}\cdot y^{1/2}}$$
$$= \frac{3\cdot x^{1-(1/6)}}{y^{(1/2)-(1/6)}} = \frac{3x^{5/6}}{y^{2/6}} = \frac{3x^{5/6}}{y^{1/3}}$$

◀

Example 6(c) was simplified with two different approaches as a reminder that sometimes you have a choice on which part of the problem to begin with when simplifying expressions that contain exponents.

Practice Exercise 4 *In Problems 1–4, simplify each expression. Express your answer so that only positive exponents appear. Assume that the variables are positive.*

1. $\left(\dfrac{2x^{3/2}}{y^{1/2}}\right)^4$ **2.** $\left(\dfrac{x^{5/3}}{y^{2/3}}\right)^6$ **3.** $(x^{3/2}y)(x^{-1}y^2)^{1/2}$ **4.** $\left(\dfrac{16x^{2/3}y^{1/4}}{x^{1/2}y^{3/4}}\right)^{1/2}$

◀

Answers to Practice Exercises

Practice Exercise 1	**1.** 2	**2.** −2	**3.** $3\sqrt{2}$	**4.** $-2\sqrt[3]{3}$	
Practice Exercise 2	**1.** 27	**2.** 4	**3.** $\frac{1}{8}$	**4.** $\frac{1}{4\sqrt{2}} = \frac{\sqrt{2}}{8}$	
Practice Exercise 3	**1.** 36	**2.** $2^{5/6}$	**3.** $3^{7/6}$ or $3\sqrt[6]{3}$	**4.** $x^{9/10}$	**5.** $y^{2/3} = \sqrt[3]{y^2}$
Practice Exercise 4	**1.** $\frac{16x^6}{y^2}$	**2.** $\frac{x^{10}}{y^4}$	**3.** xy^2	**4.** $\frac{4x^{1/12}}{y^{1/4}}$	

4.4 Assess Your Understanding

'Are You Prepared?' *Answers are given at the end of these exercises. If you get a wrong answer, read the pages listed in parentheses.*

1. Simplify: $(x^2y^3)^5$ (pp. 62–68)

2. Simplify: $(2xy^4)^{-2}(3x^2y)$ (pp. 176–183)

3. Simplify: $\sqrt[5]{32}$ (pp. 199–204)

Concepts and Vocabulary

4. *True or False:* $8^{-2/3} = -4$

5. *True or False:* $7^{2/5} = \sqrt[5]{7^2}$

6. *True or False:* $x^{1/2} \cdot x^{3/5} = x^{1/2+3/5}$

Exercises

In Problems 7–16, convert each radical expression to an equivalent expression involving rational exponents.

7. $\sqrt{3}$ 8. $\sqrt{10}$ 9. $\sqrt[3]{7}$ 10. $\sqrt[5]{9}$ 11. $\sqrt[3]{5^2}$ 12. $\sqrt[9]{2^8}$

13. $\sqrt[7]{x^4}$ 14. $\sqrt[5]{x^2}$ 15. $\sqrt{x^5}$ 16. $\sqrt{x}$

In Problems 17–50, simplify each expression.

17. $8^{2/3}$ 18. $4^{3/2}$ 19. $(-27)^{2/3}$ 20. $(-64)^{2/3}$ 21. $4^{-3/2}$ 22. $(-8)^{-5/3}$

23. $9^{-3/2}$ 24. $25^{-5/2}$ 25. $\left(\frac{9}{4}\right)^{3/2}$ 26. $\left(\frac{27}{8}\right)^{2/3}$ 27. $\left(\frac{4}{9}\right)^{-3/2}$ 28. $\left(\frac{8}{27}\right)^{-2/3}$

29. $4^{1.5}$ 30. $16^{-1.5}$ 31. $\left(\frac{1}{4}\right)^{-1.5}$ 32. $\left(\frac{1}{9}\right)^{1.5}$ 33. $(\sqrt{3})^6$ 34. $(\sqrt[3]{4})^6$

35. $(\sqrt{5})^{-2}$ 36. $(\sqrt[4]{3})^{-8}$ 37. $3^{1/2} \cdot 3^{3/2}$ 38. $5^{1/3} \cdot 5^{4/3}$ 39. $\frac{7^{1/3}}{7^{4/3}}$ 40. $\frac{6^{5/4}}{6^{1/4}}$

41. $2^{1/3} \cdot 4^{1/3}$ 42. $9^{1/3} \cdot 3^{1/3}$ 43. $\sqrt[4]{3} \cdot \sqrt[4]{27}$ 44. $\sqrt[3]{2} \cdot \sqrt[3]{4}$ 45. $(\sqrt[4]{2})^{-4}$ 46. $(\sqrt[5]{3})^{-5}$

47. $(\sqrt[3]{6})^2$ 48. $(\sqrt[4]{5})^3$ 49. $\sqrt{2}\sqrt[3]{2}$ 50. $\sqrt{5}\sqrt[3]{5}$

In Problems 51–64, simplify each expression. Express your answer so that only positive exponents occur. Assume that any variables are positive.

51. $\sqrt[8]{x^4}$ 52. $\sqrt[6]{x^3}$ 53. $\sqrt{x^3}\sqrt[4]{x}$ 54. $\sqrt[3]{x^2}\sqrt{x}$

55. $x^{3/2}x^{-1/2}$ 56. $x^{5/4}x^{-1/4}$ 57. $(x^3y^6)^{2/3}$ 58. $(x^4y^8)^{5/4}$

59. $(x^2y)^{1/3}(xy^2)^{2/3}$ 60. $(xy)^{1/4}(x^2y^2)^{1/2}$ 61. $(16x^2y^{-1/3})^{3/4}$ 62. $(4x^{-1}y^{1/3})^{3/2}$

63. $\left(\frac{x^{2/5}y^{-1/5}}{x^{-1/3}}\right)^{15}$ 64. $\left(\frac{x^{1/2}}{y^2}\right)^4\left(\frac{y^{1/3}}{x^{-2/3}}\right)^3$

'Are You Prepared?' Answers

1. $x^{10}y^{15}$ 2. $\frac{3}{4y^7}$ 3. 2

4.5 Complex Numbers

PREPARING FOR THIS SECTION *Before getting started, review the following:*

- Classification of Numbers (pp. 14–18)
- Simplifying Square Roots (pp. 189–190)

Now work the 'Are You Prepared?' problems on page 215.

OBJECTIVES

1 Identify Complex Numbers

2 Find Square Roots of Negative Numbers

In this section, we introduce complex numbers and explain their relationship to real numbers. To begin, we review sets of numbers from Chapter 1.

The *counting numbers* consist of $\{1, 2, 3, 4, 5, \dots\}$ and are used to count items. From a historical perspective, these were probably the first numbers humans had use for. Not long after devising the counting numbers, one could imagine the need for the number zero. As a result, the set of counting numbers was expanded to the set of *whole numbers*, $\{0, 1, 2, 3, 4, 5, 6, \dots\}$. Eventually, the need to express negative quantities arose and the set of whole numbers was expanded to the set of *integers*, $\{\dots, -3, -2, -1, 0, 1, 2, \dots\}$. At some point in time, humankind had a need to find a number x that would represent a portion of an item; and so, numbers such as $\frac{1}{2}, \frac{2}{3}$, etc., the *rational numbers*, were introduced. Now, if our universe were to consist only of rational numbers, there would be no number whose square equals 2. That is, there would be no number x for which $x^2 = 2$. To remedy this, mathematicians introduce numbers such as $\sqrt{2}, \sqrt[3]{5}$, etc., the *irrational numbers*. As stated in Section 1.2, the recognition of irrational numbers was accomplished among much controversy, as the discovery that $\sqrt{2}$ could not be expressed as a quotient of two integers was regarded as a fundamental flaw in the number concept. As irrational numbers became accepted, we obtained the set of *real numbers* by combining the sets of rational and irrational numbers.

One of the properties of a real number is that its square is nonnegative. For example, there is no real number x for which $x^2 = -1$. To remedy this situation, mathematicians introduce a number called the **imaginary unit,** which is denoted by i, and whose square is -1. Again, this idea of a number whose square is -1 was fraught with controversy as is evident by its name: *imaginary unit*. Eventually, the imaginary unit was accepted among mathematicians, giving us

$$i^2 = -1$$

✓1 In the progression outlined above, each time we encountered a situation that was unsuitable, we introduced a new number system to remedy that situation. And each new number system contained the earlier number system as a subset. The number system that results from introducing the number i is called the **complex number system.**

Complex numbers are numbers of the form $a + bi$, where a and b are real numbers. The real number a is called the **real part** of the number $a + bi$, the real number b is called the **imaginary part** of $a + bi$, where i is the imaginary unit and $i^2 = -1$.

Notice that the set of complex numbers contains the set of real numbers as a subset since every real number a can be expressed as $a + 0i$. The complex number $0 + bi$ is usually written as bi, and is called a **pure imaginary number.**

When a complex number is written in the form $a + bi$, where a and b are real numbers, we say it is in **standard form.** However, if the imaginary part of a complex number is negative, such as in the complex number $3 + (-2)i$, we agree to write it as $3 - 2i$.

EXAMPLE 1 Identify the real part and the imaginary part of each of the following complex numbers:

(a) $-5 + 6i$ (b) $4 - 7i$ (c) 8 (d) $-i$

Solution
(a) The real part of $-5 + 6i$ is -5 and the imaginary part is 6.
(b) Since $4 - 7i = 4 + (-7)i$, the real part of $4 - 7i$ is 4 and the imaginary part is -7.
(c) Since $8 = 8 + 0i$, the real part is 8 and the imaginary part is 0.
(d) Since $-i = 0 + (-1)i$, the real part is 0 and the imaginary part is -1. ◀

Practice Exercise 1 *Identify the real part and the imaginary part of each of the following complex numbers:*

1. $-2 + i$ **2.** $4 - 9i$ **3.** 0 **4.** $5i$

The relationship between various types of numbers is shown in the following diagram:

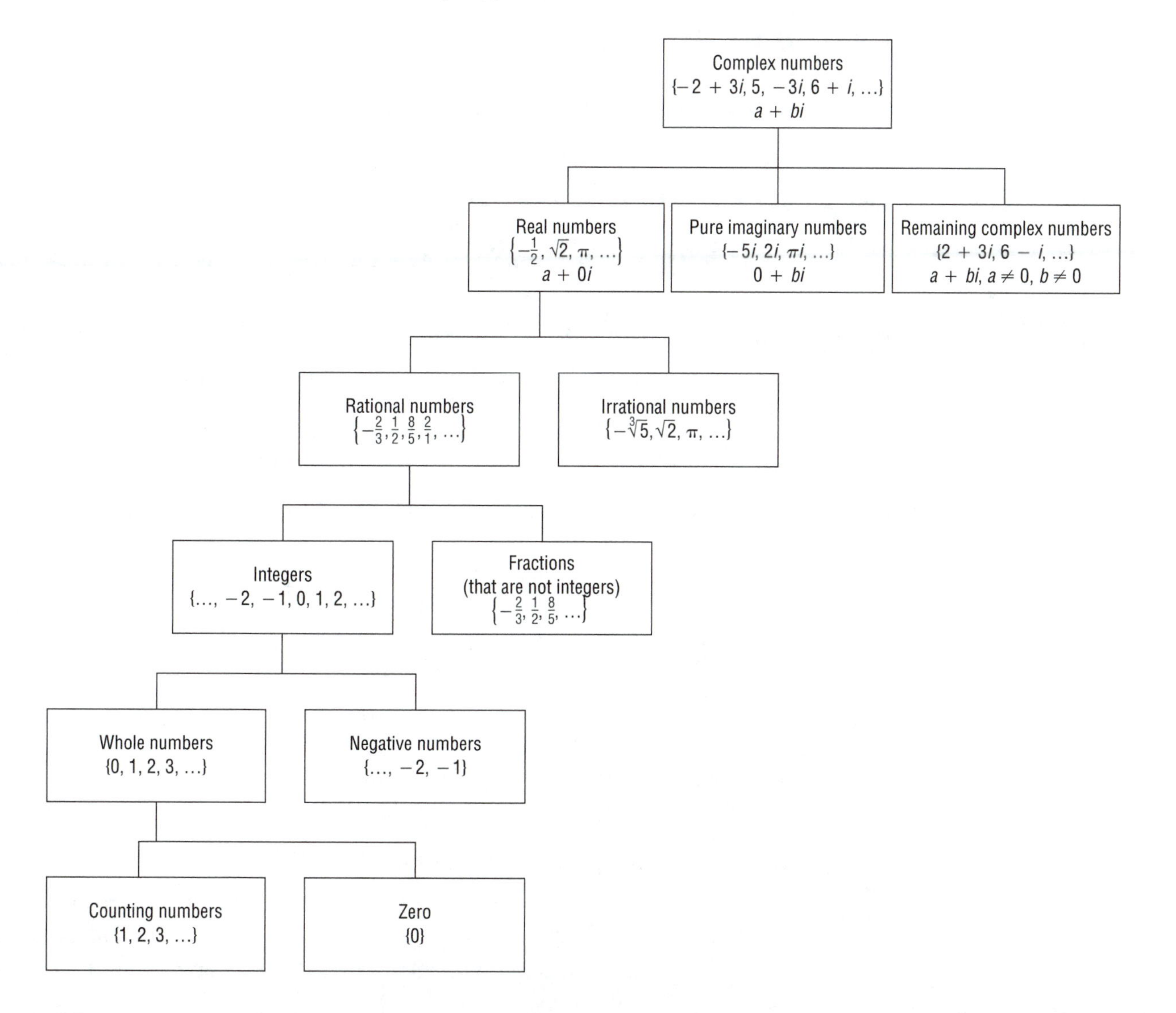

Square Roots of Negative Numbers

2 In the complex number system we can define the square root of a negative real number.

If N is a positive real number, we define the **principal square root of −N**, denoted by $\sqrt{-N}$, as.

$$\sqrt{-N} = \sqrt{-N}i$$

where i is the imaginary unit and $i^2 = -1$.

WARNING: In writing $\sqrt{-N} = \sqrt{N}i$, be sure to place i outside the $\sqrt{}$ symbol. ■

EXAMPLE 2 Simplifying Square Roots

Simplify each of the following in the complex number system.

(a) $\sqrt{-9}$ (b) $\sqrt{-49}$ (c) $\sqrt{-2}$ (d) $\sqrt{-1}$

Solution

(a) $\sqrt{-9} = \sqrt{9}i = 3i$

(b) $\sqrt{-49} = \sqrt{49}i = 7i$

(c) $\sqrt{-2} = \sqrt{2}i$

(d) $\sqrt{-1} = \sqrt{1}i = i$ ◀

Practice Exercise 2 *Simplify each of the following in the complex number system:*

1. $\sqrt{-81}$ **2.** $\sqrt{-98}$ **3.** $\sqrt{-5}$

3

EXAMPLE 3 Simplifying Complex Numbers

Simplify each of the following in the complex number system. Give your answer in standard form.

(a) $3 - \sqrt{-25}$ (b) $\sqrt{-4} - 1$ (c) $3 + \sqrt{-24}$ (d) $\dfrac{2 - \sqrt{-12}}{4}$

Solution

(a) $3 - \sqrt{-25} = 3 - \sqrt{25}i = 3 - 5i$

(b) $\sqrt{-4} - 1 = -1 + \sqrt{-4} = -1 + \sqrt{4}i = -1 + 2i$

(c) $3 + \sqrt{-24} = 3 + \sqrt{24}i = 3 + \sqrt{4 \cdot 6}i = 3 + 2\sqrt{6}i$

(d) $\dfrac{2 - \sqrt{-12}}{4} = \dfrac{2 - \sqrt{12}i}{4} = \dfrac{2 - 2\sqrt{3}i}{4}$

$$= \frac{2}{4} - \frac{2\sqrt{3}i}{4} = \frac{1}{2} - \frac{\sqrt{3}}{2}i$$ ◀

Practice Exercise 3 *Simplify each of the following in the complex number system. Give your answer in standard form.*

1. $-7 - \sqrt{-16}$ **2.** $\sqrt{-12} + 6$ **3.** $\dfrac{\sqrt{-12} + 6}{8}$

When solving an algebra problem, pay attention to the directions. Sometimes the directions will state, "Solve over the real numbers." If this is the case, we accept only real number answers and disregard any complex number answers of the form $a + bi$, with $b \neq 0$. Likewise, if the directions state, "Solve over the complex numbers," we accept all real number and all complex number answers. For certain problems, the directions might state, "Solve over the integers." If this is the case we accept only the answers that are integers and disregard any others.

Answers to Practice Exercises

Practice Exercise 1

1. The real part is -2 and the imaginary part is 1.
2. The real part is 4 and the imaginary part is -9.
3. The real part is 0 and the imaginary part is 0.
4. The real part is 0 and the imaginary part is 5.

Practice Exercise 2 **1.** $9i$ **2.** $7\sqrt{2}i$ **3.** $\sqrt{5}i$

Practice Exercise 3 **1.** $-7 - 4i$ **2.** $6 + 2\sqrt{3}i$ **3.** $\frac{3}{4} + \frac{\sqrt{3}}{4}i$

4.5 Assess Your Understanding

'Are You Prepared?' *Answers are given at the end of these exercises. If you get a wrong answer, read the pages listed in parentheses.*

1. Name the integers in the set $\left\{-3, 0, \sqrt{2}, \frac{6}{7}, \pi\right\}$. (pp. 14–18)

2. Name the irrational numbers in the set $\left\{-3, 0, \sqrt{2}, \frac{6}{7}, \pi\right\}$. (pp. 14–18)

3. The real numbers consist of all the ________ and ________ numbers. (pp. 14–18)

4. Simplify $\sqrt{48}$. (pp. 189–190)

Concepts and Vocabulary

5. Consider the complex number $-2 + 7i$: -2 is called ________ ________; 7 is called the ________ ________; and i is called the ________ ________.

6. $\sqrt{-16} =$ ________ in the complex number system.

7. Express in standard form: $3 - \sqrt{-25}$.

8. *True or False:* $\sqrt{5}i = \sqrt{5i}$

Exercises

In Problems 9–18, identify the real part and the imaginary part of each complex number.

9. $8 + 3i$ **10.** $-5 + 7i$ **11.** $6i$ **12.** $10i$ **13.** 42
14. 23 **15.** $5i - 12$ **16.** $2i + 1$ **17.** $11 - i$ **18.** $9 + i$

In Problems 19–28, simplify each of the following in the complex number system.

19. $\sqrt{-9}$ **20.** $\sqrt{-36}$ **21.** $\sqrt{-1}$ **22.** $\sqrt{-7}$ **23.** $\sqrt{-3}$
24. $\sqrt{-18}$ **25.** $\sqrt{-20}$ **26.** $\sqrt{-48}$ **27.** $\sqrt{-75}$ **28.** $\sqrt{-121}$

In Problems 29–44, simplify each of the following in the complex number system. Give your answer in standard form.

29. $4 + \sqrt{-16}$ **30.** $3 - \sqrt{-49}$ **31.** $\sqrt{-64} + 2$ **32.** $\sqrt{-144} - 10$
33. $6 + \sqrt{-15}$ **34.** $1 - \sqrt{-7}$ **35.** $8 - \sqrt{-28}$ **36.** $1 - \sqrt{-50}$

37. $\sqrt{-40} + 2$ **38.** $\sqrt{-32} + 5$ **39.** $\dfrac{2 + \sqrt{-12}}{6}$ **40.** $\dfrac{9 - \sqrt{-18}}{3}$

41. $\dfrac{4 - \sqrt{-16}}{4}$ **42.** $\dfrac{6 + \sqrt{-9}}{3}$ **43.** $\dfrac{5 - \sqrt{-8}}{6}$ **44.** $\dfrac{3 + \sqrt{-5}}{7}$

45. Suppose the directions to an algebra problem were to "Solve over the rational numbers." In your own words, explain what this means.

'Are You Prepared?' Answers

1. $-3, 0$ **2.** $\sqrt{2}, \pi$ **3.** rational; irrational **4.** $4\sqrt{3}$

Chapter Review

Things to Know

Exponents

$a^n = \underbrace{a \cdot a \cdot \cdots \cdot a}_{n \text{ factors}}$ n a positive integer

$a^0 = 1$ $a \neq 0$

$a^{-n} = \dfrac{1}{a^n}$ $a \neq 0$, n a positive integer

Laws of Exponents

$a^m \cdot a^n = a^{m+n}$ $(a^m)^n = a^{mn}$ $(a \cdot b)^n = a^n \cdot b^n$

$\dfrac{a^m}{a^n} = a^{m-n} = \dfrac{1}{a^{n-m}}$ $\left(\dfrac{a}{b}\right)^n = \dfrac{a^n}{b^n}$ $\left(\dfrac{a}{b}\right)^{-n} = \left(\dfrac{b}{a}\right)^n$

Radical

$\sqrt[n]{a} = b$ means $b^n = a$, where $a \geq 0, b \geq 0$ if n is even

$a^{m/n} = \sqrt[n]{a^m} = \left(\sqrt[n]{a}\right)^m$

Complex numbers

Numbers of the form $a + bi$, where a, b are real numbers and $i^2 = -1$

Objectives

Section		You Should Be Able To . . .	Review Exercises
4.1	1	Work with negative exponents (p. 176)	1, 2, 9, 10, 29–32
	2	Review the laws of exponents (p. 179)	21, 29–32
	3	Use scientific notation (p. 183)	37, 38
4.2	1	Evaluate square roots (p. 188)	27
	2	Simplify square roots (p. 189)	11, 12
	3	Multiply square roots (p. 190)	11, 12
	4	Add and subtract square roots (p. 191)	15, 16
	5	Multiply expressions containing square roots (p. 192)	13, 14, 20
	6	Rationalize denominators (p. 194)	41–46
	7	Simplify quotients of square roots (p. 197)	17, 18, 20

4.3	✓1	Work with nth roots (p. 199)	3, 4, 28
	✓2	Simplify nth roots (p. 201)	3, 4, 19, 28
	✓3	Approximate nth roots with a calculator (p. 202)	39, 40
	✓4	Use properties of radicals (p. 203)	11, 12, 17, 18, 20
	✓5	Combine like radicals (p. 204)	15, 16, 19
4.4	✓1	Simplify expressions containing rational exponents (p. 207)	21, 22, 24–26
	✓2	Simplify radicals using rational exponents (p. 209)	23
	✓3	Use laws of exponents with rational exponents (p. 209)	21, 22, 24, 33–36
4.5	✓1	Identify complex numbers (p. 212)	47–54
	✓2	Find square roots of negative numbers (p. 214)	47–54

Review Exercises

In Problems 1–28, perform the indicated operations.

1. $3^{-1} + 2^{-1}$

2. -3^{-2}

3. $\sqrt{25} - \sqrt[3]{-1}$

4. $\sqrt[3]{-27} + \sqrt[5]{-32}$

5. $9^{1/2} \cdot (-3)^2$

6. $(-8)^{-1/3} - \left(\frac{1}{2}\right)^{-1}$

7. $5^2 - 3^3$

8. $2^3 + 3^2 - 1$

9. $\frac{2^{-3} \cdot 5^0}{4^2}$

10. $\frac{3^{-2} - 4^1}{2^{-2}}$

11. $\sqrt{12} \cdot \sqrt{75}$

12. $\sqrt{6} \cdot \sqrt{18}$

13. $(2\sqrt{5} - 2)(2\sqrt{5} + 2)$

14. $(2\sqrt{3} + \sqrt{2})(2\sqrt{3} - \sqrt{2})$

15. $3\sqrt{24} + 2\sqrt{6}$

16. $8\sqrt{75} - 2\sqrt{27}$

17. $\frac{\sqrt{54}}{\sqrt{18}}$

18. $\sqrt{\frac{98}{8}}$

19. $4\sqrt[3]{54} - 3\sqrt[3]{2}$

20. $\frac{\sqrt{7}\sqrt{56}}{\sqrt{18}}$

21. $\left(\frac{8}{9}\right)^{-2/3}$

22. $\left(\frac{4}{27}\right)^{-3/2}$

23. $(\sqrt[3]{2})^{-3}$

24. $(4\sqrt{32})^{-1/2}$

25. $|5 - 8^{1/3}|$

26. $|9^{1/2} - 27^{2/3}|$

27. $\sqrt{|3^2 - 5^2|}$

28. $\sqrt[3]{-|4^2 - 2^3|}$

In Problems 29–36, write each expression so that all exponents are positive. Assume $x > 0$ and $y > 0$.

29. $\frac{x^{-2}}{y^{-2}}$

30. $\left(\frac{x^{-1}}{y^{-3}}\right)^2$

31. $\frac{(x^2y)^{-4}}{(xy)^{-3}}$

32. $\frac{\left(\frac{x}{y}\right)^2}{\left(\frac{x}{y}\right)^{-1}}$

33. $(25x^{-4/3}y^{-2/3})^{3/2}$

34. $(16x^{-2/3}y^{4/3})^{-3/2}$

35. $\left(\frac{2x^{-1/2}}{y^{-3/4}}\right)^{-4}$

36. $\left(\frac{8x^{-3/2}}{y^{-3}}\right)^{-2/3}$

37. Write 3.275×10^5 in standard form.

38. Write 0.00691 in scientific notation.

39. Use a calculator to approximate $\sqrt[4]{12}$. Round your answer to three decimal places.

40. Use a calculator to approximate $\frac{\sqrt{5} + 3}{\sqrt[3]{2}}$. Round your answer to three decimal places.

In Problems 41–46, rationalize the denominator of each expression.

41. $\dfrac{2}{\sqrt{3}}$

42. $\dfrac{-1}{\sqrt{5}}$

43. $\dfrac{2}{1-\sqrt{2}}$

44. $\dfrac{-4}{1+\sqrt{3}}$

45. $\dfrac{1+\sqrt{5}}{1-\sqrt{5}}$

46. $\dfrac{4\sqrt{3}+2}{2\sqrt{3}+1}$

In Problems 47–54, use the complex number system and write each expression in the standard form $a + bi$.

47. $\sqrt{-81}$

48. $\sqrt{-98}$

49. $3+\sqrt{-4}$

50. $5-\sqrt{-9}$

51. $\sqrt{-8}+10$

52. $\sqrt{-18}-4$

53. $\dfrac{12+\sqrt{-20}}{4}$

54. $\dfrac{6-\sqrt{-36}}{6}$

ANSWERS

CHAPTER 1 Numbers and Their Properties

1.1 Concepts and Vocabulary *(page 10)*

1. $x + 3 = 4 \cdot 5$ **2.** Subtraction **3.** base, exponent **4.** 9 **5.** False **6.** True

1.1 Exercises *(page 10)*

7. "The sum of 3 and 2 equals 5" is symbolized by $3 + 2 = 5$.

9. "The sum of x and 2 is the product of 3 and 4" is symbolized by $x + 2 = 3 \cdot 4$.

11. "The product of 3 and y is the sum of 1 and 2" is symbolized by $3 \cdot y = 1 + 2$.

13. "The difference x less 2 equals 6" is symbolized by $x - 2 = 6$.

15. "The quotient x divided by 2 is 6" is symbolized by $x/2 = 6$.

17. "Three times a number x is the difference twice x less 2" is symbolized by $3x = 2x - 2$.

19. "The product of x and x equals twice x" is symbolized by $x \cdot x = 2x$.

21. The sum of 2 and x is 6.

23. The product of 2 and x is the sum of 3 and 5.

25. The sum of three times a number x and 4 is 7.

27. The quotient of y divided by 3 less 6 is 4.

29. If $5 = y$, then $y = 5$ by the symmetric property.

31. If $x = y + 2$ and $y + 2 = 10$, then $x = 10$ by the transitive property.

33. If $y = 8 + 2$, then $y = 10$ by the principle of substitution. (We can also say if $y = 8 + 2$ and $8 + 2 = 10$, then $y = 10$ by the transitive property.)

35. If $2x = 10$, then $2x = 2 \cdot 5$ by the principle of substitution.

37. For $2 \cdot 2 \cdot 2$, the number 2 appears as a factor 3 times. Thus, $2 \cdot 2 \cdot 2 = 2^3$.

39. For $3 \cdot 3 \cdot 3 \cdot 3 \cdot 3$, the number 3 appears as a factor 5 times. Thus, $3 \cdot 3 \cdot 3 \cdot 3 \cdot 3 = 3^5$.

41. For $4 \cdot 4 \cdot 4$, the number 4 appears as a factor 3 times. Thus, $4 \cdot 4 \cdot 4 = 4^3$.

43. For $8 \cdot 8$, the number 8 appears as a factor 2 times. Thus, $8 \cdot 8 = 8^2$.

45. For $x \cdot x \cdot x$, the number x appears as a factor 3 times. Thus, $x \cdot x \cdot x = x^3$.

47. For 4^2, the base is 4 and the exponent is 2. The value is $4^2 = 4 \cdot 4 = 16$.

49. For 1^4, the base is 1 and the exponent is 4. The value is $1^4 = 1 \cdot 1 \cdot 1 \cdot 1 = 1$.

51. For -5^2, the base is 5 and the exponent is 2. The value is $-5^2 = -(5 \cdot 5) = -25$.

53. For -3^2, the base is 3 and the exponent is 2. The value is $-3^2 = -(3 \cdot 3) = -9$.

55. $4 + 2 \cdot 5 = 4 + 10$ Multiply before adding
$= 14$

57. $6 \cdot 2 - 1 = 12 - 1$ Multiply before subtracting
$= 11$

59. $2(3 + 4) = 2(7)$ Perform the addition within parentheses
$= 14$ Multiply

61. $5 + 4/2 = 5 + 2$ Divide before adding
$= 7$

63. $3^2 + 4 \cdot 2 = 9 + 4 \cdot 2$ Evaluate exponents $3^2 = 3 \cdot 3 = 9$
$= 9 + 8$ Multiply before adding
$= 17$

65. $8 + 2 \cdot 3^2 = 8 + 2 \cdot (9)$ Evaluate exponents $3^2 = 3 \cdot 3 = 9$
$= 8 + 18$ Multiply before adding
$= 26$

67. $8 + (2 \cdot 3)^2 = 8 + (6)^2$ Perform the multiplication within parentheses
$= 8 + 36$ Evaluate exponents $6^2 = 6 \cdot 6 = 36$
$= 44$ Add

69. $3 \cdot 4 + 2^3 = 3 \cdot 4 + 8$ Evaluate exponents $2^3 = 2 \cdot 2 \cdot 2 = 8$
$= 12 + 8$ Multiply before adding
$= 20$

71. $3 \cdot (4 + 2)^3 = 3 \cdot (6)^3$ Perform the addition within parentheses
$= 3 \cdot (216)$ Evaluate exponents $6^3 = 6 \cdot 6 \cdot 6 = 216$
$= 648$ Multiply

73. $[8 + (4 \cdot 2 + 3)^2] - (10 + 4 \cdot 2)$
$= [8 + (8 + 3)^2] - (10 + 8)$ Begin within innermost parentheses and multiply
$= [8 + (11)^2] - (10 + 8)$ Perform the addition within innermost parentheses
$= [8 + 121] - (10 + 8)$ Evaluate exponents within brackets $11^2 = 11 \cdot 11 = 121$
$= [129] - (18)$ Perform the additions within the grouping symbols
$= 111$ Perform the subtraction

75. $2^3(4 + 1)^2 = 2^3(5)^2$ Perform the addition within parentheses
$= (8) \cdot (25)$ Evaluate exponents $2^3 = 2 \cdot 2 \cdot 2 = 8,\quad 5^2 = 5 \cdot 5 = 25$
$= 200$ Multiply

77. $[(6 \cdot 2 - 3^2) \cdot 2 + 1] \cdot 2$
$= [(6 \cdot 2 - 9) \cdot 2 + 1] \cdot 2$ Evaluate exponents within parentheses $3^2 = 3 \cdot 3 = 9$
$= [(12 - 9) \cdot 2 + 1] \cdot 2$ Perform the multiplication within inner parentheses $6 \cdot 2 = 12$
$= [(3) \cdot 2 + 1] \cdot 2$ Perform the subtraction within inner parentheses $12 - 9 = 3$
$= [6 + 1] \cdot 2$ Perform the multiplication within brackets $3 \cdot 2 = 6$
$= [7] \cdot 2$ Perform the addition within brackets $6 + 1 = 7$
$= 14$ Multiply

79. $1 \cdot 2^2 + 2 \cdot 3^2 + 3 \cdot 4^2$
$= 1 \cdot (4) + 2 \cdot (9) + 3 \cdot (16)$ Evaluate exponents
$2^2 = 2 \cdot 2 = 4,\quad 3^2 = 3 \cdot 3 = 9,\quad 4^2 = 4 \cdot 4 = 16$
$= 4 + 18 + 48$ Multiply $1 \cdot 4 = 4,\quad 2 \cdot 9 = 18,\quad 3 \cdot 16 = 48$
$= 22 + 48$ Perform the addition working left to right
$= 70$

81. $\dfrac{8 + 4}{2 \cdot 3} = \dfrac{12}{6}$ Perform the addition in the numerator and the multiplication in the denominator
$= 2$ Divide

83. $\dfrac{4^2 + 2}{3^3} = \dfrac{16 + 2}{27}$ Evaluate exponents in the numerator and the denominator
$4^2 = 4 \cdot 4 = 16,\quad 3^3 = 3 \cdot 3 \cdot 3 = 27$
$= \dfrac{18}{27}$ Perform the addition in the numerator
$= \dfrac{2}{3}$ Divide

85. $4 + 10/2 - 1 = 4 + 5 - 1$ Perform the division $\dfrac{10}{2} = 5$
$= 9 - 1$ Perform the addition
$= 8$ Perform the subtraction

87. $(4 + 10)/(2 - 1) = (14)/(1)$ Perform the addition and the subtraction within parentheses
$= 14$ Divide

89. $\dfrac{2 \cdot 3}{1} + \dfrac{3 \cdot 4}{2} = \dfrac{6}{1} + \dfrac{12}{2}$ Perform the multiplication in the numerator $2 \cdot 3 = 6, 3 \cdot 4 = 12$
$= 6 + 6$ Perform the division $\dfrac{6}{1} = 6, \dfrac{12}{2} = 6$
$= 12$ Perform the addition

91. $\dfrac{3^2 + 14/2}{2^2 + 4} = \dfrac{9 + 14/2}{4 + 4}$ Evaluate the exponents in the numerator and the denominator $3^2 = 3 \cdot 3 = 9$, $2^2 = 2 \cdot 2 = 4$

$= \dfrac{9 + 7}{4 + 4}$ Perform the division in the numerator $\dfrac{14}{2} = 7$

$= \dfrac{16}{8}$ Perform the addition in the numerator and the denominator

$= 2$ Divide

93. Problem 85 does not have parentheses, whereas problems 86, 87, and 88 each have parentheses.
In Problem 85, we first perform the division, then perform the addition followed by the subtraction to get 8.
In Problem 86, we first perform the addition within parentheses, then perform the division followed by the subtraction to get 6.
In Problem 87, we first perform the addition and the subtraction within parentheses and then perform the division to get 14.
In Problem 88, we perform the subtraction within parentheses, then perform the division followed by the addition to get 14.

1.2 Concepts and Vocabulary *(page 19)*

1. roster method, set-builder notation **2.** True **3.** False **4.** infinite **5.** rational **6.** True **7.** True **8.** False

1.2 Exercises *(page 19)*

For Problems 9–18, we use the sets $A = \{1, 2, 3, 4, 5\}$ *and* $B = \{1, 2, 3\}$.

9. $1 \in A$ is true since 1 is an element of set A. **11.** $4 \in B$ is false since 4 is not an element of the set B. **13.** $A = B$ is false since each set does not contain the same elements. **15.** $B \subseteq A$ is true since every element of set B is also an element of set A. **17.** $\{1\} \subseteq A$ is true since every element of the set $\{1\}$ is also an element of set A. **19.** $\emptyset$, $\{a\}$, $\{b\}$, $\{a, b\}$ **21.** $\{1, 3, 5, 7\}$ is a finite set. **23.** $\{1, 3, 5, 7, \ldots\}$ is an infinite set. **25.** $\{x \mid x \text{ is an even integer}\} = \{\ldots, -6, -4, -2, 0, 2, 4, \ldots\}$ is an infinite set. **27.** $\{x \mid x \text{ is an even digit}\} = \{0, 2, 4, 6, 8\}$ is a finite set.

29. Divide 1 by 5, obtaining

$$\begin{array}{r} 0.2 \\ 5\overline{)1.0} \\ \underline{1.0} \\ 0 \end{array} \qquad \frac{1}{5} = 0.2$$

31. Divide 7 by 3, obtaining

$\dfrac{7}{3} = 2.33\ldots = 2.\overline{3}$ where the 3's repeat

$$\begin{array}{r} 2.333 \\ 3\overline{)7.000} \\ \underline{6} \\ 10 \\ \underline{9} \\ 10 \\ \underline{9} \\ 10 \end{array}$$

33. Divide -5 by 8, obtaining

$$\begin{array}{r} -.625 \\ 8\overline{)-5.000} \\ \underline{4\,8} \\ 20 \\ \underline{16} \\ 40 \\ \underline{40} \\ 0 \end{array} \qquad \frac{-5}{8} = -0.625$$

35. Divide 4 by 25, obtaining

$$\begin{array}{r} .16 \\ 25\overline{)4.00} \\ \underline{2\,5} \\ 1\,50 \\ \underline{1\,50} \\ 0 \end{array} \qquad \frac{4}{25} = 0.16$$

37. Divide -3 by 7, obtaining

$\dfrac{-3}{7} = -0.428571428571\ldots = -0.\overline{428571}$,

where the block 428571 repeats.

$$\begin{array}{r} -.428571 \\ 7\overline{)-3.000000} \\ \underline{2\,8} \\ 20 \\ \underline{14} \\ 60 \\ \underline{56} \\ 40 \\ \underline{35} \\ 50 \\ \underline{49} \\ 10 \\ \underline{7} \\ 30 \end{array}$$

39. **(a)** 2 and 5 are natural numbers
(b) 2, 5, and −6 are integers.
(c) $2, 5, -6, \frac{1}{2}, -1.333\ldots$ are rational numbers.
(d) $\pi + 1$ is the only irrational number.
(e) All the numbers listed are real numbers.

41. **(a)** 1, 2, 3, 4
(b) 0, 1, 2, 3, 4
(c) $0, 1, 2, 3, 4, \frac{1}{2}, \frac{3}{2}, \frac{5}{2}, \frac{7}{2}, \frac{9}{2}$
(d) None are irrational.
(e) $\{x/2 | x \text{ is a digit}\} = C = \left\{\frac{0}{2}, \frac{1}{2}, \frac{2}{2}, \frac{3}{2}, \frac{4}{2}, \frac{5}{2}, \frac{6}{2}, \frac{7}{2}, \frac{8}{2}, \frac{9}{2}\right\}$
$= \left\{0, \frac{1}{2}, 1, \frac{3}{2}, 2, \frac{5}{2}, 3, \frac{7}{2}, 4, \frac{9}{2}\right\}$

43. **(a)** $25 + 50.6 = 75.6$
The average 25-year-old-male can expect to live to be 75.6 years old.
(b) $25 + 55.4 = 80.4$
The average 25-year-old-female can expect to live to be 80.4 years old.
(c) $80.4 - 75.6 = 4.8$
The female can expect to live 4.8 years longer.

45. $\frac{4}{5} \cdot 1000 = \frac{4000}{5} = 800$
A woman has 800 red blood cells in a drop of blood.

47. $\frac{370}{1{,}000{,}000} = 0.00037$ of the atmosphere was carbon dioxide.

1.3 Concepts and Vocabulary *(page 26)*

1. truncate **2.** False **3.** False **4.** 12.596

1.3 Exercises *(page 26)*

5. **(a)** 18.953
(b) 18.952

7. **(a)** 28.653
(b) 28.653

9. **(a)** 0.063
(b) 0.062

11. **(a)** 9.999
(b) 9.998

13. $3/7 = 0.4285714286$
(a) 0.429
(b) 0.428

15. $521/15 = 34.7333\ldots$
(a) 34.733
(b) 34.733

17. $\left(\frac{4}{9}\right)^2 = \frac{16}{81} = 0.1975308642$
(a) 0.198
(b) 0.197

19. $\pi \approx 3.141592654$
(a) 3.142
(b) 3.141

21. $\pi + \frac{3}{2} \approx 4.641592654$
(a) 4.642
(b) 4.641

23. $\pi^3 \approx 31.00627668$
(a) 31.006
(b) 31.006

In Problems 25–52, the TI-83 calculator was used. For other models, the keystroke instructions may vary.

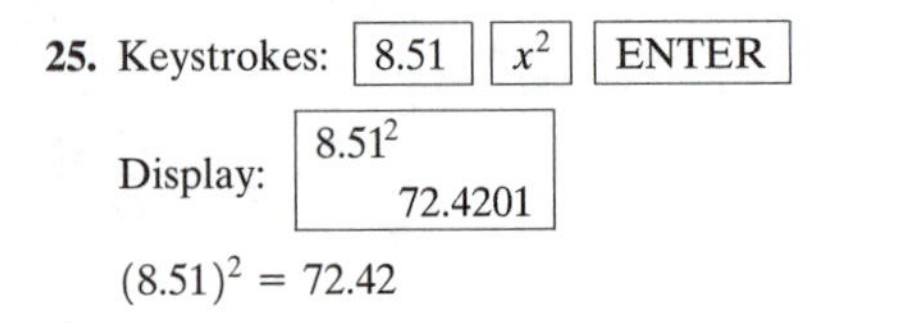

25. Keystrokes: [8.51] [x^2] [ENTER]
Display: 8.51^2 72.4201
$(8.51)^2 = 72.42$

27. Keystrokes: [4.1] [+] [3.2] [×] [8.3] [ENTER]
Display: 4.1 + 3.2*8.3 30.66
$4.1 + (3.2)(8.3) = 30.66$

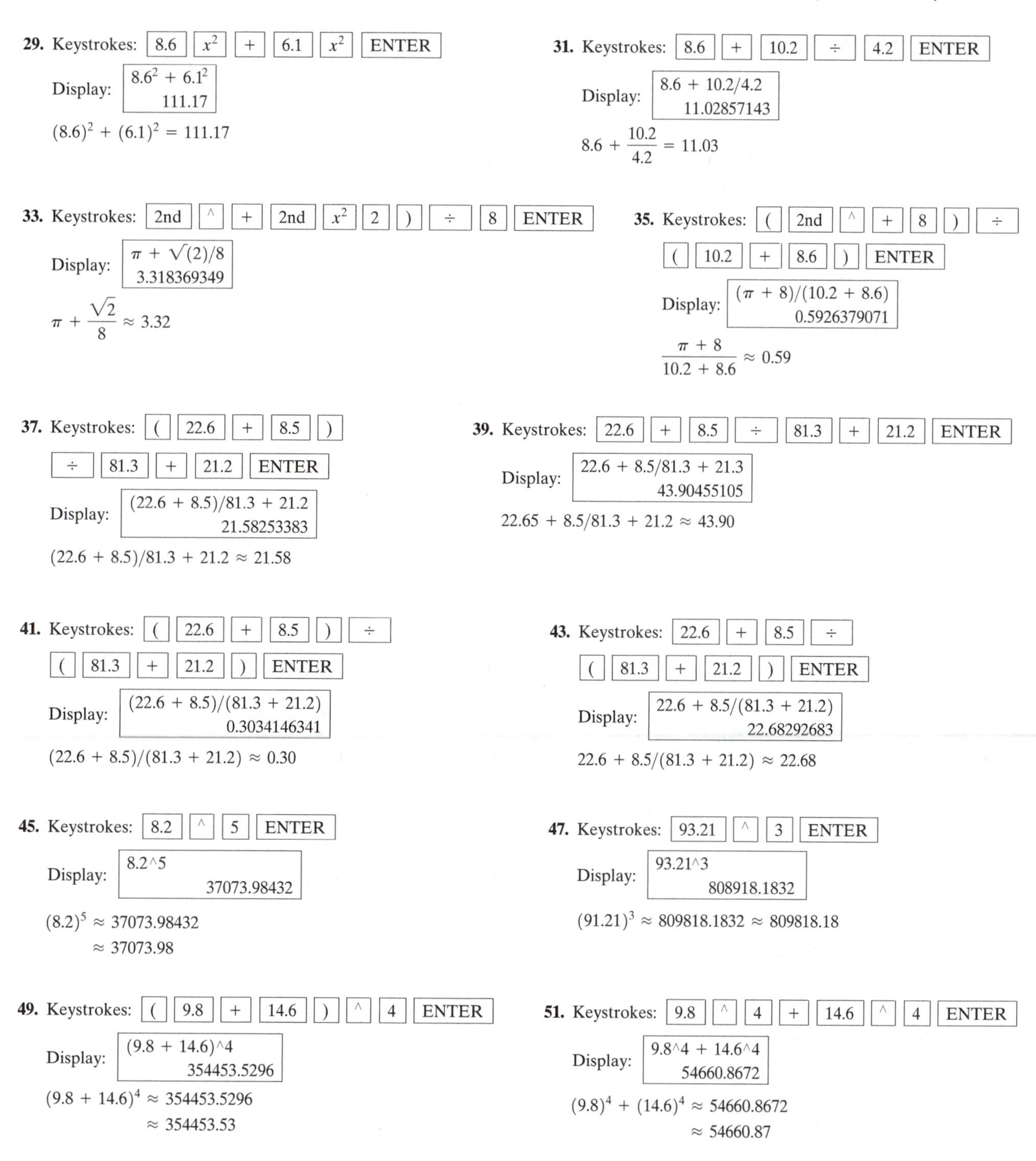

29. Keystrokes: [8.6] [x^2] [+] [6.1] [x^2] [ENTER]

Display: $8.6^2 + 6.1^2$ / 111.17

$(8.6)^2 + (6.1)^2 = 111.17$

31. Keystrokes: [8.6] [+] [10.2] [÷] [4.2] [ENTER]

Display: 8.6 + 10.2/4.2 / 11.02857143

$8.6 + \frac{10.2}{4.2} = 11.03$

33. Keystrokes: [2nd] [^] [+] [2nd] [x^2] [2] [)] [÷] [8] [ENTER]

Display: $\pi + \sqrt{}(2)/8$ / 3.318369349

$\pi + \frac{\sqrt{2}}{8} \approx 3.32$

35. Keystrokes: [(] [2nd] [^] [+] [8] [)] [÷] [(] [10.2] [+] [8.6] [)] [ENTER]

Display: $(\pi + 8)/(10.2 + 8.6)$ / 0.5926379071

$\frac{\pi + 8}{10.2 + 8.6} \approx 0.59$

37. Keystrokes: [(] [22.6] [+] [8.5] [)] [÷] [81.3] [+] [21.2] [ENTER]

Display: (22.6 + 8.5)/81.3 + 21.2 / 21.58253383

$(22.6 + 8.5)/81.3 + 21.2 \approx 21.58$

39. Keystrokes: [22.6] [+] [8.5] [÷] [81.3] [+] [21.2] [ENTER]

Display: 22.6 + 8.5/81.3 + 21.3 / 43.90455105

$22.65 + 8.5/81.3 + 21.2 \approx 43.90$

41. Keystrokes: [(] [22.6] [+] [8.5] [)] [÷] [(] [81.3] [+] [21.2] [)] [ENTER]

Display: (22.6 + 8.5)/(81.3 + 21.2) / 0.3034146341

$(22.6 + 8.5)/(81.3 + 21.2) \approx 0.30$

43. Keystrokes: [22.6] [+] [8.5] [÷] [(] [81.3] [+] [21.2] [)] [ENTER]

Display: 22.6 + 8.5/(81.3 + 21.2) / 22.68292683

$22.6 + 8.5/(81.3 + 21.2) \approx 22.68$

45. Keystrokes: [8.2] [^] [5] [ENTER]

Display: 8.2^5 / 37073.98432

$(8.2)^5 \approx 37073.98432$
≈ 37073.98

47. Keystrokes: [93.21] [^] [3] [ENTER]

Display: 93.21^3 / 808918.1832

$(91.21)^3 \approx 809818.1832 \approx 809818.18$

49. Keystrokes: [(] [9.8] [+] [14.6] [)] [^] [4] [ENTER]

Display: (9.8 + 14.6)^4 / 354453.5296

$(9.8 + 14.6)^4 \approx 354453.5296$
≈ 354453.53

51. Keystrokes: [9.8] [^] [4] [+] [14.6] [^] [4] [ENTER]

Display: 9.8^4 + 14.6^4 / 54660.8672

$(9.8)^4 + (14.6)^4 \approx 54660.8672$
≈ 54660.87

53. ERROR MESSAGE

1.4 Concepts and Vocabulary *(page 33)*

1. Distributive **2.** False. The Multiplication by Zero Property states that the product of any real number and 0 equals 0.
3. True **4.** False

1.4 Exercises *(page 33)*

5. The additive inverse of 3 is -3 because $3 + (-3) = 0$.
The reciprocal of 3 is 1/3 because $3 \cdot \frac{1}{3} = 1$.

7. The additive inverse of -4 is $-(-4) = 4$ because $-4 + 4 = 0$.
The reciprocal of -4 is $1/-4$ because $(-4) \cdot (1/-4) = 1$.

9. The additive inverse of 1/4 is $-1/4$ because $1/4 + (-1/4) = 0$.
The reciprocal of 1/4 is $\frac{4}{1}$ because $\frac{1}{4} \cdot \frac{4}{1} = \frac{1}{4} \cdot 4 = 1$.

11. The additive inverse of 1 is -1 because $1 + (-1) = 0$.
The reciprocal of 1 is $1/1 = 1$ because $1 \cdot (1/1) = 1 \cdot 1 = 1$.

13. $3(x + 4) = 3 \cdot x + 3 \cdot 4$
$= 3x + 12$

15. $x(x + 3) = x \cdot x + x \cdot 3$
$= x^2 + 3x$

17. $4x \cdot (x + 4)$
$= 4x \cdot x + 4x \cdot 4$
$= 4x^2 + 16x.$

19. Commutative Property:
$x + \frac{1}{2} = \frac{1}{2} + x$

21. Associative Property:
$x + (y + 3) = (x + y) + 3$

23. Associative Property:
$3 \cdot (2 \cdot x) = (3 \cdot 2) \cdot x$

25. Distributive Property:
$(2 + a)x = 2x + ax$

27. Distributive Property:
$8x + 5x = (8 + 5)x$

29. Cancellation Property:
If $x + \frac{3}{4} = x + y$,
then $\frac{3}{4} = y$

31. Cancellation Property:
If $3 \cdot y = 3 \cdot 20$,
then $y = 20$

33. Zero Product Property: If $ax = 0$, then either $a = 0$ or $x = 0$

35. Commutative Property

37. Distributive Property

39. Multiplicative Inverse Property

41. Additive Identity Property

43. Additive Inverse Property

45. Associative Property

47. Cancellation Property

49. Associative Property

51. Distributive Property

53. Multiplicative Identity Property

55. Zero Product Property

57. Distributive Property

59.

$3x + 8 = 20$	Given
$3x + 8 = 12 + 8$	Principle of Substitution (Substitute 12 + 8 for 20)
$3x = 12$	Cancellation Property (Cancel the 8's)
$3x = 3 \cdot 4$	Principle of Substitution (Substitute $3 \cdot 4$ for 12)
$x = 4$	Cancellation Property (Cancel the 3's)

61.

$x^2 + 14 = x(x + 2) + 4$	Given
$x^2 + 14 = (x \cdot x + x \cdot 2) + 4$	Distributive Property
$x^2 + 14 = (x^2 + x \cdot 2) + 4$	Change to Exponent Form ($x \cdot x = x^2$)
$x^2 + 14 = x^2 + (x \cdot 2 + 4)$	Associative Property
$14 = x \cdot 2 + 4$	Cancellation Property (Cancel the x^2's)
$10 + 4 = x \cdot 2 + 4$	Principle of Substitution (Substitute 10 + 4 for 14)
$10 = x \cdot 2$	Cancellation Property (Cancel the 4's)
$5 \cdot 2 = x \cdot 2$	Principle of Substitution (Substitute $5 \cdot 2$ for 10)
$5 = x$	Cancellation Property (Cancel the 2's)
$x = 5$	Symmetric Property

63. $(x + 2)(x + 3)$

$= (x + 2)\cdot x + (x + 2)\cdot 3$	Distributive Property
$= x\cdot x + 2\cdot x + x\cdot 3 + 2\cdot 3$	Distributive Property
$= x^2 + 2\cdot x + x\cdot 3 + 2\cdot 3$	Change to Exponent Form ($x\cdot x = x^2$)
$= x^2 + 2\cdot x + 3\cdot x + 2\cdot 3$	Commutative Property ($x\cdot 3 = 3\cdot x$)
$= x^2 + (2 + 3)x + 2\cdot 3$	Distributive Property
$= x^2 + 5x + 6$	Principle of Substitution (Substitute 5 for $2 + 3$, and 6 for $2\cdot 3$)

65. $\frac{a+b}{a+c} = \frac{b}{c}$ is false since if we let $a = 2, b = 8$, and $c = 2$, then $\frac{a+b}{a+c} = \frac{2+8}{2+2} = \frac{10}{4} = \frac{5}{2}$ but $\frac{b}{c} = \frac{8}{2} = 4$

67. $a\cdot a = 2a$ is false since if we let $a = 3$, then $a\cdot a = 3\cdot 3 = 9$ but $2\cdot a = 2\cdot 3 = 6$

1.5 Concepts and Vocabulary *(page 41)*

1. True **2.** negative, positive **3.** False **4.** -7 **5.** $8 + (-13)$ **6.** -11

1.5 Exercises *(page 41)*

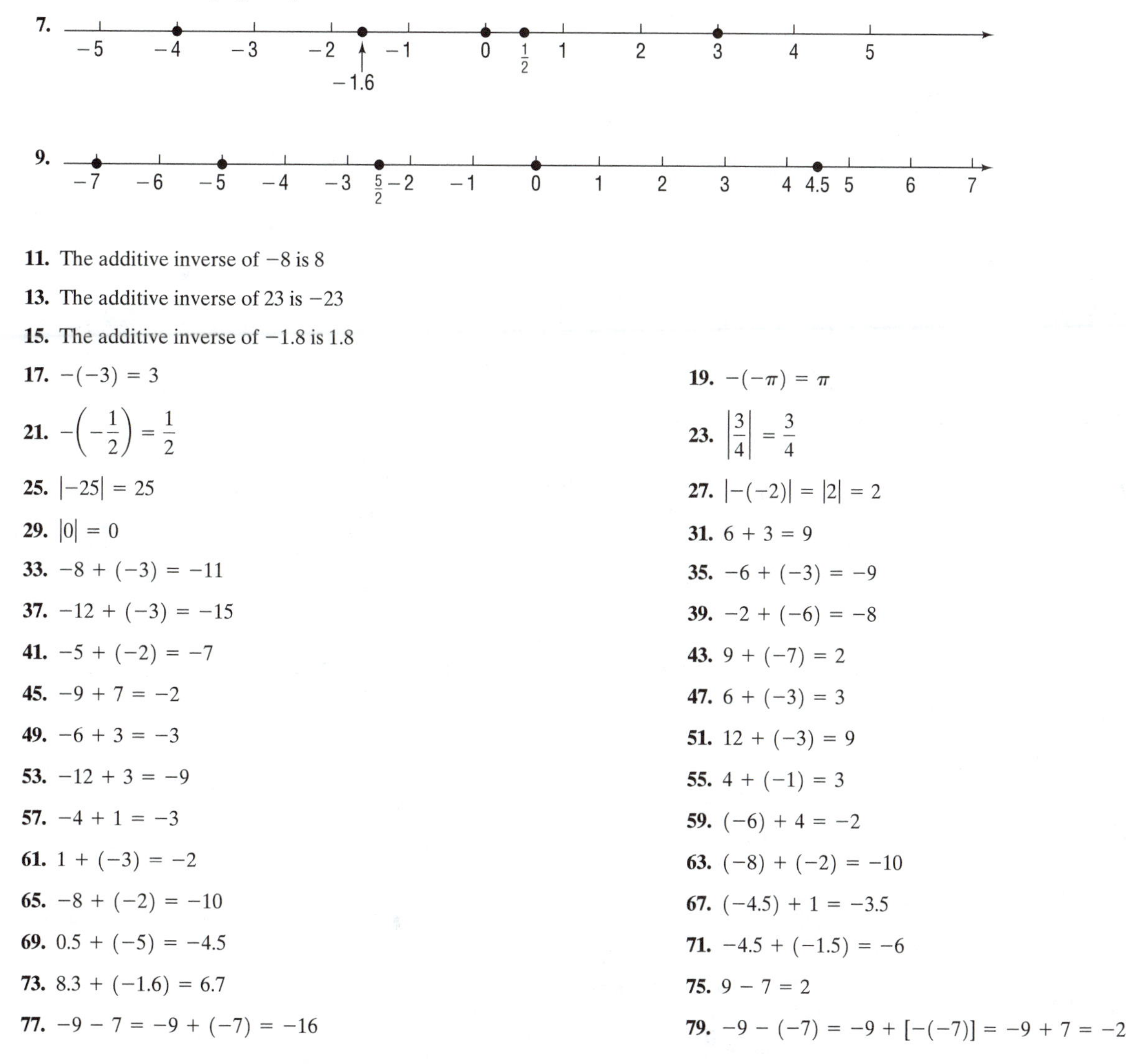

11. The additive inverse of -8 is 8

13. The additive inverse of 23 is -23

15. The additive inverse of -1.8 is 1.8

17. $-(-3) = 3$

19. $-(-\pi) = \pi$

21. $-\left(-\frac{1}{2}\right) = \frac{1}{2}$

23. $\left|\frac{3}{4}\right| = \frac{3}{4}$

25. $|-25| = 25$

27. $|-(-2)| = |2| = 2$

29. $|0| = 0$

31. $6 + 3 = 9$

33. $-8 + (-3) = -11$

35. $-6 + (-3) = -9$

37. $-12 + (-3) = -15$

39. $-2 + (-6) = -8$

41. $-5 + (-2) = -7$

43. $9 + (-7) = 2$

45. $-9 + 7 = -2$

47. $6 + (-3) = 3$

49. $-6 + 3 = -3$

51. $12 + (-3) = 9$

53. $-12 + 3 = -9$

55. $4 + (-1) = 3$

57. $-4 + 1 = -3$

59. $(-6) + 4 = -2$

61. $1 + (-3) = -2$

63. $(-8) + (-2) = -10$

65. $-8 + (-2) = -10$

67. $(-4.5) + 1 = -3.5$

69. $0.5 + (-5) = -4.5$

71. $-4.5 + (-1.5) = -6$

73. $8.3 + (-1.6) = 6.7$

75. $9 - 7 = 2$

77. $-9 - 7 = -9 + (-7) = -16$

79. $-9 - (-7) = -9 + [-(-7)] = -9 + 7 = -2$

81. $9 - (-7) = 9 + [-(-7)] = 9 + 7 = 16$

83. $10 - 17 = 10 + (-17) = -7$

85. $-10 - 17 = -10 + (-17) = -27$

87. $-10 - (-17) = -10 + [-(-17)]$
$= -10 + 17 = 7$

89. $10 - (-17) = 10 + [-(-17)]$
$= 10 + 17 = 27$

91. $4 - 7 = 4 + (-7) = -3$

93. $-4 - 7 = -4 + (-7) = -11$

95. $-4 - (-7) = -4 + [-(-7)] = -4 + 7 = 3$

97. $4 - (-7) = 4 + [-(-7)] = 4 + 7 = 11$

99. $14 - 8 = 6$

101. $-14 - 8 = -14 + (-8) = -22$

103. $14 - (-8) = 14 + [-(-8)] = 14 + 8 = 22$

105. $-14 - (-8) = -14 + [-(-8)]$
$= -14 + 8 = -6$

107. $14 - 8 + 4 = 6 + 4 = 10$

109. $14 - 8 - 4 = 6 - 4 = 2$

111. $-8 - (4 - 8) - 1 = -8 - [4 + (-8)] - 1$
$= -8 - (-4) - 1 = -8 + [-(-4)] - 1$
$= -8 + 4 - 1 = -4 + (-1) = -5$

113. $4 - (3 - 6) - (-2 - 6)$
$= 4 - [3 + (-6)] - [-2 + (-6)]$
$= 4 - (-3) - (-8)$
$= 4 + [-(-3)] + [-(-8)]$
$= 4 + 3 + 8 = 7 + 8 = 15$

1.6 Concepts and Vocabulary *(page 46)*

5. False **6.** negative **7.** -32 **8.** False

1.6 Exercises *(page 46)*

9. $5(3) = (5 \cdot 3) = 15$

11. $(-5)(-3) = (5 \cdot 3) = 15$

13. $(5)(-3) = -(5 \cdot 3) = -15$

15. $(-5)(3) = -(5 \cdot 3) = -15$

17. $9(7) = (9 \cdot 7) = 63$

19. $(-9)(-7) = (9 \cdot 7) = 63$

21. $(9)(-7) = -(9 \cdot 7) = -63$

23. $(-9)(7) = -(9 \cdot 7) = -63$

25. $(-2)^6 = (-2)(-2)(-2)(-2)(-2)(-2)$
$= (2 \cdot 2)(-2)(-2)(-2)(-2)$
$= 4(-2)(-2)(-2)(-2)$
$= -(4 \cdot 2)(-2)(-2)(-2)$
$= -8(-2)(-2)(-2)$
$= (8 \cdot 2)(-2)(-2)$
$= 16(-2)(-2)$
$= -(16 \cdot 2)(-2)$
$= -(32)(-2)$
$= (32 \cdot 2) = 64$

27. $-2^6 = -(2 \cdot 2 \cdot 2 \cdot 2 \cdot 2 \cdot 2)$
$= -[(2 \cdot 2) \cdot 2 \cdot 2 \cdot 2 \cdot 2]$
$= -[4 \cdot 2 \cdot 2 \cdot 2 \cdot 2]$
$= -[(4 \cdot 2) \cdot 2 \cdot 2 \cdot 2]$
$= -[8 \cdot 2 \cdot 2 \cdot 2]$
$= -[(8 \cdot 2) \cdot 2 \cdot 2]$
$= -[16 \cdot 2 \cdot 2]$
$= -[(16 \cdot 2) \cdot 2]$
$= -(32 \cdot 2)$
$= -64$

29. $-2 \cdot (3 + 4) + 2^3 = -2 \cdot (7) + 2^3$ — Perform addition within parentheses $3 + 4 = 7$
$= -2 \cdot (7) + 8$ — Evaluate exponents $2^3 = 2 \cdot 2 \cdot 2 = 8$
$= -14 + 8$ — Multiply $-2(7) = -(2 \cdot 7) = -14$
$= -6$ — Add

31. $4 \cdot (3 - 5) - 6 = 4 \cdot [3 + (-5)] - 6$ — Change subtraction to addition in parentheses
$= 4(-2) - 6$ — Perform addition within brackets $3 + (-5) = -2$
$= -8 - 6$ — Multiply $4(-2) = -(4 \cdot 2) = -8$
$= -8 + (-6)$ — Change subtraction to addition
$= -14$ — Perform addition

33. $-4 \cdot (-5) \cdot (-2) = 20 \cdot (-2)$ — Multiply $-4 \cdot (-5) = (4 \cdot 5) = 20$
$= -(20 \cdot 2) = -40$

35. $-5 \cdot [-7 + (-3)] = -5 \cdot (-10)$ — Perform addition within brackets $-7 + (-3) = -10$
$= 50$ — Multiply $-5 \cdot (-10) = (5 \cdot 10) = 50$

37. $[9 - (-12)] \cdot (-3) = [9 + (-(-12)] \cdot (-3)$ — Change subtraction within brackets to addition $-(-12) = 12$
$= (9 + 12) \cdot (-3)$
$= (21) \cdot (-3)$ — Perform addition in parentheses $9 + 12 = 21$
$= -63$ — Multiply $(21) \cdot (-3) = -(21 \cdot 3) = -63$

39. $(-25 - 116) \cdot 0 = [-25 + (-116)] \cdot 0$ — Change subtraction within parentheses to addition
$= (-141) \cdot 0$ — Perform addition in brackets $-25 + (-116) = -141$
$= 0$ — Multiply $(-141) \cdot 0 = -(141 \cdot 0) = 0$

41. $-15 - 3 \cdot (-4) = -15 - [3 \cdot (-4)]$ — Perform multiplication first $3 \cdot (-4) = -(3 \cdot 4) - 12$
$= -15 - (-12)$
$= -15 + [-(-12)]$ — Change subtraction to addition $-(-12) = 12$
$= -15 + 12$
$= -3$ — Add

43. $1 - 2 \cdot (-4) - 6 = 1 - (-8) - 6$ — Multiply $2 \cdot (-4) = -(2 \cdot 4) = -8$
$= 1 + [-(-8)] - 6$ — Change subtraction to addition $-(-8) = 8$
$= 1 + 8 - 6$
$= 9 - 6$ — Perform addition and subtraction
$= 3$

45. $-4^2 \cdot (4 - 8) - (6 - 8)$
$= -4^2 \cdot [4 + (-8)] - [6 + (-8)]$ — Change subtraction in parentheses to addition
$= -4^2 \cdot (-4) - (-2)$ — Perform addition within brackets $4 + (-8) = -4$; $6 + (-8) = -2$
$= -16 \cdot (-4) - (-2)$ — Evaluate exponents $-4^2 = -(4) \cdot (4) = -16$
$= 64 - (-2)$ — Multiply $-16 \cdot (-4) = (-16) \cdot (-4) = 64$
$= 64 + [-(-2)]$ — Change subtraction to addition $-(-2) = 2$
$= 64 + 2$
$= 66$ — Add

47. $(-4)^2 \cdot [4 - 8 - (6 - 8)]$
$= (-4)^2 \cdot [4 - 8 - (6 + (-8))]$ — Change subtraction in inner parentheses to addition
$= (-4)^2 \cdot [4 - 8 - (-2)]$ — Perform addition in inner parentheses $6 + (-8) = -2$
$= (-4)^2 \cdot [4 + (-8) + (-(-2))]$ — Change subtraction in brackets to addition
$= (-4)^2 \cdot [-4 + (-(-2))]$ — Perform addition $4 + (-8) = -4$
$= (-4)^2 \cdot (-4 + 2)$ — $-(-2) = 2$
$= (-4)^2 \cdot (-2)$ — Perform addition in parentheses $-4 + 2 = -2$
$= 16 \cdot (-2)$ — Evaluate exponents $(-4)^2 = (-4) \cdot (-4) = 16$
$= -32$ — Multiply $(16) \cdot (-2) = -(16 \cdot 2) = -32$

49. $1 - [1 - 3^2 + (-4)^2]$

$= 1 - [1 - 9 + 16]$ — Evaluate exponents within brackets $3^2 = 3 \cdot 3 = 9, (-4)^2 = (-4) \cdot (-4) = 16$

$= 1 - [1 + (-9) + 16]$ — Change subtraction in brackets to addition

$= 1 - (-8 + 16)$ — Perform addition within parentheses $1 + (-9) = -8$

$= 1 - (8)$ — $-8 + 16 = 8$

$= 1 + (-8)$ — Change subtraction to addition

$= -7$ — Perform addition

51. $5 \cdot (-2^3) - (-3^2)$

$= 5 \cdot (-8) - (-9)$ — Evaluate exponents $(-2^3) = -(2) \cdot (2) \cdot (2) = -8$, $(-3^2) = -(3) \cdot (3) = -(3 \cdot 3) = -9$

$= -40 - (-9)$ — Multiply $5 \cdot (-8) = -(5 \cdot 8) = -40$

$= -40 + [-(-9)]$ — Change subtraction to addition $-(-9) = 9$

$= -40 + 9$ — Perform addition

$= -31$

53. $\frac{-56}{-8} = \frac{56}{8} = 7$

55. $\frac{-12}{-4} = \frac{12}{4} = 3$

57. $\frac{35}{-7} = -\frac{35}{7} = -5$

59. $\frac{55}{-5} = -\frac{55}{5} = -11$

61. $\frac{-81}{9} = -\frac{81}{9} = -9$

63. $\frac{-6}{1} = -\frac{6}{1} = -6$

65. $\frac{14}{-1} = -\frac{14}{1} = -14$

67. $\frac{-4 - 8}{6 - 10} = \frac{-4 + (-8)}{6 + (-10)}$ — Change subtractions to additions

$= \frac{-12}{-4}$ — Perform addition in the numerator and the denominator

$= \frac{12}{4} = 3$ — Perform division $\frac{-12}{-4} = \frac{12}{4} = 3$

69. $\frac{2 - 3 + (-4)}{-2 - 3 - 4} = \frac{2 + (-3) + (-4)}{-2 + (-3) + (-4)}$ — Change all subtractions to additions

$= \frac{-5}{-9}$ — Perform addition in the numerator and the denominator working left to right $2 + (-3) + (-4) = -1 + (-4) = -5$, $-2 + (-3) + (-4) = -5 + (-4) = -9$

$= \frac{5}{9}$ — Divide $\frac{-5}{-9} = \frac{5}{9}$

71. $(-2 - 8)/(-2 + 3) = [-2 + (-8)]/(-2 + 3)$ — Change subtraction in parentheses to addition

$= -10/1$ — Perform addition within brackets and parentheses $-2 + (-8) = -10, -2 + 3 = 1$

$= -10$ — Divide $\frac{-10}{1} = -\frac{10}{1} = -10$

73. $-2 - 8/(-2) + 3 = -2 - (-4) + 3$ — Perform division first $8/(-2) = -\frac{8}{2} = -4$

$= -2 + [-(-4)] + 3$ — Change subtraction to addition $-(-4) = 4$

$= -2 + 4 + 3$

$= 5$ — Perform addition working left to right $-2 + 4 + 3 = 2 + 3 = 5$

75. The answers are different because Problem 73 does not have parentheses enclosing the numerator and the denominator.

Problems 77–84 use the formula Sharpe Ratio $= \frac{ER - RFR}{SD}$

77. $ER = .05, \quad RFR = .10, \quad SD = .01$

Sharpe Ratio $= \frac{.05 - .10}{.01} = \frac{-.05}{.01} = -5$

79. $ER = .20, \quad RFR = .05, \quad SD = .01$

Sharpe Ratio $= \frac{.20 - .05}{.01} = \frac{.15}{.01} = 15$

81. $ER = .10,\quad RFR = .06,\quad SD = .015$

$$\text{Sharpe Ratio} = \frac{.10 - .06}{.015} = \frac{.04}{.015} = 2.66\overline{6}$$

83. $ER = .05,\quad RFR = .10,\quad SD = .015$

$$\text{Sharpe Ratio} = \frac{.05 - .10}{.015} = \frac{-.05}{.015} = -3.33\overline{3}$$

85.

Quarter	Earnings
1st	+$1.20
2nd	−$0.75
3rd	−$0.30
4th	+$0.20

The gains for the year were $1.20 + $0.20 = $1.40
The losses for the year were $0.75 + $0.30 = $1.05
The annual earnings are the gains minus the losses, $1.40 − $1.05 = $0.35 per share.

87. $1 + 2 + 3 + 4 + \cdots + 99$
$= (1 + 99) + (2 + 98) + (3 + 97) + \cdots + (49 + 51) + 50$
$= \underbrace{100 + 100 + 100 + \cdots + 100}_{49\ times} + 50$
$= 49 \cdot 100 + 50 = 4900 + 50 = 4950$

89. $-2 - 4 - 6 - 8 - \cdots - 98$
$= (-2 - 98) + (-4 - 96) + (-6 - 94) + \cdots + (-48 - 52) - 50$
$= \underbrace{(-100) + (-100) + (-100) + \cdots + (-100)}_{24\ times} + (-50)$
$= 24 \cdot (-100) - 50 = -2400 - 50 = -2450$

91.

$a \cdot (b - c) = a \cdot [b + (-c)]$	Equation (3) Section 1.5
$= a \cdot b + a \cdot (-c)$	Distributive Property
$= a \cdot b + [-(a \cdot c)]$	Equation (1) Section 1.6
$= a \cdot b - a \cdot c$	Equation (3) Section 1.5

We have proved that multiplication distributes over subtraction.

1.7 Concepts and Vocabulary *(page 51)*

1. strict **2.** True **3.** True **4.** False

1.7 Exercises *(page 51)*

5. $\frac{1}{2} > 0$ **7.** $-1 > -2$ **9.** $\pi > 3.14$ **11.** $\frac{1}{2} = 0.5$

13. $\frac{2}{3} < 0.67$ **15.** $x > 2$ **17.** $x > 0$ **19.** $x < 2$

21. $x \le 1$ **23.** $x \le -3$ **25.** $|\pi| = \pi$ **27.** $\left|4 - \frac{8}{2}\right| = |4 - 4| = |0| = 0$

29. $|6 + 3(-4)| = |6 + (-12)|$
$= |-6| = 6$

31. $\left|\frac{8}{2} + \frac{2}{-1}\right| = |4 + (-2)|$
$= |2| = 2$

33. $|9 \cdot (-3)/5| = |-(9 \cdot 3)/5| = |-27/5| = 27/5$

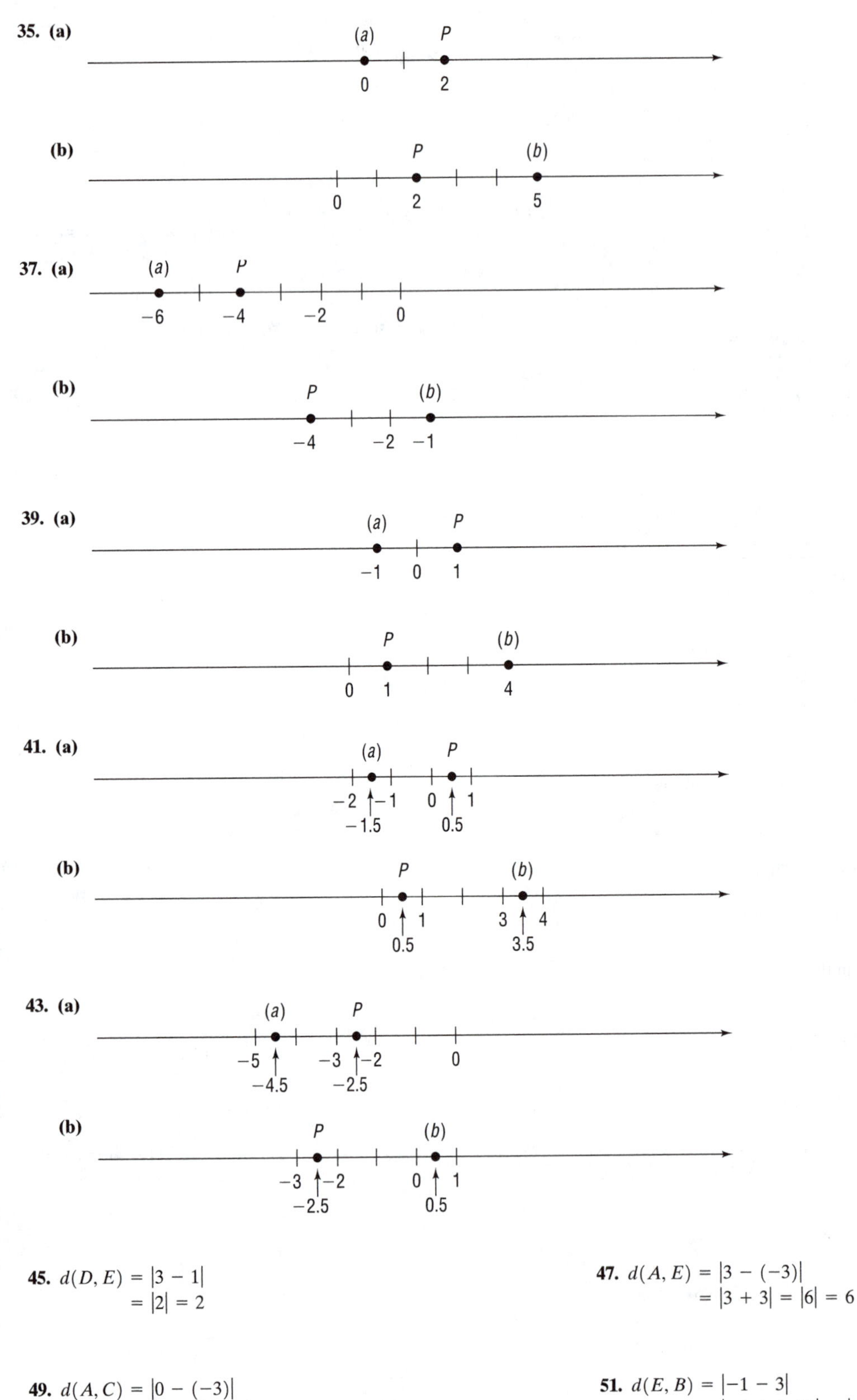

45. $d(D, E) = |3 - 1|$
$= |2| = 2$

47. $d(A, E) = |3 - (-3)|$
$= |3 + 3| = |6| = 6$

49. $d(A, C) = |0 - (-3)|$
$= |0 + 3| = |3| = 3$

51. $d(E, B) = |-1 - 3|$
$= |-1 + (-3)| = |-4| = 4$

53. $d(A, D) + d(B, E) = |1 - (-3)| + |3 - (-1)|$
$= |1 + 3| + |3 + 1|$
$= |4| + |4|$
$= 4 + 4 = 8$

1.8 Concepts and Vocabulary *(page 55)*

1. variable **2.** formula **3.** True **4.** domain **5.** -2 **6.** -2

1.8 Exercises *(page 55)*

7. $x + 2y = (-2) + 2(3) = -2 + (2\cdot 3) = -2 + 6 = 4$

9. $5xy + 2 = 5(-2)(3) + 2 = -(10\cdot 3) + 2 = -30 + 2 = -28$

11. $\frac{2x}{x - y} = \frac{2(-2)}{(-2) - 3} = \frac{-(2\cdot 2)}{-2 + (-3)} = \frac{-4}{-5} = \frac{4}{5}$

13. $\frac{3x + 2y}{2 + y} = \frac{3(-2) + 2(3)}{2 + 3} = \frac{-(3\cdot 2) + (2\cdot 3)}{2 + 3} = \frac{-6 + 6}{5} = \frac{0}{5} = 0$

15. $|x + y| = |4 + (-3)| = |1| = 1$

17. $|x| + |y| = |4| + |(-3)| = 4 + 3 = 7$

19. $\frac{|x|}{x} = \frac{|4|}{4} = \frac{4}{4} = 1$

21. $|4x - 5y| = |4(4) - 5(-3)| = |16 - (-15)| = |16 + 15| = |31| = 31$

23. $||4x| - |5y|| = ||4(4)| - |5(-3)|| = ||16| - |-15|| = |16 - 15| = |1| = 1$

25. The domain of the variable x in the expression $\frac{5}{x - 3}$ is $\{x|x \neq 3\}$, since if $x = 3$, then the denominator becomes zero, which is not allowed.

27. The domain of the variable x in the expression $\frac{x}{x + 4}$ is $\{x|x \neq -4\}$, since if $x = -4$, then the denominator becomes zero, which is not allowed.

29. The domain of the variable x in the expression $\frac{1}{x} + \frac{1}{x - 1}$ is $\{x|x \neq 0, x \neq 1\}$, since if $x = 0$, the denominator becomes zero in the first term and if $x = 1$, the denominator becomes zero in the second term.

31. The domain of the variable x in the expression $\frac{x + 1}{x(x + 2)}$ is $\{x|x \neq 0, x \neq -2\}$, since if $x = 0$ or $x = -2$, the denominator becomes zero, which is not allowed.

33. In the formula for the area A of a square with side of length x, $A = x^2$, the domain of the variable x is the set of positive real numbers. The domain of the variable A is also the set of positive real numbers.

35. $A = lw$. The domain of the variables A, l, and w are the set of positive real numbers.

37. **(a)** $C = \pi d$. The domain of the variables C and d are the set of positive real numbers.

(b) $C = 2\pi r$. The domain of the variables C and r are the set of positive real numbers.

39. $A = \frac{\sqrt{3}}{4}x^2$. The domain of the variables A and x are the set of positive real numbers.

41. $V = \frac{4}{3}\pi r^3$. The domain of the variables V and r are the set of positive real numbers.

43. $V = x^3$. The domain of the variables V and x^3 are the set of positive real numbers.

In Problems 45–48 use the formulas $A = x^2$ for the area A and $P = 4x$ for the perimeter P.

45. $x = 2$ inches, then

$A = x^2 = (2)^2 = 4$ square inches

$P = 4x = 4(2) = 8$ inches

47. $x = 3$ meters, then

$A = x^2 = (3)^2 = 9$ square meters

$P = 4x = 4(3) = 12$ meters

In Problems 49–52 use the formulas $V = \frac{4}{3}\pi r^3$ for the volume of a sphere and $S = 4\pi r^2$ for the surface area of a sphere.

49. $r = 2$ meters, then

$V = \frac{4}{3}\pi(2)^3 = \frac{4}{3}(8)\pi = \frac{32}{3}\pi \approx 33.51$ cubic meters

$S = 4\pi(2)^2 = 4 \cdot 4\pi = 16\pi \approx 50.27$ square meters

51. $r = 4$ inches, then

$V = \frac{4}{3}\pi(4)^3 = \frac{4}{3}(64)\pi = \frac{256}{3}\pi \approx 268.08$ cubic inches

$S = 4\pi(4)^2 = 4(16)\pi = 64\pi \approx 201.06$ square inches

In Problems 53–56 use the formula $C = \frac{5}{9}(F - 32)$

53. $F = 32°$, then

$C = \frac{5}{9}(32 - 32) = \frac{5}{9}(0) = 0°$

55. $F = 68°$, then

$C = \frac{5}{9}(68 - 32) = \frac{5}{9}(36) = 5 \cdot 4 = 20°$

57. The distance s an object will fall after t seconds is given by the formula $s = \frac{1}{2}gt^2$, where g is approximately 32 ft/sec^2.

After 1 second, $s = \frac{1}{2}(32)(1)^2 = 16 \cdot 1 = 16$ feet

After 2 seconds, $s = \frac{1}{2}(32)(2)^2 = 16 \cdot 4 = 64$ feet

In Problems 59–64 we use the formula $h = vt - \frac{1}{2}(32)t^2$, where h is the height of the object after t seconds and v is the initial velocity.

59. If $v = 50$ ft/sec and $t = 1$ second, then

$$h = 50(1) - \frac{1}{2}(32)(1)^2$$
$$= 50 - 16$$
$$= 34 \text{ feet}$$

61. If $v = 50$ ft/sec and $t = 3$ seconds, then

$$h = 50(3) - \frac{1}{2}(32)(3)^2$$
$$= 150 - 16 \cdot 9$$
$$= 150 - 144$$
$$= 6 \text{ feet}$$

63. If $v = 80$ ft/sec and $t = 2$ seconds, then

$$h = 80(2) - \frac{1}{2}(32)(2)^2$$
$$= 160 - 16 \cdot 4$$
$$= 160 - 64$$
$$= 96 \text{ feet}$$

65. (a) If the weekly cost C of manufacturing x watches is given by the formula $C = 4000 + 2x$, then the domain of the variable x is the set of nonnegative integers.

(b) The domain of the variable C is the set of even integers greater than or equal to 4000.

(c) If $x = 1000$, then

$$C = 4000 + 2(1000) = 4000 + 2000 = \$6000$$

(d) If $x = 2000$, then

$$C = 4000 + 2(2000) = 4000 + 4000 = \$8000$$

Chapter 1 Review Exercises *(page 59)*

1. $x + 2 = 5$

3. $3 - x = y + 2$

5. $4x = 2 - 3x$

7. $(2^3 + 5)\cdot 2 = (8 + 5)\cdot 2 = (13)\cdot 2 = 26$

9. $7 + (3\cdot 4 + 1)^2 + 3^2\cdot 4$
$= 7 + (12 + 1)^2 + 9\cdot 4$
$= 7 + (13)^2 + 9\cdot 4$
$= 7 + (13)\cdot(13) + 9\cdot 4$
$= 7 + 169 + 36 = 212$

11. $\dfrac{3 + 5}{3 + 1} = \dfrac{8}{4} = 2$

Problems 13–20 use the sets $A = \{1, 3, 7, 9\}$, $B = \{1, 2, 3, 4\}$

13. $3 \in A$ is true since 3 is an element of the set A.

15. $4 \notin A$ is true since 4 is not an element of the set A.

17. $A \subseteq B$ is false since there are elements of set A which are not also elements of set B.

19. "A is a finite set" is true since set A contains a finite number of elements.

21. (a) 8 is the only natural number.
(b) 8 and 0 are integers.
(c) $8, 0, \frac{1}{2}, 1.25, 8.333\ldots$ are rational numbers.
(d) $-\sqrt{2}$ and $\frac{\pi}{2}$ are irrational numbers.
(e) All the numbers listed are real numbers.

23. 0.375

25. $0.\overline{63}$ where 63 repeats

27. Distributive Property

29. Principle of Substitution

31. Multiplicative Inverse Property

33. Cancellation Property

35. Additive Identity Property

37. $3 - (8 - 10) - (-3 - 4)$
$= 3 - (-2) - (-7)$
$= 3 + 2 + 7$
$= 5 + 7 = 12$

39. $-3^2\cdot(-2)^3 = (-9)\cdot(-8) = 72$

41. $0^2 + (-5 - 2) = 0 + (-7) = -7$

43. $-4 - [2 - (-2)^3 + 3^2]$
$= -4 - [2 - (-8) + 9]$
$= -4 - [2 + 8 + 9]$
$= -4 - (10 + 9)$
$= -4 - 19 = -23$

45. $1 - (-6 + 2 - 1) - (1 - 2^3)$
$= 1 - (-4 - 1) - (1 - 8)$
$= 1 - (-5) - (-7)$
$= 1 + 5 + 7$
$= 6 + 7 = 13$

47. $(-2)^2(3 - 4^2)$
$= 4 \cdot (3 - 16)$
$= 4(-13) = -52$

49. $-2 - (-3) - (-4)$
$= -2 + 3 + 4$
$= 1 + 4 = 5$

51. $1 - 2 - 3 - 4 = -1 - 3 - 4 = -4 - 4 = -8$

53. $(1^2 - 2^2 + 3^2) - [(-1)^2 - (-2)^2 - (-3)^2]$
$= (1 - 4 + 9) - [1 - 4 - 9]$
$= (-3 + 9) - [-3 - 9]$
$= (6) - [-12]$
$= 6 + 12 = 18$

55. $|-2 - 3^2| = |-2 - 9|$
$= |-11| = 11$

57. $3^2 - |-4| = 9 - |-4|$
$= 9 - 4 = 5$

59. $|3 - 4| - |4 - 3|$
$|-1| - |1| = 1 - 1 = 0$

61. $|-2| - |-3| = 2 - 3 = -1$

63. $1 - |2^2 - (-2)^3| = 1 - |4 - (-8)| = 1 - |4 + 8|$
$= 1 - |12| = 1 - 12 = -11$

65. $\dfrac{-5 - 13}{1 - 5} = \dfrac{-18}{-4} = \dfrac{9}{2}$

In Problems 67–74 we are given $x = 3$ and $y = -5$.

67. $2x + 3y = 2 \cdot (3) + 3 \cdot (-5)$
$= 6 + (-15) = -9$

69. $|x - y| = |3 - (-5)|$
$= |3 + 5| = |8| = 8$

71. $xy + x = (3)(-5) + 3$
$= -15 + 3 = -12$

73. $y^2 - x^2 = (-5)^2 - (3)^2$
$= 25 - 9 = 16$

75. 183.44

77. $\pi + 3.213 \approx 6.354592654 \approx 6.35$

79. $-5 > -8$

81. $-6 < \dfrac{11}{2}$

83. $d(C, B) = |0 - (-1)| = |0 + 1| = |1| = 1$

85. $d(D, E) = |5 - 2| = |3| = 3$

87. The domain of the variable x in $\dfrac{3 + x}{3 - x}$ is $\{x | x \neq 3\}$ since if $x = 3$, the denominator becomes zero.

89. The domain of the variable x in $\dfrac{1}{x^2}$ is $\{x | x \neq 0\}$ since if $x = 0$, the denominator becomes zero.

91. The domain of the variable x in $\dfrac{x + 7}{x - 5}$ is $\{x | x \neq 5\}$ since if $x = 5$, the denominator becomes zero.

93. Using the formula

$$h = 144t - \left(\frac{1}{2}\right)(32)t^2$$

(a) After 1 second,

$$h = 144(1) - \left(\frac{1}{2}\right)(32)(1)^2$$
$$= 144 - 16$$
$$= 128 \text{ feet}$$

(b) After 4 seconds,

$$h = 144(4) - \left(\frac{1}{2}\right)(32)(4)^2$$
$$= 576 - 16 \cdot 16$$
$$= 576 - 256 = 320 \text{ feet}$$

(c) After 9 seconds,

$$h = 144(9) - \left(\frac{1}{2}\right)(32)(9)^2$$
$$= 1296 - 16 \cdot 81$$
$$= 1296 - 1296 = 0 \text{ feet}$$

(d) After 4.5 seconds,

$$h = 144(4.5) - \left(\frac{1}{2}\right)(32)(4.5)^2$$
$$= 648 - 16 \cdot (20.25)$$
$$= 648 - 324 = 324 \text{ feet}$$

95. Dan's earnings: $(6.25)(30) = \$187.50$
Deductions: $18 + 6.50 + 11.63 + 2.73 + 10 = \$48.86$
Dan's paycheck: $\$187.50 - \$48.86 = \$138.64$

97.

$2x - 8 = 10$	Given
$2x + (-8) = 10$	Equation (3), Section 1.5
$2x + (-8) = 18 + (-8)$	Principle of Substitution (Substitute $18 + (-8)$ for 10)
$2x = 18$	Cancellation Property (Cancel the -8's)
$2x = 2 \cdot 9$	Principle of Substitution (Substitute $2 \cdot 9$ for 18)
$x = 9$	Cancellation Property (Cancel the 2's)

CHAPTER 2 Polynomials

2.1 Concepts and Vocabulary *(page 69)*

3. 1 **4.** −1 **5.** True **6.** False **7.** True **8.** True

2.1 Exercises *(page 69)*

9. $2^3 \cdot 2^2 = 2^{3+2} = 2^5 = 32$

11. $5^2 \cdot 5^4 = 5^{2+4} = 5^6$

13. $(-2)^3 \cdot (-2) = (-2)^{3+1} = (-2)^4 = 16$

15. $(-2)^2 \cdot (-2)^3 = (-2)^{2+3} = (-2)^5 = -32$

17. $2^4 \cdot 2^0 = 2^4 \cdot 1 = 2^4 = 16$ or $2^4 \cdot 2^0 = 2^{4+0} = 2^4 = 16$

19. $-2^4 \cdot 2 = -1 \cdot 2^4 \cdot 2 = -1 \cdot 2^{4+1} = -1 \cdot 2^5 = -1 \cdot 32 = -32$

21. $x^3 \cdot x^9 = x^{3+9} = x^{12}$

23. $y \cdot y^3 = y^{1+3} = y^4$

25. $(3^4)^5 = 3^{4 \cdot 5} = 3^{20}$

27. $(2^3)^2 = 2^{3 \cdot 2} = 2^6 = 64$

29. $(5^2)^1 = 5^{2 \cdot 1} = 5^2 = 25$

31. $[(-6)^8]^2 = (-6)^{8 \cdot 2} = (-6)^{16}$

33. $-(6^8)^2 = -1 \cdot 6^{8 \cdot 2} = -1 \cdot 6^{16} = -6^{16}$

35. $(8^2)^0 = 8^{2 \cdot 0} = 8^0 = 1$

37. $(x^3)^9 = x^{3 \cdot 9} = x^{27}$

39. $(z^5)^0 = z^{5 \cdot 0} = z^0 = 1$

41. $(2x)^5 = 2^5 \cdot x^5 = 32x^5$

43. $(-3x)^4 = (-3)^4 \cdot x^4 = 81x^4$

45. $(3xy)^3 = 3^3 \cdot x^3 \cdot y^3 = 27x^3y^3$

47. $(2x^3)^0 = 2^0 \cdot x^{3 \cdot 0} = 1 \cdot x^0 = 1 \cdot 1 = 1$

49. $(-3x^5)^4 = (-3)^4 \cdot x^{5 \cdot 4} = 81x^{20}$

51. $(3x^4y)^3 = 3^3 \cdot x^{4 \cdot 3} \cdot y^3 = 27x^{12}y^3$

53. $\frac{2^{10}}{2^3} = 2^{10-3} = 2^7 = 128$

55. $\frac{(-5)^6}{(-5)^0} = (-5)^{6-0} = (-5)^6$

57. $\frac{4^3}{4} = 4^{3-1} = 4^2 = 16$

59. $\frac{y^5}{y^4} = y^{5-4} = y^1 = y$

61. $\frac{5^4}{5^{10}} = \frac{1}{5^{10-4}} = \frac{1}{5^6}$

63. $\frac{6^8}{6^9} = \frac{1}{6^{9-8}} = \frac{1}{6}$

65. $\frac{(-2)^3}{(-2)^7} = \frac{1}{(-2)^{7-3}} = \frac{1}{(-2)^4} = \frac{1}{16}$

67. $\frac{x^2}{x^9} = \frac{1}{x^{9-2}} = \frac{1}{x^7}$

69. $\left(\frac{3}{5}\right)^2 = \frac{3^2}{5^2} = \frac{9}{25}$

71. $\left(\frac{1}{2}\right)^4 = \frac{1^4}{2^4} = \frac{1}{16}$

73. $\left(\frac{x}{-3}\right)^2 = \frac{x^2}{(-3)^2} = \frac{x^2}{9}$

75. $\left(\frac{x}{5y}\right)^2 = \frac{x^2}{(5y)^2} = \frac{x^2}{5^2y^2} = \frac{x^2}{25y^2}$

77. $\left(\frac{x^3}{-3y}\right)^2 = \frac{(x^3)^2}{(-3y)^2} = \frac{x^{3\cdot2}}{(-3)^2y^2} = \frac{x^6}{9y^2}$

79. $\left(\frac{3x^2}{5y^4}\right)^2 = \frac{(3x^2)^2}{(5y^4)^2} = \frac{3^2x^{2\cdot2}}{5^2y^{4\cdot2}} = \frac{9x^4}{25y^8}$

81. $3^0 = 1$

83. $-3^0 = -1\cdot3^0 = -1\cdot1 = -1$

2.2 Concepts and Vocabulary *(page 76)*

3. False **4.** 3, leading **5.** trinomial **6.** False **7.** $-3x^5$ **8.** -6

2.2 Exercises *(page 76)*

9. monomial; variable: x; coefficient: 5; degree: 3

11. not monomial because of variable in denominator

13. not monomial because exponent is negative

15. monomial; variables: x and y; coefficient: 5; degree: 7

17. not monomial because of variable in denominator

19. monomial; variable: x; coefficient: $-\frac{1}{3}$; degree: 3

21. not monomial because of rational exponent

23. monomial; constant, no variables; coefficient: -4; degree: 0

25. not monomial because expression has two terms

27. monomial; variables: x and y; coefficient: 2; degree: 10

29. $8x^3 + 4x^3 = (8+4)\cdot x^3 = 12x^3$; this expression is a monomial.

31. $5x - 2x + 4 = (5-2)\cdot x + 4 = 3x + 4$; this expression is a binomial.

33.
$$\begin{aligned}10x^2 - 21x^2 + x + 1 &= (10-21)x^2 + x + 1\\ &= (10 + (-21))x^2 + x + 1\\ &= -11x^2 + x + 1;\end{aligned}$$
this expression is a trinomial.

35.
$$\begin{aligned}4x^5 - 5x^2 + 5x - 3x^2 &= 4x^5 - 5x^2 - 3x^2 + 5x\\ &= 4x^5 + (-5-3)\cdot x^2 + 5x\\ &= 4x^5 + (-5 + (-3))\cdot x^2 + 5x\\ &= 4x^5 + (-8)\cdot x^2 + 5x\\ &= 4x^5 - 8x^2 + 5x;\end{aligned}$$
this expression is a trinomial.

37.
$$\begin{aligned}4(x^2-5) + 5(x^2+1) &= 4\cdot x^2 + 4(-5) + 5\cdot x^2 + 5(1)\\ &= 4x^2 + (-20) + 5x^2 + 5\\ &= (4+5)\cdot x^2 + (-20) + 5\\ &= 9x^2 + (-15)\\ &= 9x^2 - 15;\end{aligned}$$
this expression is a binomial.

39.
$$\begin{aligned}-3(2x-1) - 5(3-2x) &= -6x + 3 - 15 + 10x\\ &= -6x + 10x + 3 - 15\\ &= (-6+10)x + (3-15)\\ &= 4x - 12\end{aligned}$$
this expression is a binomial.

41. $3x^2 + 4x + 1$ is in standard form. The coefficients are 3, 4, 1; the degree is 2.

43. $1 - x^2 = -x^2 + 1 = (-1) \cdot x^2 + 0 \cdot x + 1$ is in standard form. The coefficients are $-1, 0, 1$; the degree is 2.

45. $9y^2 + 8y - 5 = 9y^2 + 8y + (-5)$ is in standard form. The coefficients are $9, 8, -5$; the degree is 2.

47. $1 - x + x^2 - x^3 = -x^3 + x^2 - x + 1 = (-1) \cdot x^3 + 1 \cdot x^2 + (-1) \cdot x + 1$ is in standard form. The coefficients are $-1, 1, -1, 1$; the degree is 3.

49. $\sqrt{2} - x = -x + \sqrt{2} = (-1) \cdot x + \sqrt{2}$ is in standard form. The coefficients are $-1, \sqrt{2}$; the degree is 1.

51. $xy^2 - 1 + x$; two variables, degree is 3.

53. $x^2y + y^2z + z^2x - 3xyz$; three variables, degree is 3.

55. $2x^4y^3z + 5x^2y^2 - 7z^5 + 10$; three variables, degree is 8.

57. $x^2 - 1$ for $x = 1$

Replace x by 1 in $x^2 - 1$. The result is

$$x^2 - 1 = (1)^2 - 1 = 1 - 1 = 0$$
$$\uparrow$$
$$x = 1$$

59. $x^2 + x + 1$ for $x = -1$

Replace x by -1 in $x^2 + x + 1$. The result is

$$x^2 + x + 1 = (-1)^2 + (-1) + 1 = 1 + (-1) + 1 = 1$$
$$\uparrow$$
$$x = -1$$

61. $3x - 6$ for $x = 2$

Replace x by 2 in $3x - 6$. The result is

$$3x - 6 = 3(2) - 6 = 6 - 6 = 6 + (-6) = 0$$
$$\uparrow$$
$$x = 2$$

63. $5y^3 - 3y^2 + 4$ for $y = 2$

Replace y by 2 in $5y^3 - 3y^2 + 4$. The result is

$$5y^3 - 3y^2 + 4 = 5(2)^3 - 3(2)^2 + 4 = 5 \cdot (8) - 3 \cdot (4) + 4$$
$$\uparrow$$
$$y = 2$$
$$= 40 - 12 + 4 = 40 + (-12) + 4 = 28 + 4 = 32$$

65. $5x^2 - 3y^3 + z^2 - 4$ for $x = 2, y = -1, z = 3$

Replace x by 2, y by -1, z by 3 in $5x^2 - 3y^3 + z^2 - 4$. The result is

$$5x^2 - 3y^3 + z^2 - 4 = 5 \cdot (2)^2 - 3 \cdot (-1)^3 + (3)^2 - 4$$
$$\uparrow$$
$$x = 2,\ y = -1,\ z = 3$$
$$= 5 \cdot 4 - 3 \cdot (-1) + 9 - 4 = 20 - (-3) + 9 - 4$$
$$= 20 + 3 + 9 + (-4) = 23 + 5 = 28$$

67. $x^5 - x^4 + x^3 - x^2 + x - 1$ for $x = 1$

Replace x by 1 in $x^5 - x^4 + x^3 - x^2 + x - 1$. The result is

$$x^5 - x^4 + x^3 - x^2 + x - 1 = (1)^5 - (1)^4 + (1)^3 - (1)^2 + (1) - 1$$
$$\uparrow$$
$$x = 1$$
$$= 1 - 1 + 1 - 1 + 1 - 1$$
$$= 1 + (-1) + 1 + (-1) + 1 + (-1)$$
$$= 0$$

69. $x^3 + x^2 + x + 1$ for $x = -1.5$

Replace x by -1.5 in $x^3 + x^2 + x + 1$. The result is

$$x^3 + x^2 + x + 1 = (-1.5)^3 + (-1.5)^2 + (-1.5) + 1$$
$$\uparrow$$
$$x = -1.5$$
$$= -3.375 + 2.25 + (-1.5) + 1 = -1.625$$

71. If initial velocity is 160 feet/sec., the height h after t seconds is given by the formula

$$h = -16t^2 + 160t$$

(a) The height of the object after 3 seconds is

$$h = -16 \cdot (3)^2 + 160 \cdot (3)$$
$$= -16 \cdot 9 + 160 \cdot (3)$$
$$= -144 + 480 = 336 \text{ feet}$$

(b) The height of the object after 4 seconds is

$$h = -16 \cdot (4)^2 + 160 \cdot (4)$$
$$= -16 \cdot 16 + 160 \cdot (4)$$
$$= -256 + 640 = 384 \text{ feet}$$

2.3 Concepts and Vocabulary *(page 82)*

5. like terms **6.** False

2.3 Exercises *(page 82)*

7. $(3x + 2) + (5x - 3) = 3x + 2 + 5x - 3 = (3x + 5x) + (2 - 3)$
$= 8x - 1$

9. $(5x^2 - x + 4) + (x^2 + x + 1) = 5x^2 - x + 4 + x^2 + x + 1$
$= (5x^2 + x^2) + (-x + x) + (4 + 1)$
$= 6x^2 + 5$

11. $(8x^3 + x^2 - x) + (3x^2 + 2x + 1) = 8x^3 + x^2 - x + 3x^2 + 2x + 1$
$= 8x^3 + (x^2 + 3x^2) + (-x + 2x) + 1$
$= 8x^3 + 4x^2 + x + 1$

13. $(2x + 3y + 6) + (x - y + 4) = 2x + 3y + 6 + x - y + 4$
$= (2x + x) + (3y - y) + (6 + 4)$
$= 3x + 2y + 10$

15. $3(x^2 + 3x + 4) + 2(x - 3)$
$= (3 \cdot x^2 + 3 \cdot 3x + 3 \cdot 4) + (2 \cdot x - 3 \cdot 2)$
$= (3x^2 + 9x + 12) + (2x - 6)$
$= 3x^2 + 9x + 12 + 2x - 6$
$= 3x^2 + (9x + 2x) + (12 - 6)$
$= 3x^2 + 11x + 6$

17. $5(2x^3 - x^2 + 3x + 4) + 3(x^2 - 4x + 1) = 10x^3 - 5x^2 + 15x + 20 + 3x^2 - 12x + 3$
$= 10x^3 + (-5x^2 + 3x^2) + (15x - 12x) + (20 + 3)$
$= 10x^3 - 2x^2 + 3x + 23$

19. $(x^2 + x + 5) - (3x - 2) = x^2 + x + 5 - 3x + 2$
$= x^2 + (x - 3x) + (5 + 2)$
$= x^2 - 2x + 7$

21. $(x^3 - 3x^2 + 5x + 10) - (2x^2 - x + 1)$
$= x^3 - 3x^2 + 5x + 10 - 2x^2 + x - 1$
$= x^3 + (-3x^2 - 2x^2) + (5x + x) + (10 - 1)$
$= x^3 - 5x^2 + 6x + 9$

23. $(6x^5 + x^3 + x) - (5x^4 - x^3 + 3x^2) = 6x^5 + x^3 + x - 5x^4 + x^3 - 3x^2$
$= 6x^5 - 5x^4 + (x^3 + x^3) - 3x^2 + x$
$= 6x^5 - 5x^4 + 2x^3 - 3x^2 + x$

25. $(2x^3 + x^2 + 3x + 4) - (x^2 - 4x + 1) = 2x^3 + x^2 + 3x + 4 - x^2 + 4x - 1$
$= 2x^3 + (x^2 - x^2) + (3x + 4x) + (4 - 1)$
$= 2x^3 + 7x + 3$

27. $3(x^2 - 3x + 1) - 2(3x^2 + x - 4)$
$= 3x^2 - 9x + 3 - 6x^2 - 2x + 8$
$= (3x^2 - 6x^2) + (-9x - 2x) + (3 + 8)$
$= -3x^2 - 11x + 11$

29. $6(x^3 + x^2 - 3) - 4(2x^3 - 3x^2) = 6x^3 + 6x^2 - 18 - 8x^3 + 12x^2$
$= (6x^3 - 8x^3) + (6x^2 + 12x^2) - 18$
$= -2x^3 + 18x^2 - 18$

31. $(x^2 - x + 2) + (2x^2 - 3x + 5) - (x^2 + 1)$
$= x^2 - x + 2 + 2x^2 - 3x + 5 - x^2 - 1$
$= (x^2 + 2x^2 - x^2) + (-x - 3x) + (2 + 5 - 1)$
$= 2x^2 - 4x + 6$

33. $9(y^2 - 2y + 1) - 6(1 - y^2) = 9y^2 - 18y + 9 - 6 + 6y^2$
$= (9y^2 + 6y^2) - 18y + (9 - 6)$
$= 15y^2 - 18y + 3$

35. $4(x^3 - x^2 + 1) - 3(x^2 + 1) + 2(x^3 + x + 2)$
$= 4x^3 - 4x^2 + 4 - 3x^2 - 3 + 2x^3 + 2x + 4$
$= (4x^3 + 2x^3) + (-4x^2 - 3x^2) + 2x + (4 - 3 + 4)$
$= 6x^3 - 7x^2 + 2x + 5$

37. $(2x^2 + 3y^2 - xy) + (x^2 + 2y^2 - 3xy)$
$= 2x^2 + 3y^2 - xy + x^2 + 2y^2 - 3xy$
$= (2x^2 + x^2) + (3y^2 + 2y^2) + (-xy - 3xy)$
$= 3x^2 - 4xy + 5y^2$

39. $(4x^2 - 3xy + 2y^2) - (x^2 - y^2 + 4xy)$
$= 4x^2 - 3xy + 2y^2 - x^2 + y^2 - 4xy$
$= (4x^2 - x^2) + (-3xy - 4xy) + (2y^2 + y^2)$
$= 3x^2 - 7xy + 3y^2$

41. $(4x^2y + 5xy^2 + xy) - (3x^2y - 4xy^2 - xy)$
$= 4x^2y + 5xy^2 + xy - 3x^2y + 4xy^2 + xy$
$= (4x^2y - 3x^2y) + (5xy^2 + 4xy^2) + (xy + xy)$
$= x^2y + 9xy^2 + 2xy$

43. $3(x^2 + 2xy - 4y^2) + 4(3x^2 - 2xy + 4y^2)$
$= 3x^2 + 6xy - 12y^2 + 12x^2 - 8xy + 16y^2$
$= (3x^2 + 12x^2) + (6xy - 8xy) + (-12y^2 + 16y^2)$
$= 15x^2 - 2xy + 4y^2$

45. $-3(2x^2 - xy - 2y^2) - 5(-x^2 - xy + y^2)$
$= -6x^2 + 3xy + 6y^2 + 5x^2 + 5xy - 5y^2$
$= (-6x^2 + 5x^2) + (3xy + 5xy) + (6y^2 - 5y^2)$
$= -x^2 + 8xy + y^2$

47. $(5xy + 3yz - 4xz) + (4xy - 2yz - xz)$
$= 5xy + 3yz - 4xz + 4xy - 2yz - xz$
$= (5xy + 4xy) + (3yz - 2yz) + (-4xz - xz)$
$= 9xy + yz - 5xz$

49. $(4xy - 2yz - xz) - (xy - yz - xz) = 4xy - 2yz - xz - xy + yz + xz$
$= (4xy - xy) + (-2yz + yz) + (-xz + xz)$
$= 3xy - yz$

51. $3(2xy - yz + 3xz) + 4(-xy + 2yz - 3xz)$
$= 6xy - 3yz + 9xz - 4xy + 8yz - 12xz$
$= (6xy - 4xy) + (-3yz + 8yz) + (9xz - 12xz)$
$= 2xy + 5yz - 3xz$

53. $4(xy - 2yz + 3xz) - 2(-2xy + yz - 3xz)$
$= 4xy - 8yz + 12xz + 4xy - 2yz + 6xz$
$= (4xy + 4xy) + (-8yz - 2yz) + (12xz + 6xz)$
$= 8xy - 10yz + 18xz$

55. $(x^3 + 3x^2 - x - 2) + (-4x^3 - x^2 + x + 4)$
$= x^3 + 3x^2 - x - 2 - 4x^3 - x^2 + x + 4$
$= (x^3 - 4x^3) + (3x^2 - x^2) + (-x + x) + (-2 + 4)$
$= -3x^3 + 2x^2 + 2$

57. $(x^3 + 3x^2 - x - 2) - (-4x^3 - x^2 + x + 4)$
$= x^3 + 3x^2 - x - 2 + 4x^3 + x^2 - x - 4$
$= (x^3 + 4x^3) + (3x^2 + x^2) + (-x - x) + (-2 - 4)$
$= 5x^3 + 4x^2 - 2x - 6$

59. $2(4x^3 + x^2 - x - 2) + 3(x^3 - 2x^2 + x - 2)$
$= 8x^3 + 2x^2 - 2x - 4 + 3x^3 - 6x^2 + 3x - 6$
$= (8x^3 + 3x^3) + (2x^2 - 6x^2) + (-2x + 3x) + (-4 - 6)$
$= 11x^3 - 4x^2 + x - 10$

61. $R = 10x$ and $C = 100 + 5x$
$R - C = 10x - (100 + 5x) = 10x - 100 - 5x = (10x - 5x) - 100$
$= 5x - 100$

63. The profit P is the difference $R - C$.
$P = R - C = 3x - (2x + 100) = 3x - 2x - 100$
$= x - 100$

2.4 Concepts and Vocabulary *(page 93)*

4. True **5.** False **6.** Distributive **7.** First, Outer, Inner, Last **8.** add

2.4 Exercises *(page 93)*

9. $x^2 \cdot x^3 = x^5$

11. $3x^2 \cdot 4x^4 = 3 \cdot 4 \cdot x^2 \cdot x^4$
$= 12x^6$

13. $-7x \cdot 8x^2 = -7 \cdot 8 \cdot x \cdot x^2$
$= -56x^3$

15. $3x(2x^3) = 3 \cdot 2 \cdot x \cdot x^3$
$= 6x^4$

17. $(-3x^2y^4)(-2xy^2) = -3 \cdot -2 \cdot x^2 \cdot x \cdot y^4 \cdot y^2 = 6x^3y^6$

19. $(3xy)(4xy^3) = 3 \cdot 4 \cdot x \cdot x \cdot y \cdot y^3 = 12x^2y^4$

21. $(1 - 2x) \cdot 3x^2 = 1 \cdot 3x^2 - 2x \cdot 3x^2$
$= 1 \cdot 3 \cdot x^2 - 2 \cdot 3 \cdot x \cdot x^2$
$= 3x^2 - 6x^3$

23. $3xy(2x^2 - xy^2) = 3xy \cdot 2x^2 - 3xy \cdot xy^2 = 6x^3y - 3x^2y^3$

25. $x(x^2 + x - 4) = x \cdot x^2 + x \cdot x - x \cdot 4 = x^3 + x^2 - 4x$

27. $(x + 4)(x^2 - 2x + 3)$
$= x(x^2 - 2x + 3) + 4(x^2 - 2x + 3)$
$= x \cdot x^2 - x \cdot 2x + x \cdot 3 + 4 \cdot x^2 - 4 \cdot 2x + 4 \cdot 3$
$= x^3 - 2x^2 + 3x + 4x^2 - 8x + 12$
$= x^3 + 2x^2 - 5x + 12$

29. $(3x - 2)(x^2 - 2x + 3)$
$= 3x(x^2 - 2x + 3) - 2(x^2 - 2x + 3)$
$= 3x \cdot x^2 - 3x \cdot 2x + 3x \cdot 3 - 2 \cdot x^2 - 2 \cdot (-2x) - 2 \cdot 3$
$= 3x^3 - 6x^2 + 9x - 2x^2 + 4x - 6$
$= 3x^3 - 8x^2 + 13x - 6$

31. $(-2x + 3)(-x^2 - x - 1)$
$= -2x(-x^2 - x - 1) + 3(-x^2 - x - 1)$
$= -2x \cdot (-x^2) - 2x \cdot (-x) - 2x \cdot (-1) + 3 \cdot (-x^2) + 3 \cdot (-x) + 3 \cdot (-1)$
$= 2x^3 + 2x^2 + 2x - 3x^2 - 3x - 3$
$= 2x^3 - x^2 - x - 3$

33. $(x + 4)(x - 3) = x(x - 3) + 4(x - 3)$
$= x^2 - 3x + 4x - 12$
$= x^2 + x - 12$

35. $(x + 8)(2x + 1) = x(2x + 1) + 8(2x + 1)$
$= x \cdot 2x + x \cdot 1 + 8 \cdot 2x + 8 \cdot 1$
$= 2x^2 + x + 16x + 8$
$= 2x^2 + 17x + 8$

37. $(-3x + 1)(x + 4) = -3x(x + 4) + 1(x + 4)$
$= -3x \cdot x - 3x \cdot 4 + 1 \cdot x + 1 \cdot 4$
$= -3x^2 - 12x + x + 4$
$= -3x^2 - 11x + 4$

39. $(1 - 4x)(2 - 3x) = 1 \cdot (2 - 3x) - 4x(2 - 3x)$
$= 1 \cdot 2 - 1 \cdot 3x - 4x \cdot 2 - 4x \cdot (-3x)$
$= 2 - 3x - 8x + 12x^2$
$= 2 - 11x + 12x^2$
$= 12x^2 - 11x + 2$

41. $(2x + 3)(x^2 - 2) = 2x(x^2 - 2) + 3(x^2 - 2)$
$= 2x \cdot x^2 - 2x \cdot 2 + 3 \cdot x^2 - 3 \cdot 2$
$= 2x^3 - 4x + 3x^2 - 6$
$= 2x^3 + 3x^2 - 4x - 6$

43. $(3x - 5)(2x + 3) = 3x(2x + 3) - 5(2x + 3)$
$= 3x \cdot 2x + 3x \cdot 3 - 5 \cdot 2x - 5 \cdot 3$
$= 6x^2 + 9x - 10x - 15$
$= 6x^2 - x - 15$

45. $(x - 7)(x + 7) = x^2 - 49$

47. $(2x + 3)(2x - 3) = 4x^2 - 9$

49. $(3x + 4)(3x - 4) = (3x)^2 - (4)^2 = 9x^2 - 16$

51. $(x + 4)^2 = x^2 + 8x + 16$

53. $(x - 4)^2 = x^2 - 8x + 16$

55. $(2x - 3)^2 = (2x - 3) \cdot (2x - 3) = (2x)^2 - 2 \cdot 2x \cdot 3 + (3)^2$
$= 4x^2 - 12x + 9$

57. $(1 + x)^2 - (1 - x)^2 = 1^2 + 2\cdot 1\cdot x + x^2 - (1^2 - 2\cdot 1\cdot x + x^2)$
$= 1 + 2x + x^2 - 1 + 2x - x^2$
$= 4x$

59. $(x + a)^2 - x^2 = x^2 + 2ax + a^2 - x^2 = 2ax + a^2$

61. $(x - 1)(x^2 + x + 1)$
$= x(x^2 + x + 1) - 1(x^2 + x + 1)$
$= x\cdot x^2 + x\cdot x + x\cdot 1 - 1\cdot x^2 - 1\cdot x - 1\cdot 1$
$= x^3 + x^2 + x - x^2 - x - 1$
$= x^3 - 1$

63. $(x + 3)(x^2 - 3x + 9) = x(x^2 - 3x + 9) + 3(x^2 - 3x + 9)$
$= x\cdot x^2 + x\cdot(-3x) + x\cdot 9 + 3\cdot x^2 + 3\cdot(-3x) + 3\cdot 9$
$= x^3 - 3x^2 + 9x + 3x^2 - 9x + 27$
$= x^3 + 27$

65. $(x + 5)^3 = x^3 + 3\cdot 5x^2 + 3\cdot 5^2\cdot x + 5^3 = x^3 + 15x^2 + 75x + 125$

67. $(x + a)^3 - x^3 = x^3 + 3ax^2 + 3a^2x + a^3 - x^3 = 3ax^2 + 3a^2x + a^3$

69. $(2x + 5)(x^2 + x + 1)$
$= 2x(x^2 + x + 1) + 5(x^2 + x + 1)$
$= 2x\cdot x^2 + 2x\cdot x + 2x\cdot 1 + 5\cdot x^2 + 5x + 5\cdot 1$
$= 2x^3 + 2x^2 + 2x + 5x^2 + 5x + 5$
$= 2x^3 + 7x^2 + 7x + 5$

71. $(3x - 1)(2x^2 - 3x + 2)$
$= 3x(2x^2 - 3x + 2) - 1(2x^2 - 3x + 2)$
$= 3x\cdot 2x^2 - 3x\cdot 3x + 3x\cdot 2 - 1\cdot 2x^2 - 1\cdot(-3x) - 1\cdot 2$
$= 6x^3 - 9x^2 + 6x - 2x^2 + 3x - 2$
$= 6x^3 - 11x^2 + 9x - 2$

73. $(x^2 + x - 1)(x^2 - x + 1)$
$= x^2(x^2 - x + 1) + x(x^2 - x + 1) - 1(x^2 - x + 1)$
$= x^2\cdot x^2 - x^2\cdot x + x^2\cdot 1 + x\cdot x^2 - x\cdot x + x\cdot 1 - 1\cdot x^2 - 1\cdot(-x) - 1\cdot 1$
$= x^4 - x^3 + x^2 + x^3 - x^2 + x - x^2 + x - 1$
$= x^4 - x^2 + 2x - 1$

75. $(x + 2)^2(x - 1)$
$= (x^2 + 4x + 4)(x - 1)$
$= x^2(x - 1) + 4x(x - 1) + 4(x - 1)$
$= x^2\cdot x - x^2\cdot 1 + 4x\cdot x - 4x\cdot 1 + 4\cdot x - 4\cdot 1$
$= x^3 - x^2 + 4x^2 - 4x + 4x - 4$
$= x^3 + 3x^2 - 4$

77. $(x - 1)^2(2x + 3)$
$= (x^2 - 2x + 1)(2x + 3)$
$= x^2(2x + 3) - 2x(2x + 3) + 1(2x + 3)$
$= 2x^3 + 3x^2 - 4x^2 - 6x + 2x + 3$
$= 2x^3 - x^2 - 4x + 3$

79. $(x - 1)(x - 2)(x - 3)$
$= [x(x - 2) - 1(x - 2)](x - 3)$
$= [x\cdot x - x\cdot 2 - 1\cdot x - 1\cdot(-2)](x - 3)$
$= [x^2 - 2x - x + 2](x - 3)$
$= [x^2 - 3x + 2](x - 3)$
$= x^2(x - 3) - 3x(x - 3) + 2(x - 3)$
$= x^2\cdot x - x^2\cdot 3 - 3x\cdot x - 3x\cdot(-3) + 2\cdot x - 2\cdot 3$
$= x^3 - 3x^2 - 3x^2 + 9x + 2x - 6$
$= x^3 - 6x^2 + 11x - 6$

81. $(x + 1)^3 - (x - 1)^3$
$= x^3 + 3\cdot 1\cdot x^2 + 3\cdot 1^2\cdot x + 1^3 - (x^3 - 3\cdot 1\cdot x^2 + 3\cdot 1^2\cdot x - 1^3)$
$= x^3 + 3x^2 + 3x + 1 - (x^3 - 3x^2 + 3x - 1)$
$= x^3 + 3x^2 + 3x + 1 - x^3 + 3x^2 - 3x + 1$
$= 6x^2 + 2$

83. $(x + y)(x - 2y) = x\cdot x + x\cdot(-2y) + y\cdot x + y\cdot(-2y)$
$= x^2 - 2xy + xy - 2y^2$
$= x^2 - xy - 2y^2$

85. $(2x - y)(3x + 4y) = 2x\cdot 3x + 2x\cdot 4y + (-y)\cdot 3x + (-y)\cdot 4y$
$= 6x^2 + 8xy - 3xy - 4y^2$
$= 6x^2 + 5xy - 4y^2$

87. $(x - 2y)(x^2 + 2xy + 4y^2) = x\cdot x^2 + x\cdot 2xy + x\cdot 4y^2 + (-2y)\cdot x^2 + (-2y)\cdot 2xy + (-2y)\cdot 4y^2$
$= x^3 + 2x^2y + 4xy^2 - 2x^2y - 4xy^2 - 8y^3$
$= x^3 - 8y^3$

89. $(x-y)^2-(x+y)^2 = x^2-2xy+y^2-(x^2+2xy+y^2)$
$= x^2-2xy+y^2-x^2-2xy-y^2$
$= -4xy$

91. $(x+2y)^2+(x-3y)^2$
$= x^2+2\cdot x\cdot 2y+(2y)^2+x^2-2\cdot x\cdot 3y+(3y)^2$
$= x^2+4xy+4y^2+x^2-6xy+9y^2$
$= 2x^2-2xy+13y^2$

93. $[(x+y)^2+z^2]+[x^2+(y+z)^2]$
$= [x^2+2xy+y^2+z^2]+[x^2+y^2+2yz+z^2]$
$= x^2+2xy+y^2+z^2+x^2+y^2+2yz+z^2$
$= 2x^2+2y^2+2z^2+2xy+2yz$

95. $(x-y)(x^2+xy+y^2) = x(x^2+xy+y^2)-y(x^2+xy+y^2)$
$= x\cdot x^2+x\cdot xy+x\cdot y^2-y\cdot x^2-y\cdot xy-y\cdot y^2$
$= x^3+x^2y+xy^2-x^2y-xy^2-y^3$
$= x^3-y^3$

97. $(2x^2+xy+y^2)(3x^2-xy+2y^2)$
$= 2x^2(3x^2-xy+2y^2)+xy(3x^2-xy+2y^2)+y^2(3x^2-xy+2y^2)$
$= 2x^2\cdot 3x^2-2x^2\cdot xy+2x^2\cdot 2y^2+xy\cdot 3x^2-xy\cdot xy+xy\cdot 2y^2+y^2\cdot 3x^2-y^2\cdot xy+y^2\cdot 2y^2$
$= 6x^4-2x^3y+4x^2y^2+3x^3y-x^2y^2+2xy^3+3x^2y^2-xy^3+2y^4$
$= 6x^4-2x^3y+3x^3y^2+4x^2y-x^2y^2+3x^2y^3+2xy^3-xy^3+2y^4$
$= 6x^4+x^3y+6x^2y^2+xy^3+2y^4$

99. $(x+y+z)(x-y-z)$
$= x(x-y-z)+y(x-y-z)+z(x-y-z)$
$= x\cdot x-x\cdot y-x\cdot z+y\cdot x-y\cdot y-y\cdot z+z\cdot x-z\cdot y-z\cdot z$
$= x^2-xy-xz+xy-y^2-yz+xz-yz-z^2$
$= x^2-xy+xy-xz+xz-y^2-yz-yz-z^2$
$= x^2-y^2-z^2-2yz$

101. $(x-a)^2 = (x-a)(x-a) = x(x-a)-a(x-a)$
$= x\cdot x-x\cdot a-a\cdot x-a(-a)$
$= x^2-ax-ax+a^2$
$= x^2-2ax+a^2$

103. $(x-a)^3 = (x-a)^2(x-a) = (x^2-2ax+a^2)(x-a)$
$= x^2(x-a)-2ax(x-a)+a^2(x-a)$
$= x^3-ax^2-2ax^2+2a^2x+a^2x-a^3$
$= x^3-3ax^2+3a^2x-a^3$

105. The product of two polynomials equals the sum of their respective degree since the degree of the product equals the degree of the product of the leading terms:

$$(a_nx^n+a_{n-1}x^{n-1}+\cdots+a_1x+a_0)\cdot(b_mx^m+b_{m-1}x^{m-1}+\cdots+b_1x+b_0)$$
$$= a_nb_mx^{n+m}+(a_nb_{m-1}+a_{n-1}b_m)x^{n+m-1}+\cdots$$

2.5 Concepts and Vocabulary *(page 100)*

3. zero, less than **4.** quotient, divisor, remainder **5.** False **6.** $5x^2$

2.5 Exercises *(page 100)*

7. $\dfrac{3^5}{3^2} = 3^{5-2} = 3^3 = 27$

9. $\dfrac{x^6}{x^2} = x^{6-2} = x^4$

11. $\dfrac{25x^4}{5x^2} = 5x^2$

13. $\dfrac{20y^4}{4y^3} = 5y$

15. $\dfrac{45x^2y^3}{9xy} = 5xy^2$

17. $\dfrac{5x^3 - 3x^2 + x}{x} = \dfrac{5x^3}{x} - \dfrac{3x^2}{x} + \dfrac{x}{x} = 5x^2 - 3x + 1$

19. $\dfrac{10x^5 - 5x^4 + 15x^2}{5x^2} = \dfrac{10x^5}{5x^2} - \dfrac{5x^4}{5x^2} + \dfrac{15x^2}{5x^2} = 2x^3 - x^2 + 3$

21. $\dfrac{-21x^3 + x^2 - 3x + 4}{x} = \dfrac{-21x^3}{x} + \dfrac{x^2}{x} - \dfrac{3x}{x} + \dfrac{4}{x} = -21x^2 + x - 3 + \dfrac{4}{x}$

23. $\dfrac{8x^3 - x^2 + 1}{x^2} = \dfrac{8x^3}{x^2} - \dfrac{x^2}{x^2} + \dfrac{1}{x^2} = 8x - 1 + \dfrac{1}{x^2}$

25. $\dfrac{2x^3 - x^2 + 1}{x^3} = 2 - \dfrac{1}{x} + \dfrac{1}{x^3}$

27.

$$
\begin{array}{r|l}
 & 4x^2 - 3x + 1 \\
\hline
x & 4x^3 - 3x^2 + x + 1 \\
 & \underline{4x^3} \\
 & \quad -3x^2 \\
 & \quad \underline{-3x^2} \\
 & \qquad\quad x \\
 & \qquad\quad \underline{x} \\
 & \qquad\qquad 1
\end{array}
$$

The quotient is $4x^2 - 3x + 1$; the remainder is 1.
Check: $x(4x^2 - 3x + 1) + 1 = 4x^3 - 3x^2 + x + 1$

$\dfrac{4x^3 - 3x^2 + x + 1}{x} = 4x^2 - 3x + 1 + \dfrac{1}{x}$

29.

$$
\begin{array}{r|l}
 & 4x^2 - 11x + 23 \\
\hline
x + 2 & 4x^3 - 3x^2 + x + 1 \\
 & \underline{4x^3 + 8x^2} \\
 & \quad -11x^2 + x \\
 & \quad \underline{-11x^2 - 22x} \\
 & \qquad\qquad 23x + 1 \\
 & \qquad\qquad \underline{23x + 46} \\
 & \qquad\qquad\quad -45
\end{array}
$$

The quotient is $4x^2 - 11x + 23$; the remainder is -45.
Check: $(x + 2)(4x^2 - 11x + 23) + (-45)$
$= 4x^3 - 11x^2 + 23x + 8x^2 - 22x + 46 + (-45)$
$= 4x^3 - 3x^2 + x + 1$

$\dfrac{4x^3 - 3x^2 + x + 1}{x + 2} = 4x^2 - 11x + 23 + \dfrac{-45}{x + 2}$

31.

$$
\begin{array}{r|l}
 & 4x^2 + 13x + 53 \\
\hline
x - 4 & 4x^3 - 3x^2 + x + 1 \\
 & \underline{4x^3 - 16x^2} \\
 & \qquad 13x^2 + x \\
 & \qquad \underline{13x^2 - 52x} \\
 & \qquad\qquad 53x + 1 \\
 & \qquad\qquad \underline{53x - 212} \\
 & \qquad\qquad\quad 213
\end{array}
$$

The quotient is $4x^2 + 13x + 53$; the remainder is 213.
Check: $(x - 4)(4x^2 + 13x + 53) + 213$
$= 4x^3 + 13x^2 + 53x - 16x^2 - 52x - 212 + 213$
$= 4x^3 - 3x^2 + x + 1$

$\dfrac{4x^3 - 3x^2 + x + 1}{x - 4} = 4x^2 + 13x + 53 + \dfrac{213}{x - 4}$

33.

$$
\begin{array}{r|l}
 & 4x - 3 \\
\hline
x^2 & 4x^3 - 3x^2 + x + 1 \\
 & \underline{4x^3} \\
 & \quad -3x^2 \\
 & \quad \underline{-3x^2} \\
 & \qquad\quad x + 1
\end{array}
$$

The quotient is $4x - 3$; the remainder is $x + 1$.
Check: $x^2(4x - 3) + (x + 1) = 4x^3 - 3x^2 + x + 1$

$\dfrac{4x^3 - 3x^2 + x + 1}{x^2} = 4x - 3 + \dfrac{x + 1}{x^2}$

35.

$$
\begin{array}{r|l}
 & 4x - 3 \\
\hline
x^2 + 2 & 4x^3 - 3x^2 + x + 1 \\
 & \underline{4x^3 \qquad\quad + 8x} \\
 & \quad -3x^2 - 7x + 1 \\
 & \quad \underline{-3x^2 \qquad - 6} \\
 & \qquad\quad -7x + 7
\end{array}
$$

The quotient is $4x - 3$; the remainder is $-7x + 7$.
Check: $(x^2 + 2)(4x - 3) + (-7x + 7) = 4x^3 - 3x^2 + 8x - 6 - 7x + 7$
$= 4x^3 - 3x^2 + x + 1$

$\dfrac{4x^3 - 3x^2 + x + 1}{x^2 + 2} = 4x - 3 + \dfrac{-7x + 7}{x^2 + 2}$

37.
$$\begin{array}{r} 4 \\ x^3 - 1\overline{)4x^3 - 3x^2 + x + 1} \\ \underline{4x^3 \qquad\qquad\quad - 4} \\ -3x^2 + x + 5 \end{array}$$

The quotient is 4; the remainder is $-3x^2 + x + 5$.
Check:
$$(x^3 - 1)(4) + (-3x^2 + x + 5) = 4x^3 - 4 - 3x^2 + x + 5$$
$$= 4x^3 - 3x^2 + x + 1$$
$$\frac{4x^3 - 3x^2 + x + 1}{x^3 - 1} = 4 + \frac{-3x^2 + x + 5}{x^3 - 1}$$

39.
$$\begin{array}{r} 4x - 7 \\ x^2 + x + 1\overline{)4x^3 - 3x^2 + x + 1} \\ \underline{4x^3 + 4x^2 + 4x\phantom{{}+1}} \\ -7x^2 - 3x + 1 \\ \underline{-7x^2 - 7x - 7} \\ 4x + 8 \end{array}$$

The quotient is $4x - 7$; the remainder is $4x + 8$.
Check: $(4x - 7)(x^2 + x + 1) + (4x + 8)$
$$= 4x^3 + 4x^2 + 4x - 7x^2 - 7x - 7 + 4x + 8$$
$$= 4x^3 - 3x^2 + x + 1$$
$$\frac{4x^3 - 3x^2 + x + 1}{x^2 + x + 1} = 4x - 7 + \frac{4x + 8}{x^2 + x + 1}$$

41.
$$\begin{array}{r} 4x + 9 \\ x^2 - 3x - 4\overline{)4x^3 - 3x^2 + x + 1} \\ \underline{4x^3 - 12x^2 - 16x\phantom{{}+1}} \\ 9x^2 + 17x + 1 \\ \underline{9x^2 - 27x - 36} \\ 44x + 37 \end{array}$$

The quotient is $4x + 9$; the remainder is $44x + 37$
Check: $(4x + 9)(x^2 - 3x - 4) + 44x + 37$
$$= 4x^3 - 12x^2 - 16x + 9x^2 - 27x - 36 + 44x + 37$$
$$= 4x^3 - 3x^2 + x + 1$$
$$\frac{4x^3 - 3x^2 + x + 1}{x^2 - 3x - 4} = 4x + 9 + \frac{44x + 37}{x^2 - 3x - 4}$$

43.
$$\begin{array}{r} x^3 + x^2 + x + 1 \\ x - 1\overline{)x^4 + 0x^3 + 0x^2 + 0x - 1} \\ \underline{x^4 - x^3} \\ x^3 \\ \underline{x^3 - x^2} \\ x^2 \\ \underline{x^2 - x} \\ x - 1 \\ \underline{x - 1} \\ 0 \end{array}$$

The quotient is $x^3 + x^2 + x + 1$; the remainder is 0.
Check: $(x - 1)(x^3 + x^2 + x + 1)$
$$= x^4 + x^3 + x^2 + x - x^3 - x^2 - x - 1 = x^4 - 1$$
$$\frac{x^4 - 1}{x - 1} = x^3 + x^2 + x + 1$$

45.
$$\begin{array}{r} x^2 + 1 \\ x^2 - 1\overline{)x^4 + 0x^3 + 0x^2 + 0x - 1} \\ \underline{x^4 \qquad\quad - x^2} \\ x^2 \qquad - 1 \\ \underline{x^2 \qquad - 1} \\ 0 \end{array}$$

The quotient is $x^2 + 1$; the remainder is 0.
Check: $(x^2 - 1)(x^2 + 1) + 0 = x^4 + x^2 - x^2 - 1 = x^4 - 1$
$$\frac{x^4 - 1}{x^2 - 1} = x^2 + 1$$

47.
$$\begin{array}{r} -4x^2 - 3x - 3 \\ x - 1\overline{)-4x^3 + x^2 + 0x - 4} \\ \underline{-4x^3 + 4x^2} \\ -3x^2 + 0x \\ \underline{-3x^2 + 3x} \\ -3x - 4 \\ \underline{-3x + 3} \\ -7 \end{array}$$

The quotient is $-4x^2 - 3x - 3$; the remainder is -7.
Check:
$$(x - 1)(-4x^2 - 3x - 3) + (-7)$$
$$= -4x^3 - 3x^2 - 3x + 4x^2 + 3x + 3 - 7 = -4x^3 + x^2 - 4$$
$$\frac{-4x^3 + x^2 - 4}{x - 1} = -4x^2 - 3x - 3 + \frac{-7}{x - 1}$$

49. $1 - x^2 + x^4 = x^4 - x^2 + 1$

$$\begin{array}{r}
x^2 - x - 1 \\
x^2 + x + 1\overline{)x^4 + 0x^3 - x^2 + 0x + 1} \\
\underline{x^4 + x^3 + x^2} \\
-x^3 - 2x^2 + 0x \\
\underline{-x^3 - x^2 - x} \\
-x^2 + x + 1 \\
\underline{-x^2 - x - 1} \\
2x + 2
\end{array}$$

The quotient is $x^2 - x - 1$; the remainder is $2x + 2$.
Check: $(x^2 + x + 1)(x^2 - x - 1) + 2x + 2$
$= x^4 - x^3 - x^2 + x^3 - x^2 - x + x^2 - x - 1 + 2x + 2$
$= x^4 - x^2 + 1$

$$\frac{x^4 - x^2 + 1}{x^2 + x + 1} = x^2 - x - 1 + \frac{2x + 2}{x^2 + x + 1}$$

51. $1 - x^2 = -x^2 + 1; 1 - x^2 + x^4 = x^4 + 0x^3 - x^2 + 0x + 1$

$$\begin{array}{r}
-x^2 \\
-x^2 + 1\overline{)x^4 + 0x^3 - x^2 + 0x + 1} \\
\underline{x^4 \qquad - x^2 \qquad\qquad} \\
1
\end{array}$$

The quotient is $-x^2$; the remainder is 1.
Check: $-x^2(-x^2 + 1) + 1 = x^4 - x^2 + 1$

$$\frac{x^4 - x^2 + 1}{-x^2 + 1} = -x^2 + \frac{1}{-x^2 + 1}$$

53.

$$\begin{array}{r}
x^2 + ax + a^2 \\
x - a\overline{)x^3 + 0x^2 + 0x - a^3} \\
\underline{x^3 - ax^2} \\
ax^2 + 0x \\
\underline{ax^2 - a^2x} \\
a^2x - a^3 \\
\underline{a^2x - a^3} \\
0
\end{array}$$

The quotient is $x^2 + ax + a^2$; the remainder is 0.
Check:
$(x - a)(x^2 + ax + a^2) = x^3 + ax^2 + a^2x - ax^2 - a^2x - a^3$
$= x^3 - a^3$

$$\frac{x^3 - a^3}{x - a} = x^2 + ax + a^2$$

55.

$$\begin{array}{r}
x^3 + ax^2 + a^2x + a^3 \\
x - a\overline{)x^4 + 0x^3 + 0x^2 + 0x - a^4} \\
\underline{x^4 - ax^3} \\
ax^3 + 0x^2 \\
\underline{ax^3 - a^2x^2} \\
a^2x^2 + 0x \\
\underline{a^2x^2 - a^3x} \\
a^3x - a^4 \\
\underline{a^3x - a^4} \\
0
\end{array}$$

The quotient is $x^3 + ax^2 + a^2x + a^3$; the remainder is 0.
Check: $(x - a)(x^3 + ax^2 + a^2x + a^3)$
$= x^4 + ax^3 + a^2x^2 + a^3x - ax^3 - a^2x^2 - a^3x - a^4$
$= x^4 - a^4$

$$\frac{x^4 - a^4}{x - a} = x^3 + ax^2 + a^2x + a^3$$

2.6 Concepts and Vocabulary *(page 106)*

1. prime **2.** $5x(x^3 - 4)$ **3.** $6x - 1$ **4.** $2x^3 + 5x^2 - 8$

2.6 Exercises *(page 106)*

5. $3x + 6 = 3(x + 2)$

7. $ax^2 + a = a(x^2 + 1)$

9. $x^3 + x^2 + x = x(x^2 + x + 1)$

11. $2x^2 + 2x + 2 = 2(x^2 + x + 1)$

13. $3x^2y - 6xy^2 + 12xy = 3xy(x - 2y + 4)$

15. $-5x^2y^3 - 45xy^2 - 50xy = -5xy(xy^2 + 9y + 10)$

17. $-3x^2 - 12x = -3x(x + 4)$

19. $81x + 24y = 3(27x + 8y)$

21. $54x^3 + 17x^2 = x^2(54x + 17)$

23. $18x^4 - 36x^2 + 63x = 9x(2x^3 - 4x + 7)$

25. $-7x^2 - 7x = -7x(x + 1)$

27. $x^4 + 81x^2 = x^2(x^2 + 81)$

29. $x(x + 3) - 6(x + 3) = (x - 6)(x + 3)$

31. $x(x^2 + 4) + 5(x^2 + 4) = (x + 5)(x^2 + 4)$

33. $9x(x + 5) - 2(x + 5) = (9x - 2)(x + 5)$

35. $3x(2x + 7) + (2x + 7) = (3x + 1)(2x + 7)$

37. $12x(x - 4) - (x - 4) = (12x - 1)(x - 4)$

39. $3x^2(4x + 5) - 6x(4x + 5) = 3x(4x + 5) \cdot x + 3x(4x + 5) \cdot (-2) = 3x(4x + 5)(x - 2)$

41. $10x^2(3x + 11) - 5x(3x + 11) = 5x(3x + 11) \cdot 2x + 5x(3x + 11) \cdot (-1) = 5x(3x + 11)(2x - 1)$

43. $(x + 2)^2 - 5(x + 2) = (x + 2)(x + 2 - 5) = (x + 2)(x - 3)$

45. $x^3 + x^2 + 2x + 2 = (x^3 + x^2) + (2x + 2) = x^2(x + 1) + 2(x + 1) = (x^2 + 2)(x + 1)$

47. $x^3 - x^2 + 3x - 3 = (x^3 - x^2) + (3x - 3) = x^2(x - 1) + 3(x - 1) = (x^2 + 3)(x - 1)$

49. $2x^3 - 10x^2 + 3x - 15 = (2x^3 - 10x^2) + (3x - 15) = 2x^2(x - 5) + 3(x - 5) = (2x^2 + 3)(x - 5)$

51. $3x^3 + 12x^2 + 2x + 8 = (3x^3 + 12x^2) + (2x + 8) = 3x^2(x + 4) + 2(x + 4) = (3x^2 + 2)(x + 4)$

53. $6x^3 - 15x^2 - 8x + 20 = (6x^3 - 15x^2) - (8x - 20) = 3x^2(2x - 5) - 4(2x - 5) = (3x^2 - 4)(2x - 5)$

55. $5x^3 + 2x^2 - 15x - 6 = (5x^3 + 2x^2) - (15x + 6) = x^2(5x + 2) - 3(5x + 2) = (x^2 - 3)(5x + 2)$

57. $6x^3 - 12x^2 - 5x + 10 = (6x^3 - 12x^2) - (5x - 10) = 6x^2(x - 2) - 5(x - 2) = (6x^2 - 5)(x - 2)$

59. $x^3 + 2x^2 + x + 2 = (x^3 + 2x^2) + (x + 2) = x^2(x + 2) + 1(x + 2) = (x^2 + 1)(x + 2)$

61. $3x^3 - 3x^2 + x - 1 = (3x^3 - 3x^2) + (x - 1) = 3x^2(x - 1) + 1(x - 1) = (3x^2 + 1)(x - 1)$

63. $3x^3 - 15x^2 - x + 5 = (3x^3 - 15x^2) - (x - 5) = 3x^2(x - 5) - 1(x - 5) = (3x^2 - 1)(x - 5)$

65. $3x^3 - 6x^2 - x + 2 = (3x^3 - 6x^2) - (x - 2) = 3x^2(x - 2) - 1(x - 2) = (3x^2 - 1)(x - 2)$

67. $x^2 + 2x - x - 2 = (x^2 + 2x) - (x + 2) = x(x + 2) - 1(x + 2) = (x - 1)(x + 2)$

2.7 Concepts and Vocabulary *(page 116)*

3. $2x - 1$ **4.** True

2.7 Exercises *(page 116)*

5. $x^2 + 5x + 6$

Integers whose product is 6	1, 6	−1, −6	2, 3	−2, −3
Sum	7	−7	5	−5

Since the coefficient of the middle term is 5: $x^2 + 5x + 6 = (x + 2)(x + 3)$

7. $x^2 + 7x + 6$

Integers whose product is 6	1, 6	−1, −6	2, 3	−2, −3
Sum	7	−7	5	−5

Since the coefficient of the middle term is 7: $x^2 + 7x + 6 = (x + 1)(x + 6)$

9. $x^2 + 7x + 10$

Integers whose product is 10	1, 10	−1, −10	2, 5	−2, −5
Sum	11	−11	7	−7

Since the coefficient of the middle term is 7: $x^2 + 7x + 10 = (x + 2)(x + 5)$

11. $x^2 + 17x + 16$

Integers whose product is 16	1, 16	−1, −16	2, 8	−2, −8	4, 4	−4, −4
Sum	17	−17	10	−10	8	−8

Since the coefficient of the middle term is 17: $x^2 + 17x + 16 = (x + 1)(x + 16)$

13. $x^2 + 12x + 20$

Integers whose product is 20	1, 20	−1, −20	2, 10	−2, −10	4, 5	−4, −5
Sum	21	−21	12	−12	9	−9

Since the coefficient of the middle term is 12: $x^2 + 12x + 20 = (x + 2)(x + 10)$

15. $x^2 + 12x + 11$

Integers whose product is 11	1, 11	−1, −11
Sum	12	−12

Since the coefficient of the middle term is 12: $x^2 + 12x + 11 = (x + 1)(x + 11)$

17. $x^2 - 10x + 21$

Integers whose product is 21	1, 21	−1, −21	3, 7	−3, −7
Sum	22	−22	10	−10

Since the coefficient of the middle term is −10: $x^2 - 10x + 21 = (x - 3)(x - 7)$

19. $x^2 - 11x + 10$

Integers whose product is 10	1, 10	−1, −10	2, 5	−2, −5
Sum	11	−11	7	−7

Since the coefficient of the middle term is −11: $x^2 - 11x + 10 = (x - 1)(x - 10)$

21. $x^2 - 9x + 20$

Integers whose product is 20	1, 20	−1, −20	2, 10	−2, −10	4, 5	−4, −5
Sum	21	−21	12	−12	9	−9

Since the coefficient of the middle term is −9: $x^2 - 9x + 20 = (x - 4)(x - 5)$

23. $x^2 - 10x + 16$

Integers whose product is 16	1, 16	−1, −16	2, 8	−2, −8	4, 4	−4, −4
Sum	17	−17	10	−10	8	−8

Since the coefficient of the middle term is −10: $x^2 - 10x + 16 = (x - 2)(x - 8)$

25. $x^2 - 7x - 8$

Integers whose product is −8	1, −8	−1, 8	2, −4	−2, 4
Sum	−7	7	−2	2

Since the coefficient of the middle term is −7: $x^2 - 7x - 8 = (x + 1)(x - 8)$

27. $x^2 + 7x - 8$

Integers whose product is −8	1, −8	−1, 8	2, −4	−2, 4
Sum	−7	7	−2	2

Since the coefficient of the middle term is 7: $x^2 - 7x - 8 = (x - 1)(x + 8)$

29. $x^2 - 6x + 5$

Integers whose product is 5	1, 5	−1, −5
Sum	6	−6

Since the coefficient of the middle term is −6: $x^2 - 6x + 5 = (x - 1)(x - 5)$

31. $x^2 - 4x - 5$

Integers whose product is −5	1, −5	−1, 5
Sum	−4	4

Since the coefficient of the middle term is −4: $x^2 - 4x - 5 = (x + 1)(x - 5)$

33. $x^2 - 2x - 15$

Integers whose product is −15	1, −15	−1, 15	3, −5	−3, 5
Sum	−14	14	−2	2

Since the coefficient of the middle term is −2: $x^2 - 2x - 15 = (x + 3)(x - 5)$

35. $x^2 - 8x + 15$

Integers whose product is 15	1, 15	−1, −15	3, 5	−3, −5
Sum	16	−16	8	−8

Since the coefficient of the middle term is −8: $x^2 - 8x + 15 = (x - 3)(x - 5)$

37. $2x^2 - 12x - 32 = 2(x^2 - 6x - 16)$

Integers whose product is −16	1, −16	−1, 16	2, −8	−2, 8	4, −4
Sum	−15	15	−6	6	0

Since the coefficient of the middle term is −6: $x^2 - 6x - 16 = (x + 2)(x - 8)$
Thus, $2x^2 - 12x - 32 = 2(x + 2)(x - 8)$

39. $x^3 + 10x^2 + 16x = x(x^2 + 10x + 16)$

Integers whose product is 16	1, 16	−1, −16	2, 8	−2, −8	4, 4	−4, −4
Sum	17	−17	10	−10	8	−8

Since the coefficient of the middle term is 10: $x^2 + 10x + 16 = (x + 2)(x + 8)$
Thus, $x^3 + 10x^2 + 16x = x(x + 2)(x + 8)$

41. $x^3 - x^2 - 6x = x(x^2 - x - 6)$

Integers whose product is −6	1, −6	−1, 6	2, −3	−2, 3
Sum	−5	5	−1	1

Since the coefficient of the middle term is −1: $x^2 - x - 6 = (x + 2)(x - 3)$
Thus, $x^3 - x^2 - 6x = x(x + 2)(x - 3)$

43. $x^3 - x^2 - 2x = x(x^2 - x - 2)$

Integers whose product is −2	1, −2	−1, 2
Sum	−1	1

Since the coefficient of the middle term is −1: $x^2 - x - 2 = (x + 1)(x - 2)$
Thus, $x^3 - x^2 - 2x = x(x + 1)(x - 2)$

45. $x^3 + x^2 - 6x = x(x^2 + x - 6)$

Integers whose product is −6	1, −6	−1, 6	2, −3	−2, 3
Sum	−5	5	−1	1

Since the coefficient of the middle term is 1: $x^2 + x - 6 = (x - 2)(x + 3)$
Thus, $x^3 + x^2 - 6x = x(x - 2)(x + 3)$

47. $x^3 + x^2 - 2x = x(x^2 + x - 2)$

Integers whose product is −2	1, −2	−1, 2
Sum	−1	1

Since the coefficient of the middle term is 1: $x^2 + x - 2 = (x - 1)(x + 2)$
Thus, $x^3 + x^2 - 2x = x(x - 1)(x + 2)$

49. $3x^2 + 4x + 1$
$AC = 3 \cdot 1 = 3$

Integers whose product is 3	1, 3	−1, −3
Sum	4	−4

Since the coefficient of the middle term is 4: $3x^2 + 4x + 1 = 3x^2 + x + 3x + 1 = (3x^2 + x) + (3x + 1)$
$= x(3x + 1) + 1(3x + 1)$
$= (x + 1)(3x + 1)$

51. $2z^2 + 5z + 3$
$AC = 2 \cdot 3 = 6$

Integers whose product is 6	1, 6	−1, −6	2, 3	−2, −3
Sum	7	−7	5	−5

Since the coefficient of the middle term is 5: $2z^2 + 5z + 3 = 2z^2 + 2z + 3z + 3 = (2z^2 + 2z) + (3z + 3)$
$= 2z(z + 1) + 3(z + 1)$
$= (2z + 3)(z + 1)$

53. $3x^2 + 2x - 8$
$AC = 3 \cdot (-8) = -24$

Integers whose product is −24	1, −24	−1, 24	2, −12	−2, 12	3, −8	−3, 8	4, −6	−4, 6
Sum	−23	23	−10	10	−5	5	−2	2

Since the coefficient of the middle term is 2: $3x^2 + 2x - 8 = 3x^2 - 4x + 6x - 8 = (3x^2 - 4x) + (6x - 8)$
$= x(3x - 4) + 2(3x - 4)$
$= (x + 2)(3x - 4)$

55. $3x^2 - 2x - 8$
$AC = 3 \cdot (-8) = -24$

Integers whose product is −24	1, −24	−1, 24	2, −12	−2, 12	3, −8	−3, 8	4, −6	−4, 6
Sum	−23	23	−10	10	−5	5	−2	2

Since the coefficient of the middle term is −2: $3x^2 - 2x - 8 = 3x^2 + 4x - 6x - 8 = (3x^2 + 4x) - (6x + 8)$
$= x(3x + 4) - 2(3x + 4)$
$= (x - 2)(3x + 4)$

57. $3x^2 + 14x + 8$
$AC = 3 \cdot 8 = 24$

Integers whose product is 24	1, 24	−1, −24	2, 12	−2, −12	3, 8	−3, −8	4, 6	−4, −6
Sum	25	−25	14	−14	11	−11	10	−10

Since the coefficient of the middle term is 14: $3x^2 + 14x + 8 = 3x^2 + 2x + 12x + 8$
$= (3x^2 + 2x) + (12x + 8)$
$= x(3x + 2) + 4(3x + 2)$
$= (x + 4)(3x + 2)$

59. $3x^2 + 10x - 8$
$AC = 3 \cdot (-8) = -24$

Integers whose product is −24	1, −24	−1, 24	2, −12	−2, 12	3, −8	−3, 8	4, −6	−4, 6
Sum	−23	23	−10	10	−5	5	−2	2

Since the coefficient of the middle term is 10: $3x^2 + 10x - 8 = 3x^2 - 2x + 12x - 8$
$= (3x^2 - 2x) + (12x - 8)$
$= x(3x - 2) + 4(3x - 2)$
$= (x + 4)(3x - 2)$

61. $12x^2 - x - 1$
$AC = 12 \cdot (-1) = -12$

Integers whose product is −12	1, −12	−1, 12	2, −6	−2, 6	3, −4	−3, 4
Sum	−11	11	−4	4	−1	1

Since the coefficient of the middle term is −1: $12x^2 - x - 1 = 12x^2 + 3x - 4x - 1 = (12x^2 + 3x) - (4x + 1)$
$= 3x(4x + 1) - 1(4x + 1)$
$= (3x - 1)(4x + 1)$

63. $2x^2 - 7x + 3$
$AC = 2 \cdot 3 = 6$

Integers whose product is 6	1, 6	−1, −6	2, 3	−2, −3
Sum	7	−7	5	−5

Since the coefficient of the middle term is −7: $2x^2 - 7x + 3 = 2x^2 - x - 6x + 3 = (2x^2 - x) - (6x - 3)$
$= x(2x - 1) - 3(2x - 1)$
$= (x - 3)(2x - 1)$

65. $-4x^2 - 7x - 5 = -1(4x^2 + 7x + 5)$
$AC = 4 \cdot 5 = 20$

Integers whose product is 20	1, 20	−1, −20	2, 10	−2, −10	4, 5	−4, −5
Sum	21	−21	12	−12	9	−9

Since the coefficient of the middle term is 7 and no sum equals 7, this polynomial cannot be factored any further.

67. $-2x^2 + 3x + 2 = -1(2x^2 - 3x - 2)$
$AC = 2 \cdot (-2) = -4$

Integers whose product is −4	1, −4	−1, 4	2, −2
Sum	−3	3	0

Since the coefficient of the middle term is −3: $2x^2 - 3x - 2 = 2x^2 - 4x + 1x - 2 = (2x^2 - 4x) + (1x - 2)$
$= 2x(x - 2) + 1(x - 2)$
$= (2x + 1)(x - 2)$

So, $-2x^2 + 3x + 2 = -1(x - 2)(2x + 1)$

69. $10x^2 - 13x - 3$
$AC = 10 \cdot (-3) = -30$

Integers whose product is −30	1, −30	−1, 30	2, −15	−2, 15	3, −10	−3, 10	5, −6	−5, 6
Sum	−29	29	−13	13	−7	7	−1	1

Since the coefficient of the middle term is −13: $10x^2 - 13x - 3 = 10x^2 - 15x + 2x - 3$
$= (10x^2 - 15x) + (2x - 3)$
$= 5x(2x - 3) + 1(2x - 3)$
$= (5x + 1)(2x - 3)$

71. $16x^2 + 16x + 3$
$AC = 16 \cdot 3 = 48$

Integers whose product is 48	1, 48	−1, −48	2, 24	−2, −24	3, 16	−3, −16	4, 12	−4, −12	6, 8	−6, −8
Sum	49	−49	26	−26	19	−19	16	−16	14	−14

Since the coefficient of the middle term is −16:
$$\begin{aligned} 16x^2 + 16x + 3 &= 16x^2 + 4x + 12x + 3 \\ &= (16x^2 + 4x) + (12x + 3) \\ &= 4x(4x + 1) + 3(4x + 1) \\ &= (4x + 3)(4x + 1) \end{aligned}$$

73. $12x^2 - 5x - 2$
$AC = 12 \cdot (-2) = -24$

Integers whose product is −24	1, −24	−1, 24	2, −12	−2, 12	3, −8	−3, 8	4, −6	−4, 6
Sum	−23	23	−10	10	−5	5	−2	2

Since the coefficient of the middle term is −5:
$$\begin{aligned} 12x^2 - 5x - 2 &= 12x^2 - 8x + 3x - 2 = (12x^2 - 8x) + (3x - 2) \\ &= 4x(3x - 2) + 1(3x - 2) \\ &= (4x + 1)(3x - 2) \end{aligned}$$

75. $2x^2 + 3x - 2$
$AC = 2 \cdot (-2) = -4$

Integers whose product is −4	1, −4	−1, 4	2, −2
Sum	−3	3	0

Since the coefficient of the middle term is 3:
$$\begin{aligned} 2x^2 + 3x - 2 &= 2x^2 - 1x + 4x - 2 = (2x^2 - 1x) + (4x - 2) \\ &= x(2x - 1) + 2(2x - 1) \\ &= (x + 2)(2x - 1) \end{aligned}$$

77. $8x^2 - 2x - 15$
$AC = 8 \cdot (-15) = -120$

Integers whose product is −120	1, −120	−1, 120	2, −60	−2, 60	3, −40	−3, 40	4, −30	−4, 30
Sum	−119	119	−58	58	−37	37	−26	26
Integers whose product is −120	5, −24	−5, 24	6, −20	−6, 20	8, −15	−8, 15	10, −12	−10, 12
Sum	−19	19	−14	14	−7	7	−2	2

Since the coefficient of the middle term is −2:
$$\begin{aligned} 8x^2 - 2x - 15 &= 8x^2 - 12x + 10x - 15 \\ &= (8x^2 - 12x) + (10x - 15) \\ &= 4x(2x - 3) + 5(2x - 3) \\ &= (4x + 5)(2x - 3) \end{aligned}$$

79. $8x^2 - 26x + 15$
$AC = 8 \cdot 15 = 120$

Integers whose product is 120	1, 120	−1, −120	2, 60	−2, −60	3, 40	−3, −40	4, 30	−4, −30
Sum	121	−121	62	−62	43	−43	34	−34
Integers whose product is 120	5, 24	−5, −24	6, 20	−6, −20	8, 15	−8, −15	10, 12	−10, −12
Sum	29	−29	26	−26	23	−23	22	−22

Since the coefficient of the middle term is −26: $8x^2 - 26x + 15 = 8x^2 - 6x - 20x + 15$
$$= (8x^2 - 6x) - (20x - 15)$$
$$= 2x(4x - 3) - 5(4x - 3)$$
$$= (2x - 5)(4x - 3)$$

81. $-3x^2 + x + 14 = -1(3x^2 - x - 14)$
$AC = 3 \cdot (-14) = -42$

Integers whose product is −42	1, −42	−1, 42	2, −21	−2, 21	3, −14	−3, 14	6, −7	−6, 7
Sum	−41	41	−19	19	−11	11	−1	1

Since the coefficient of the middle term is −1: $3x^2 - 1x - 10 = 3x^2 + 6x - 7x - 14$
$$= (3x^2 + 6x) - (7x + 14)$$
$$= 3x(x + 2) - 7(x + 2)$$
$$= (3x - 7)(x + 2)$$

So, $-3x^2 + x + 14 = -1(x + 2)(3x - 7)$

83. $12x^2 - 11x + 2$
$AC = 12 \cdot 2 = 24$

Integers whose product is 24	1, 24	−1, −24	2, 12	−2, −12	3, 8	−3, −8	4, 6	−4, −6
Sum	25	−25	14	−14	11	−11	10	−10

Since the coefficient of the middle term is −11: $12x^2 - 11x + 2 = 12x^2 - 3x - 8x + 2$
$$= (12x^2 - 3x) - (8x - 2)$$
$$= 3x(4x - 1) - 2(4x - 1)$$
$$= (3x - 2)(4x - 1)$$

85. $3x^2 + 5x - 2$
$AC = 3 \cdot (-2) = -6$

Integers whose product is −6	1, −6	−1, 6	2, −3	−2, 3
Sum	−5	5	−1	1

Since the coefficient of the middle term is 5: $3x^2 + 5x - 2 = 3x^2 - 1x + 6x - 2 = (3x^2 - 1x) + (6x - 2)$
$= x(3x - 1) + 2(3x - 1)$
$= (x + 2)(3x - 1)$

87. $8x^2 + 23x + 15$
$AC = 8 \cdot 15 = 120$

Integers whose product is 120	1, 120	−1, −120	2, 60	−2, −60	3, 40	−3, −40	4, 30	−4, −30
Sum	121	−121	62	−62	43	−43	34	−34
Integers whose product is 120	5, 24	−5, −24	6, 20	−6, −20	8, 15	−8, −15	10, 12	−10, −12
Sum	29	−29	26	−26	23	−23	22	−22

Since the coefficient of the middle term is 23: $8x^2 + 23x + 15 = 8x^2 + 8x + 15x + 15$
$= (8x^2 + 8x) + (15x + 15)$
$= 8x(x + 1) + 15(x + 1)$
$= (8x + 15)(x + 1)$

89. $10x^2 + 9x - 2$
$AC = 10 \cdot (-2) = -20$

Integers whose product is −20	1, −20	−1, 20	2, −10	−2, 10	4, −5	−4, 5
Sum	−19	19	−8	8	−1	1

Since the coefficient of the middle term is 9 and no sum equals 9, this polynomial is prime.

91. $18x^2 - 9x - 27 = 9(2x^2 - x - 3)$
$AC = 2 \cdot (-3) = -6$

Integers whose product is −6	1, −6	−1, 6	2, −3	−2, 3
Sum	−5	5	−1	1

Since the coefficient of the middle term is −1: $2x^2 - 1x - 3 = 2x^2 + 2x - 3x - 3 = (2x^2 + 2x) - (3x + 3)$
$= 2x(x + 1) - 3(x + 1)$
$= (2x - 3)(x + 1)$

So, $18x^2 - 9x - 27 = 9(2x - 3)(x + 1)$

93. $8x^2 + 2x + 6 = 2(4x^2 + x + 3)$
$AC = 4 \cdot 3 = 12$

Integers whose product is 12	1, 12	−1, −12	2, 6	−2, −6	3, 4	−3, −4
Sum	13	−13	8	−8	7	−7

Since the coefficient of the middle term is 1 and no sum equals 1, this polynomial cannot be factored any further.

95. $x^2 - x + 4$

Integers whose product is 4	1, 4	−1, −4	2, 2	−2, −2
Sum	5	−5	4	−4

Since the coefficient of the middle term is −1 and no sum equals −1, this polynomial is prime.

97. $4x^3 - 10x^2 - 6x = 2x(2x^2 - 5x - 3)$
$AC = 2 \cdot (-3) = -6$

Integers whose product is −6	1, −6	−1, 6	2, −3	−2, 3
Sum	−5	5	−1	1

Since the coefficient of the middle term is −5: $2x^2 - 5x - 3 = 2x^2 + 1x - 6x - 3 = (2x^2 + 1x) - (6x + 3)$
$= x(2x + 1) - 3(2x + 1)$
$= (x - 3)(2x + 1)$

So, $4x^3 - 10x^2 - 6x = 2x(x - 3)(2x + 1)$

99. $x^4 + 2x^2 + 1$

Integers whose product is 1	1, 1	−1, −1
Sum	2	−2

Since the coefficient of the middle term is 2: $x^4 + 2x^2 + 1 = (x^2 + 1)(x^2 + 1) = (x^2 + 1)^2$

101. $x^4 + 8x^2 + 7$

Integers whose product is 7	1, 7	−1, −7
Sum	8	−8

Since the coefficient of the middle term is 8: $x^4 + 8x^2 + 7 = (x^2 + 1)(x^2 + 7)$

103. $x^2 + 4 = x^2 + 0 \cdot x + 4$

Integers whose product is 4	1, 4	−1, −4	2, 2	−2, −2
Sum	5	−5	4	−4

Since the coefficient of the middle term is 0 and no sum equals 0, this polynomial is prime.

2.8 Concepts and Vocabulary *(page 120)*

4. $2x - 3$ **5.** False **6.** False **7.** True **8.** True

2.8 Exercises *(page 120)*

9. $x^2 - 1 = (x + 1)(x - 1)$

11. $4x^2 - 1 = (2x + 1)(2x - 1)$

13. $x^2 - 16 = (x + 4)(x - 4)$

15. $25x^2 - 4 = (5x + 2)(5x - 2)$

17. $9x^2 - 16 = (3x + 4)(3x - 4)$

19. $x^2 + 2x + 1 = (x + 1)(x + 1) = (x + 1)^2$

21. $x^2 + 4x + 4 = (x + 2)(x + 2) = (x + 2)^2$

23. $x^2 - 10x + 25 = (x - 5)(x - 5) = (x - 5)^2$

25. $x^2 + 6x + 9 = (x + 3)(x + 3) = (x + 3)^2$

27. $4x^2 + 4x + 1 = (2x + 1)(2x + 1) = (2x + 1)^2$

29. $16x^2 + 8x + 1 = (4x + 1)(4x + 1) = (4x + 1)^2$

31. $4x^2 + 12x + 9 = (2x + 3)(2x + 3) = (2x + 3)^2$

33. $x^3 - x = x(x^2 - 1) = x(x + 1)(x - 1)$

35. $x^3 - 27 = (x - 3)(x^2 + 3x + 9)$

37. $8x^3 + 27 = (2x + 3)(4x^2 - 6x + 9)$

39. $x^3 + 6x^2 + 9x = x(x^2 + 6x + 9) = x(x + 3)(x + 3) = x(x + 3)^2$

41. $x^4 - 81 = (x^2 - 9)(x^2 + 9) = (x - 3)(x + 3)(x^2 + 9)$

43. $x^6 - 2x^3 + 1 = (x^3 - 1)(x^3 - 1) = (x^3 - 1)^2 = [(x - 1)(x^2 + x + 1)]^2$

45. $x^7 - x^5 = x^5(x^2 - 1) = x^5(x - 1)(x + 1)$

47. $2z^3 + 8z^2 + 8z = 2z(z^2 + 4z + 4) = 2z(z + 2)(z + 2) = 2z(z + 2)^2$

49. $16x^2 - 24x + 9 = (4x - 3)(4x - 3) = (4x - 3)^2$

51. $2x^4 - 2x = 2x(x^3 - 1) = 2x(x - 1)(x^2 + x + 1)$

53. $x^4 + 2x^2 + 1 = (x^2 + 1)(x^2 + 1) = (x^2 + 1)^2$

55. $16x^4 - 1 = (4x^2 - 1)(4x^2 + 1) = (2x - 1)(2x + 1)(4x^2 + 1)$

57. $x^4 + 81x^2 = x^2(x^2 + 81)$

Chapter 2 Review Exercises *(page 122)*

1. $3^2 \cdot 3^0 = 3^{2+0} = 3^2 = 9$

3. $\frac{4^4}{4^2} = 4^{4-2} = 4^2 = 16$

5. $[(-2)^2]^3 = (-2)^{2\cdot 3} = (-2)^6 = 64$

7. $(8^0)^4 = 8^{0\cdot 4} = 8^0 = 1$

9. $-5x^3 + 3x^2 - 9x^2 + 10 = -5x^3 - 6x^2 + 10$; trinomial. The leading coefficient is -5; the degree is 3.

11. $9x^5 + x^5 + 12x^3 = 10x^5 + 12x^3$; binomial. The leading coefficient is 10; the degree is 5.

13. $7x + x = 8x$; monomial. The leading coefficient is 8; the degree is 1.

15. $3x^3 + x - 2$ for $x = 2$

$3(2)^3 + 2 - 2$
$= 3 \cdot 8 + 2 - 2$
$= 24 + 2 - 2$
$= 26 - 2 = 24$

17. $-x^2 + 1$ for $x = 5$

$-(5)^2 + 1 = -25 + 1 = -24$

19. $5z^3 - z^2 + z$ for $z = -2$

$5(-2)^3 - (-2)^2 + (-2) = 5(-8) - (4) + (-2)$
$= -40 + (-4) + (-2) = -46$

21. $x^2y - xy^2 + 2$ for $x = -1, y = 1$

$(-1)^2(1) - (-1)(1)^2 + 2 = 1 \cdot 1 - (-1)(1) + 2 = 1 + 1 + 2 = 4$

23. $x^3 - x^2 + x - 1$ for $x = 1.2$

$(1.2)^3 - (1.2)^2 + (1.2) - 1 = 0.488$

25. $(5x^2 - x + 2) + (-2x^2 + x + 1) = 5x^2 - x + 2 - 2x^2 + x + 1$
$= 3x^2 + 3$

27. $3(x^2 + x - 2) + 2(3x^2 - x + 2)$
$= 3x^2 + 3x - 3(2) + 2 \cdot 3x^2 - 2x + 2(2)$
$= 3x^2 + 3x - 6 + 6x^2 - 2x + 4$
$= 9x^2 + x - 2$

29. $(3x^4 - x^2 + 1) - (2x^4 + x^3 - x + 2)$
$= 3x^4 - x^2 + 1 - 2x^4 - x^3 + x - 2$
$= x^4 - x^3 - x^2 + x - 1$

31. $2(-x^2 + x + 1) - 4(2x^2 + 3x - 2)$
$= 2(-x^2) + 2x + 2(1) - 4(2x^2) - 4(3x) - 4(-2)$
$= -2x^2 + 2x + 2 - 8x^2 - 12x + 8$
$= -10x^2 - 10x + 10$

33. $(4x^3 - x^2 - 2x + 3) + (-8x^3 + x^2 + 3x + 2) - (2x^3 - x^2 + 2x)$
$= 4x^3 - x^2 - 2x + 3 - 8x^3 + x^2 + 3x + 2 - 2x^3 + x^2 - 2x$
$= -6x^3 + x^2 - x + 5$

35. $2(3x + 4y - 2) + 5(x + y - 1)$
$= 2(3x) + 2(4y) + (2)(-2) + 5x + 5y - 5(1)$
$= 6x + 8y - 4 + 5x + 5y - 5$
$= 11x + 13y - 9$

37. $(2x^2 + 3xy + 4y^2) - (x^2 - y^2) = 2x^2 + 3xy + 4y^2 - x^2 + y^2$
$= x^2 + 3xy + 5y^2$

39. $4x(x^2 + x + 2) + 8(x^3 - x^2 + 1)$
$= 4x(x^2) + 4x(x) + 4x(2) + 8x^3 - 8x^2 + 8(1)$
$= 4x^3 + 4x^2 + 8x + 8x^3 - 8x^2 + 8$
$= 12x^3 - 4x^2 + 8x + 8$

41. $(9x^2)(-4x) = -36x^{2+1} = -36x^3$

43. $(-5xy^3)(3x^2y^4) = -15x^{1+2}y^{3+4} = -15x^3y^7$

45. $(2x + 5)(x^2 - x - 1) = 2x(x^2) - 2x(x) - 2x(1) + 5x^2 - 5x - 5(1)$
$= 2x^3 - 2x^2 - 2x + 5x^2 - 5x - 5$
$= 2x^3 + 3x^2 - 7x - 5$

47. $(4x - 1)^2 \cdot (x + 2) = (16x^2 - 8x + 1)(x + 2)$
$= 16x^2(x) + 16x^2(2) - 8x(x) - 8x(2) + 1x + 1(2)$
$= 16x^3 + 32x^2 - 8x^2 - 16x + x + 2$
$= 16x^3 + 24x^2 - 15x + 2$

49. $$\begin{aligned} x(2x+1)^3 &= x(8x^3+12x^2+6x+1) \\ &= 8x^3(x)+12x^2(x)+6x(x)+x \\ &= 8x^4+12x^3+6x^2+x \end{aligned}$$

51. $$\begin{aligned} (x+1)^2+x(x+1) &= x^2+2x+1+x(x)+1(x) \\ &= x^2+2x+1+x^2+x \\ &= 2x^2+3x+1 \end{aligned}$$

53. $$\begin{aligned} (2x+3y)(3x-4y) &= 2x(3x)-(2x)(4y)+3y(3x)-(3y)(4y) \\ &= 6x^2-8xy+9xy-12y^2 \\ &= 6x^2+xy-12y^2 \end{aligned}$$

55. $$\begin{aligned} &3x(2x-1)^2-4x(x+2)-3(x+2)(5x-1) \\ &= 3x(4x^2-4x+1)-4x(x)-4x(2)-3[5x(x)-1(x)+2(5x)-(2)(1)] \\ &= 3x(4x^2)-3x(4x)+3x(1)-4x^2-8x-3[5x^2-x+10x-2] \\ &= 12x^3-12x^2+3x-4x^2-8x-15x^2+3x-30x+6 \\ &= 12x^3-31x^2-32x+6 \end{aligned}$$

57. $$\frac{45x^5}{-5x} = -9x^{5-1} = -9x^4$$

59. $$\frac{-x^4y^3}{xy^3} = -x^{4-1}y^{3-3} = -x^3y^0 = -x^3\cdot 1 = -x^3$$

61. $$\frac{3x^3+x^2-x+4}{x} = \frac{3x^3}{x}+\frac{x^2}{x}-\frac{x}{x}+\frac{4}{x} = 3x^2+x-1+\frac{4}{x}$$

63. $$\frac{4x^3-8x^2+8}{2x^2} = \frac{4x^3}{2x^2}-\frac{8x^2}{2x^2}+\frac{8}{2x^2} = 2x-4+\frac{4}{x^2}$$

65. $$\frac{1-x^2+x^4}{x^4} = \frac{1}{x^4}-\frac{x^2}{x^4}+\frac{x^4}{x^4} = \frac{1}{x^4}-\frac{1}{x^2}+1$$

67.
$$\begin{array}{r} 3x^2+8x+25 \\ x-3\overline{)3x^3-x^2+x+4} \\ \underline{3x^3-9x^2} \qquad\quad \\ 8x^2+x \qquad \\ \underline{8x^2-24x} \qquad \\ 25x+4 \\ \underline{25x-75} \\ 79 \end{array}$$

The quotient is $3x^2+8x+25$; the remainder is 79.
Check: $(x-3)(3x^2+8x+25)+79$
$= 3x^3+8x^2+25x-9x^2-24x-75+79$
$= 3x^3-x^2+x+4$

69.
$$\begin{array}{r} -3x^2+4 \qquad\qquad \\ x^2+1\overline{)-3x^4+0x^3+x^2+0x+2} \\ \underline{-3x^4 \qquad\quad -3x^2} \qquad\qquad \\ 4x^2+0x+2 \\ \underline{4x^2 \qquad\quad +4} \\ -2 \end{array}$$

The quotient is $-3x^2+4$; the remainder is -2.
Check: $(x^2+1)(-3x^2+4)-2 = -3x^4+4x^2-3x^2+4-2$
$= -3x^4+x^2+2$

71.
$$\begin{array}{r} 8x^2+24x+62 \qquad\qquad \\ x^2-3x+1\overline{)8x^4+0x^3-2x^2+5x+1} \\ \underline{8x^4-24x^3+8x^2} \qquad\qquad\quad \\ 24x^3-10x^2+5x \qquad \\ \underline{24x^3-72x^2+24x} \qquad \\ 62x^2-19x+1 \\ \underline{62x^2-186x+62} \\ 167x-61 \end{array}$$

The quotient is $8x^2+24x+62$; the remainder is $167x-61$.
Check: $(x^2-3x+1)(8x^2+24x+62)+167x-61$
$= (8x^4+24x^3+62x^2)-24x^3-72x^2-186x+8x^2+24x+62+167x-61$
$= 8x^4-2x^2+5x+1$

73.

$$\begin{array}{r}
x^4 - x^3 + x^2 - x + 1 \\
x + 1\overline{)x^5 + 0x^4 + 0x^3 + 0x^2 + 0x + 1} \\
\underline{x^5 + x^4} \\
-x^4 + 0x^3 \\
\underline{-x^4 - x^3} \\
x^3 + 0x^2 \\
\underline{x^3 + x^2} \\
-x^2 + 0x \\
\underline{-x^2 - x} \\
x + 1 \\
\underline{x + 1} \\
0
\end{array}$$

The quotient is $x^4 - x^3 + x^2 - x + 1$; the remainder is 0.

Check: $(x + 1)(x^4 - x^3 + x^2 - x + 1)$

$= x^5 - x^4 + x^3 - x^2 + x + x^4 - x^3 + x^2 - x + 1$

$= x^5 + 1$

So, $\dfrac{x^5 + 1}{x + 1} = x^4 - x^3 + x^2 - x + 1$

75.

$$\begin{array}{r}
3x^4 - 2x^2 + 1 \\
2x + 1\overline{)6x^5 + 3x^4 - 4x^3 - 2x^2 + 2x + 1} \\
\underline{6x^5 + 3x^4} \\
-4x^3 - 2x^2 \\
\underline{-4x^3 - 2x^2} \\
2x + 1 \\
\underline{2x + 1} \\
0
\end{array}$$

The quotient is $3x^4 - 2x^2 + 1$; the remainder is 0.

Check: $(2x + 1)(3x^4 - 2x^2 + 1) = 6x^5 - 4x^3 + 2x + 3x^4 - 2x^2 + 1$

$= 6x^5 + 3x^4 - 4x^3 - 2x^2 + 2x + 1$

So, $\dfrac{6x^5 + 3x^4 - 4x^3 - 2x^2 + 2x + 1}{2x + 1} = 3x^4 - 2x^2 + 1$

77. $3x^2 - 6x = 3x(x - 2)$

79. $9x^3 - x = x(9x^2 - 1)$

$= x(3x + 1)(3x - 1)$

81. $9x^3 + x = x(9x^2 + 1)$

83. $8x^3 - 1 = (2x - 1)(4x^2 + 2x + 1)$

85. $x^3 - 6x^2 + 9x = x(x^2 - 6x + 9) = x(x - 3)^2$

87. $x^2 + 8x + 12$

Integers whose product is 12	1, 12	−1, −12	2, 6	−2, −6	3, 4	−3, −4
Sum	13	−13	8	−8	7	−7

Since the coefficient of the middle term is 8: $x^2 + 8x + 12 = (x + 2)(x + 6)$

89. $x^2 + 4x - 12$

Integers whose product is −12	1, −12	−1, 12	2, −6	−2, 6	3, −4	−3, 4
Sum	−11	11	−4	4	−1	1

Since the coefficient of the middle term is 4: $x^2 + 4x - 12 = (x - 2)(x + 6)$

91. $x^2 - 6x + 9 = (x - 3)(x - 3) = (x - 3)^2$

93. $4x^2 - 8x - 5$

Integers whose product is −20	1, −20	−1, 20	2, −10	−2, 10	4, −5	−4, 5
Sum	−19	19	−8	8	−1	1

Since the coefficient of the middle term is −8:

$$4x^2 - 8x - 5 = 4x^2 - 10x + 2x - 5 = (4x^2 - 10x) + (2x - 5)$$
$$= 2x(2x - 5) + 1(2x - 5) = (2x + 1)(2x - 5)$$

95. $2x(3x + 1) - 3(3x + 1) = (2x - 3)(3x + 1)$

97. $x^3 - x^2 + x - 1 = (x^3 - x^2) + (x - 1)$
$= x^2(x - 1) + (x - 1)$
$= (x^2 + 1)(x - 1)$

99. $x^2 + x + 1$

Integers whose product is 1	1, 1	−1, −1
Sum	2	−2

Since the coefficient of the middle term is 1 and no sum equals 1, the polynomial is prime.

101. $12x^2 - 11x - 15$
$AC = 12 \cdot (-15) = -180$

Integers whose product is −180	1, −180	−1, 180	2, −90	−2, 90	3, −60	−3, 60	4, −45	−4, 45	5, −36
Sum	−179	179	−88	88	−57	57	−41	41	−31
Integers whose product is −180	−5, 36	6, −30	−6, 30	9, −20	−9, 20	10, −18	−10, 18	12, −15	−12, −5
Sum	31	−24	24	−11	11	−8	8	−3	3

Since the coefficient of the middle term is −11: $12x^2 - 11x - 15 = 12x^2 + 9x - 20x - 15$
$= (12x^2 + 9x) - (20x + 15)$
$= 3x(4x + 3) - 5(4x + 3)$
$= (3x - 5)(4x + 3)$

103. $12x^2 - 27x + 15$
$AC = 12 \cdot 15 = 180$

Integers whose product is −180	1, 180	−1, −180	2, 90	−2, −90	3, 60	−3, −60	4, 45	−4, −45	5, 36
Sum	181	−181	92	−92	63	−63	49	−49	41
Integers whose product is −180	−5, −36	6, 30	−6, −30	9, 20	−9, −20	10, 18	−10, −18	12, 15	−12, −15
Sum	−41	36	−36	29	−29	28	−28	27	−27

Since the coefficient of the middle term is −27: $12x^2 - 27x + 15 = 12x^2 - 12x - 15x + 15$
$= (12x^2 - 12x) - (15x - 15)$
$= 12x(x - 1) - 15(x - 1)$
$= (12x - 15)(x - 1)$

CHAPTER 3 Rational Expressions

3.1 Concepts and Vocabulary *(page 139)*

3. True **4.** $\frac{-5}{9}$ **5.** numerators, denominators **6.** False **7.** 180 **8.** lowest terms **9.** True **10.** False

3.1 Exercises *(page 140)*

11. $\frac{5}{8}$ and $\frac{30}{48}$

(a)

$$\begin{array}{r} 0.625 \\ 8\overline{)5.000} \\ \underline{4\,8} \\ 20 \\ \underline{16} \\ 40 \\ \underline{40} \\ 0 \end{array} \qquad \begin{array}{r} 0.625 \\ 48\overline{)30.000} \\ \underline{288} \\ 120 \\ \underline{96} \\ 240 \\ \underline{240} \\ 0 \end{array}$$

Each fraction has the same decimal form, 0.625, so they are equivalent.

(b) $\frac{5}{8} = \frac{30}{48}$ since $5 \cdot 48 = 240$ and $8 \cdot 30 = 240$

(c) $\frac{30}{48} = \frac{6 \cdot 5}{6 \cdot 8} = \frac{5}{8}$ Cancellation Property, cancel the 6's.

13. $\frac{10}{8}$ and $\frac{5}{4}$

(a)

$$\begin{array}{r} 1.25 \\ 8\overline{)10.00} \\ \underline{8} \\ 20 \\ \underline{16} \\ 40 \\ \underline{40} \\ 0 \end{array} \qquad \begin{array}{r} 1.25 \\ 4\overline{)5.00} \\ \underline{4} \\ 10 \\ \underline{8} \\ 20 \\ \underline{20} \\ 0 \end{array}$$

Each fraction has the same decimal form, 1.25, so they are equivalent.

(b) $\frac{10}{8} = \frac{5}{4}$ since $10 \cdot 4 = 40$ and $8 \cdot 5 = 40$

(c) $\frac{10}{8} = \frac{2 \cdot 5}{2 \cdot 4} = \frac{5}{4}$ Cancellation Property, cancel the 2's

15. $\frac{-1}{8}$ and $\frac{-14}{112}$

(a)

$$\begin{array}{r} -0.125 \\ 8\overline{)-1.000} \\ \underline{8} \\ 20 \\ \underline{16} \\ 40 \\ \underline{40} \\ 0 \end{array} \qquad \begin{array}{r} -0.125 \\ 112\overline{)-14.000} \\ \underline{112} \\ 280 \\ \underline{224} \\ 560 \\ \underline{560} \\ 0 \end{array}$$

Each fraction has the same decimal form, −0.125, so they are equivalent.

(b) $\frac{-1}{8} = \frac{-14}{112}$ since $-1 \cdot 112 = -112$ and $-14 \cdot 8 = -112$

(c) $\frac{-14}{112} = \frac{(-1) \cdot 14}{8 \cdot 14} = \frac{-1}{8}$ Cancellation Property, cancel the 14's.

17. $\frac{48}{14}$ and $\frac{-72}{-21}$

(a)

$$\begin{array}{r} 3.428571 \\ 14\overline{)48.000000} \\ \underline{42} \\ 60 \\ \underline{56} \\ 40 \\ \underline{28} \\ 120 \\ \underline{112} \\ 80 \\ \underline{70} \\ 100 \\ \underline{98} \\ 20 \\ \underline{14} \\ 6 \end{array} \qquad \begin{array}{r} 3.428571 \\ -21\overline{)-72.000000} \\ \underline{63} \\ 90 \\ \underline{84} \\ 60 \\ \underline{42} \\ 180 \\ \underline{168} \\ 120 \\ \underline{105} \\ 150 \\ \underline{147} \\ 30 \\ \underline{21} \\ 9 \end{array}$$

Each fraction has the same decimal form, 3.438571, so they are equivalent.

(b) $\frac{48}{14} = \frac{-72}{-21}$ since $48 \cdot (-21) = -1008$ and $14 \cdot (-72) = -1008$

(c) $\frac{-72}{-21} = \frac{24(-3)}{7(-3)} = \frac{24}{7}$ and $\frac{48}{14} = \frac{2 \cdot 24}{2 \cdot 7} = \frac{24}{7}$ Cancellation Property, cancel the -3's and the 2's

19. $\frac{-40}{32}$ and $\frac{-60}{48}$

(a)

$$\begin{array}{r} -1.25 \\ 32\overline{)-40.00} \\ \underline{32} \\ 80 \\ \underline{64} \\ 160 \\ \underline{160} \\ 0 \end{array} \qquad \begin{array}{r} -1.25 \\ 48\overline{)-60.00} \\ \underline{48} \\ 120 \\ \underline{96} \\ 240 \\ \underline{240} \\ 0 \end{array}$$

Each fraction has the decimal form -1.25, so they are equivalent.

(b) $\frac{-40}{32} = \frac{-60}{48}$ since $(-40) \cdot 48 = -1920$ and $32 \cdot (-60) = -1920$

(c) $\frac{-40}{32} = \frac{(-5) \cdot 8}{4 \cdot 8} = \frac{-5}{4}$ and $\frac{-60}{48} = \frac{(-5) \cdot 12}{4 \cdot 12} = \frac{-5}{4}$ Cancellation Property, cancel the 8's and the 12's

21. $\frac{30}{54} = \frac{2 \cdot 3 \cdot 5}{2 \cdot 3 \cdot 3 \cdot 3}$

$= \frac{5}{3 \cdot 3}$ Cancel the 2's and the 3's

$= \frac{5}{9}$

23. $\frac{82}{18} = \frac{2 \cdot 41}{2 \cdot 3 \cdot 3}$

$= \frac{41}{3 \cdot 3}$ Cancel the 2's

$= \frac{41}{9}$

25. $\frac{-21}{36} = \frac{3\cdot(-7)}{2\cdot 2\cdot 3\cdot 3}$
$= \frac{-7}{2\cdot 2\cdot 3}$ Cancel the 3's
$= \frac{-7}{12}$

27. $\frac{-28}{-36} = \frac{(-2)\cdot 2\cdot 7}{(-2)\cdot 2\cdot 3\cdot 3}$
$= \frac{7}{3\cdot 3}$ Cancel the −2's and the 2's
$= \frac{7}{9}$

29. $\frac{5}{4} + \frac{1}{4} = \frac{5+1}{4} = \frac{6}{4} = \frac{2\cdot 3}{2\cdot 2} = \frac{3}{2}$

31. $\frac{13}{2} + \frac{9}{2} + \frac{11}{2} = \frac{13+9+11}{2} = \frac{33}{2}$

33. $\frac{13}{5} - \frac{8}{5} = \frac{13-8}{5} = \frac{5}{5} = 1$

35. $\frac{9}{2} - \frac{7}{2} = \frac{9-7}{2} = \frac{2}{2} = 1$

37. $\frac{8}{3} - \frac{1}{3} + \frac{2}{3} = \frac{8-1}{3} + \frac{2}{3} = \frac{7}{3} + \frac{2}{3} = \frac{9}{3} = \frac{3\cdot 3}{3} = \frac{3}{1} = 3$

39. $\frac{9}{5} + \frac{6}{5} - \frac{4}{5} = \frac{9+6}{5} - \frac{4}{5} = \frac{15}{5} - \frac{4}{5} = \frac{15-4}{5} = \frac{11}{5}$

41. $\frac{3}{2}\cdot\frac{8}{9} = \frac{3\cdot 8}{2\cdot 9} = \frac{24}{18} = \frac{2\cdot 2\cdot 2\cdot 3}{2\cdot 3\cdot 3} = \frac{2\cdot 2}{3} = \frac{4}{3}$

43. $\frac{16}{25}\cdot\frac{5}{8} = \frac{16\cdot 5}{25\cdot 8} = \frac{80}{200} = \frac{2\cdot 2\cdot 2\cdot 2\cdot 5}{2\cdot 2\cdot 2\cdot 5\cdot 5} = \frac{2}{5}$

45. $\frac{25}{18}\cdot\frac{-6}{5} = \frac{25\cdot(-6)}{18\cdot 5} = \frac{-150}{90} = \frac{(-2)\cdot 3\cdot 5\cdot 5}{2\cdot 3\cdot 3\cdot 5} = \frac{(-2)\cdot 5}{2\cdot 3} = \frac{(-1)\cdot 2\cdot 5}{2\cdot 3} = \frac{(-1)\cdot 5}{3} = \frac{-5}{3}$

47. $\frac{8}{9}\cdot\frac{3}{2}\cdot\frac{18}{7} = \frac{8\cdot 3\cdot 18}{9\cdot 2\cdot 7} = \frac{432}{126} = \frac{2\cdot 2\cdot 2\cdot 2\cdot 3\cdot 3\cdot 3}{2\cdot 3\cdot 3\cdot 7} = \frac{2\cdot 2\cdot 2\cdot 3}{7} = \frac{24}{7}$

49. $\frac{1}{3} + \frac{2}{3}\cdot\frac{1}{2} = \frac{1}{3} + \frac{2\cdot 1}{3\cdot 2}$ Perform the multiplication before adding
$= \frac{1}{3} + \frac{1}{3}$ Cancel the 2's
$= \frac{1+1}{3} = \frac{2}{3}$

51. $\frac{4}{5}\cdot\frac{10}{3} - \frac{8}{3} = \frac{4\cdot 10}{5\cdot 3} - \frac{8}{3}$ Perform multiplication before subtraction
$= \frac{40}{15} - \frac{8}{3}$
$= \frac{8\cdot 5}{3\cdot 5} - \frac{8}{3}$ Reduce $\frac{40}{15}$
$= \frac{8}{3} - \frac{8}{3}$ Perform subtraction
$= \frac{8-8}{3} = \frac{0}{3} = 0$

53. $\frac{-5}{8}\cdot\frac{12}{25} = \frac{(-5)\cdot 12}{8\cdot 25} = \frac{-2\cdot 2\cdot 3\cdot 5}{2\cdot 2\cdot 2\cdot 5\cdot 5} = \frac{-3}{2\cdot 5} = \frac{-3}{10}$

55. $\frac{-16}{-3}\cdot\frac{9}{20} = \frac{(-16)\cdot 9}{(-3)\cdot 20} = \frac{-2\cdot 2\cdot 2\cdot 2\cdot 3\cdot 3}{-3\cdot 2\cdot 2\cdot 5} = \frac{2\cdot 2\cdot 3}{5} = \frac{12}{5}$

57. $\frac{\frac{15}{8}}{\frac{25}{2}} = \frac{15}{8} \div \frac{25}{2} = \frac{15}{8}\cdot\frac{2}{25} = \frac{15\cdot 2}{8\cdot 25} = \frac{2\cdot 3\cdot 5}{2\cdot 2\cdot 2\cdot 5\cdot 5} = \frac{3}{2\cdot 2\cdot 5} = \frac{3}{20}$

59. $\frac{\frac{-15}{32}}{\frac{25}{24}} = \frac{-15}{32} \div \frac{25}{24} = \frac{-15}{32}\cdot\frac{24}{25} = \frac{(-15)\cdot 24}{32\cdot 25} = \frac{-2\cdot 2\cdot 2\cdot 3\cdot 3\cdot 5}{2\cdot 2\cdot 2\cdot 2\cdot 2\cdot 5\cdot 5} = \frac{-3\cdot 3}{2\cdot 2\cdot 5} = \frac{-9}{20}$

61. $\frac{\frac{12}{5}}{\frac{24}{15}} = \frac{12}{5} \div \frac{24}{15} = \frac{12}{5}\cdot\frac{15}{24} = \frac{12\cdot 15}{5\cdot 24} = \frac{2\cdot 2\cdot 3\cdot 3\cdot 5}{2\cdot 2\cdot 2\cdot 3\cdot 5} = \frac{3}{2}$

63. $\dfrac{\frac{22}{9}}{\frac{33}{12}} = \dfrac{22}{9} \div \dfrac{33}{12} = \dfrac{22}{9}\cdot\dfrac{12}{33} = \dfrac{22\cdot 12}{9\cdot 33} = \dfrac{2\cdot 2\cdot 2\cdot 3\cdot 11}{3\cdot 3\cdot 3\cdot 11} = \dfrac{2\cdot 2\cdot 2}{3\cdot 3} = \dfrac{8}{9}$

65. $6 = 2\cdot 3$ and $4 = 2\cdot 2$
LCM $= 2\cdot 2\cdot 3 = 12$

67. $6 = 2\cdot 3$ and $9 = 3\cdot 3$
LCM $= 2\cdot 3\cdot 3 = 18$

69. $12 = 2\cdot 2\cdot 3$ and $18 = 2\cdot 3\cdot 3$
LCM $= 2\cdot 2\cdot 3\cdot 3\cdot = 36$

71. $15 = 3\cdot 5$ and $18 = 2\cdot 3\cdot 3$
LCM $= 2\cdot 3\cdot 3\cdot 5 = 90$

73. $4 = 2\cdot 2, 6 = 2\cdot 3$ and $15 = 3\cdot 5$
LCM $= 2\cdot 2\cdot 3\cdot 5 = 60$

75. $\dfrac{4}{3} + \dfrac{3}{2}$
LCM $= 2\cdot 3 = 6$
$\dfrac{4\cdot 2}{3\cdot 2} + \dfrac{3\cdot 3}{2\cdot 3} = \dfrac{8}{6} + \dfrac{9}{6} = \dfrac{8+9}{6} = \dfrac{17}{6}$

77. $\dfrac{1}{5} + \dfrac{3}{4}$
$5 = 5$ and $4 = 2\cdot 2$
LCM $= 2\cdot 2\cdot 5 = 20$
$\dfrac{1\cdot 4}{5\cdot 4} + \dfrac{3\cdot 5}{4\cdot 5} = \dfrac{4}{20} + \dfrac{15}{20} = \dfrac{4+15}{20} = \dfrac{19}{20}$

79. $\dfrac{4}{9} + \dfrac{3}{8}$
$9 = 3\cdot 3$ and $8 = 2\cdot 2\cdot 2$
LCM $= 2\cdot 2\cdot 2\cdot 3\cdot 3 = 72$
$\dfrac{4\cdot 8}{9\cdot 8} + \dfrac{3\cdot 9}{8\cdot 9} = \dfrac{32}{72} + \dfrac{27}{72} = \dfrac{32+27}{72} = \dfrac{59}{72}$

81. $\dfrac{8}{3} - \dfrac{1}{2}$
LCM $= 2\cdot 3 = 6$
$\dfrac{8\cdot 2}{3\cdot 2} - \dfrac{1\cdot 3}{2\cdot 3} = \dfrac{16}{6} - \dfrac{3}{6} = \dfrac{16-3}{6} = \dfrac{13}{6}$

83. $\dfrac{1}{5} - \dfrac{3}{4}$
$5 = 5$ and $4 = 2\cdot 2$
LCM $= 2\cdot 2\cdot 5 = 20$
$\dfrac{1\cdot 4}{5\cdot 4} - \dfrac{3\cdot 5}{4\cdot 5} = \dfrac{4}{20} - \dfrac{15}{20} = \dfrac{4-15}{20} = \dfrac{4+(-15)}{20} = \dfrac{-11}{20}$

85. $\dfrac{4}{9} - \dfrac{3}{8}$
LCM $= 9\cdot 8 = 72$
$\dfrac{4}{9}\cdot\dfrac{8}{8} - \dfrac{9}{9}\cdot\dfrac{3}{8} = \dfrac{32}{72} - \dfrac{27}{72} = \dfrac{32-27}{72} = \dfrac{5}{72}$

87. $\dfrac{2}{7} + \dfrac{5}{14}$
$7 = 7, 14 = 2\cdot 7$
LCM $= 2\cdot 7 = 14$
$\dfrac{2}{7}\cdot\dfrac{2}{2} + \dfrac{5}{14} = \dfrac{4}{14} + \dfrac{5}{14} = \dfrac{4+5}{14} = \dfrac{9}{14}$

89. $\dfrac{9}{13} + \dfrac{11}{26}$
$13 = 13$ and $26 = 2\cdot 13$
LCM $= 2\cdot 13 = 26$
$\dfrac{9\cdot 2}{13\cdot 2} + \dfrac{11}{26} = \dfrac{18}{26} + \dfrac{11}{26} = \dfrac{18+11}{26} = \dfrac{29}{26}$

91. $\dfrac{9}{10} - \dfrac{1}{5}$
$10 = 2\cdot 5$ and $5 = 5$
LCM $= 2\cdot 5 = 10$
$\dfrac{9}{10} - \dfrac{1\cdot 2}{5\cdot 2} = \dfrac{9}{10} - \dfrac{2}{10} = \dfrac{9-2}{10} = \dfrac{7}{10}$

93. $\dfrac{5}{12} - \dfrac{11}{24}$
$12 = 2\cdot 2\cdot 3$ and $24 = 2\cdot 2\cdot 2\cdot 3$
LCM $= 2\cdot 2\cdot 2\cdot 3 = 24$
$\dfrac{5\cdot 2}{12\cdot 2} - \dfrac{11}{24} = \dfrac{10}{24} - \dfrac{11}{24} = \dfrac{10-11}{24} = \dfrac{10+(-11)}{24} = \dfrac{-1}{24}$

95. $\dfrac{5}{9} + \dfrac{5}{6}$
$9 = 3\cdot 3$ and $6 = 2\cdot 3$
LCM $= 2\cdot 3\cdot 3 = 18$
$\dfrac{5\cdot 2}{9\cdot 2} + \dfrac{5\cdot 3}{6\cdot 3} = \dfrac{10}{18} + \dfrac{15}{18} = \dfrac{10+15}{18} = \dfrac{25}{18}$

97. $\dfrac{1}{6} + \dfrac{11}{15}$
$6 = 2\cdot 3$ and $15 = 3\cdot 5$
LCM $= 2\cdot 3\cdot 5 = 30$
$\dfrac{1\cdot 5}{6\cdot 5} + \dfrac{11\cdot 2}{15\cdot 2} = \dfrac{5}{30} + \dfrac{22}{30} = \dfrac{5+22}{30} = \dfrac{27}{30} = \dfrac{3\cdot 3\cdot 3}{2\cdot 3\cdot 5} = \dfrac{3\cdot 3}{2\cdot 5} = \dfrac{9}{10}$

99. $\frac{3}{8}+\frac{5}{12}$

$8 = 2\cdot 2\cdot 2$ and $12 = 2\cdot 2\cdot 3$

$\text{LCM} = 2\cdot 2\cdot 2\cdot 3 = 24$

$$\frac{3\cdot 3}{8\cdot 3}+\frac{5\cdot 2}{12\cdot 2}=\frac{9}{24}+\frac{10}{24}=\frac{9+10}{24}=\frac{19}{24}$$

101. $\frac{5}{6}+\frac{3}{4}$

$6 = 2\cdot 3$ and $4 = 2\cdot 2$

$\text{LCM} = 2\cdot 2\cdot 3 = 12$

$$\frac{5\cdot 2}{6\cdot 2}+\frac{3\cdot 3}{4\cdot 3}=\frac{10}{12}+\frac{9}{12}=\frac{9+10}{12}=\frac{19}{12}$$

103. $\frac{5}{9}-\frac{5}{6}$

$9 = 3\cdot 3$ and $6 = 2\cdot 3$

$\text{LCM} = 2\cdot 3\cdot 3 = 18$

$$\frac{5\cdot 2}{9\cdot 2}-\frac{5\cdot 3}{6\cdot 3}=\frac{10}{18}-\frac{15}{18}=\frac{10-15}{18}=\frac{10+(-15)}{18}=\frac{-5}{18}$$

105. $\frac{1}{6}-\frac{11}{15}$

$6 = 2\cdot 3$ and $15 = 3\cdot 5$

$\text{LCM} = 2\cdot 3\cdot 5 = 30$

$$\frac{1\cdot 5}{6\cdot 5}-\frac{11\cdot 2}{15\cdot 2}= =\frac{5}{30}-\frac{22}{30}=\frac{5-22}{30}=\frac{5+(-22)}{30}=\frac{-17}{30}$$

107. $\frac{3}{8}-\frac{5}{12}$

$8 = 2\cdot 2\cdot 2$ and $12 = 2\cdot 2\cdot 3$

$\text{LCM} = 2\cdot 2\cdot 2\cdot 3 = 24$

$$\frac{3\cdot 3}{8\cdot 3}-\frac{5\cdot 2}{12\cdot 2}=\frac{9}{24}-\frac{10}{24}=\frac{9-10}{24}=\frac{9+(-10)}{24}=\frac{-1}{24}$$

109. $\frac{5}{6}-\frac{3}{4}$

$6 = 2\cdot 3$ and $4 = 2\cdot 2$

$\text{LCM} = 2\cdot 2\cdot 3 = 12$

$$\frac{5\cdot 2}{6\cdot 2}-\frac{3\cdot 3}{4\cdot 3}=\frac{10}{12}-\frac{9}{12}=\frac{10-9}{12}=\frac{1}{12}$$

111. $\frac{5}{2}-\frac{8}{3}+\frac{1}{6}$

$\text{LCM} = 2\cdot 3 = 6$

$$\frac{5\cdot 3}{2\cdot 3}-\frac{8\cdot 2}{3\cdot 2}+\frac{1}{6}=\frac{15}{6}-\frac{16}{6}+\frac{1}{6}$$

$$=\frac{15-16}{6}+\frac{1}{6}=\frac{15+(-16)}{6}+\frac{1}{6}=\frac{-1}{6}+\frac{1}{6}=\frac{-1+1}{6}=\frac{0}{6}=0$$

113. $\frac{3}{4}+\frac{8}{3}-\frac{11}{12}$

$4 = 2\cdot 2, 3 = 3$ and $12 = 2\cdot 2\cdot 3$

$\text{LCM} = 2\cdot 2\cdot 3 = 12$

$$\frac{3\cdot 3}{4\cdot 3}+\frac{8\cdot 4}{3\cdot 4}-\frac{11}{12}=\frac{9}{12}+\frac{32}{12}-\frac{11}{12}$$

$$=\frac{9+32}{12}-\frac{11}{12}=\frac{41}{12}-\frac{11}{12}=\frac{41-11}{12}=\frac{30}{12}=\frac{2\cdot 3\cdot 5}{2\cdot 2\cdot 3}=\frac{5}{2}$$

115. $\frac{1}{12}+\frac{4}{3}+\frac{3}{4}$

$12 = 2\cdot 2\cdot 3, 3 = 3$ and $4 = 2\cdot 2$

$\text{LCM} = 2\cdot 2\cdot 3 = 12$

$$\frac{1}{12}+\frac{4\cdot 4}{3\cdot 4}+\frac{3\cdot 3}{4\cdot 3}=\frac{1}{12}+\frac{16}{12}+\frac{9}{12}=\frac{1+16+9}{12}=\frac{26}{12}=\frac{2\cdot 13}{2\cdot 2\cdot 3}=\frac{13}{2\cdot 3}=\frac{13}{6}$$

117. $\frac{1}{2}+\frac{1}{3}+\frac{1}{4}$

$2 = 2, 3 = 3$ and $4 = 2\cdot 2$

$\text{LCM} = 2\cdot 2\cdot 3 = 12$

$$\frac{1\cdot 6}{2\cdot 6}+\frac{1\cdot 4}{3\cdot 4}+\frac{1\cdot 3}{4\cdot 3}=\frac{6}{12}+\frac{4}{12}+\frac{3}{12}=\frac{6+4+3}{12}=\frac{13}{12}$$

119. $\frac{3}{8} + \frac{1}{4} - \frac{1}{6}$

$8 = 2 \cdot 2 \cdot 2, 4 = 2 \cdot 2$ and $6 = 2 \cdot 3$

$\text{LCM} = 2 \cdot 2 \cdot 2 \cdot 3 = 24$

$$\frac{3 \cdot 3}{8 \cdot 3} + \frac{1 \cdot 6}{4 \cdot 6} - \frac{1 \cdot 4}{6 \cdot 4} = \frac{9}{24} + \frac{6}{24} - \frac{4}{24} = \frac{9+6}{24} - \frac{4}{24} = \frac{15}{24} - \frac{4}{24} = \frac{15-4}{24} = \frac{11}{24}$$

121. $\frac{3}{4} \cdot \left(\frac{3}{2} + \frac{2}{3}\right)$

$2 = 2$ and $3 = 3$

$\text{LCM} = 2 \cdot 3 = 6$

$$\frac{3}{4} \cdot \left(\frac{3 \cdot 3}{2 \cdot 3} + \frac{2 \cdot 2}{3 \cdot 2}\right) = \frac{3}{4} \cdot \left(\frac{9}{6} + \frac{4}{6}\right) = \frac{3}{4} \cdot \left(\frac{9+4}{6}\right) = \frac{3}{4} \cdot \left(\frac{13}{6}\right)$$

$$= \frac{3 \cdot 13}{4 \cdot 6} = \frac{3 \cdot 13}{2 \cdot 2 \cdot 2 \cdot 3} = \frac{13}{2 \cdot 2 \cdot 2} = \frac{13}{8}$$

123. $\frac{5}{6} + \frac{3}{4} - \left(\frac{1}{2}\right)^2 = \frac{5}{6} + \frac{3}{4} - \left(\frac{1}{4}\right)$ Evaluate exponents $\left(\frac{1}{2}\right)^2 = \left(\frac{1}{2}\right) \cdot \left(\frac{1}{2}\right) = \frac{1 \cdot 1}{2 \cdot 2} = \frac{1}{4}$

$6 = 2 \cdot 3$ and $4 = 2 \cdot 2$

$\text{LCM} = 2 \cdot 2 \cdot 3 = 12$

$$\frac{5 \cdot 2}{6 \cdot 2} + \frac{3 \cdot 3}{4 \cdot 3} - \frac{1 \cdot 3}{4 \cdot 3} = \frac{10}{12} + \frac{9}{12} - \frac{3}{12} = \frac{10+9}{12} - \frac{3}{12}$$

$$= \frac{19}{12} - \frac{3}{12} = \frac{19-3}{12} = \frac{16}{12} = \frac{2 \cdot 2 \cdot 2 \cdot 2}{2 \cdot 2 \cdot 3} = \frac{2 \cdot 2}{3} = \frac{4}{3}$$

125. $\frac{5}{12} - \frac{1}{18} + \frac{7}{30}$

$12 = 2 \cdot 2 \cdot 3, 18 = 2 \cdot 3 \cdot 3$ and $30 = 2 \cdot 3 \cdot 5$

$\text{LCM} = 2 \cdot 2 \cdot 3 \cdot 3 \cdot 5 = 180$

$$\frac{5 \cdot 15}{12 \cdot 15} - \frac{1 \cdot 10}{18 \cdot 10} + \frac{7 \cdot 6}{30 \cdot 6} = \frac{75}{180} - \frac{10}{180} + \frac{42}{180} = \frac{75-10}{180} + \frac{42}{180}$$

$$= \frac{65}{180} + \frac{42}{180} = \frac{65+42}{180} = \frac{107}{180}$$

127. $\frac{4}{5} \cdot \frac{25}{8} + \frac{5}{2} \cdot \frac{3}{25} = \frac{4 \cdot 25}{5 \cdot 8} + \frac{5 \cdot 3}{2 \cdot 25} = \frac{100}{40} + \frac{15}{50}$*

$40 = 2 \cdot 2 \cdot 2 \cdot 5$ and $50 = 2 \cdot 5 \cdot 5$

$\text{LCM} = 2 \cdot 2 \cdot 2 \cdot 5 \cdot 5 = 200$

$$\frac{100 \cdot 5}{40 \cdot 5} + \frac{15 \cdot 4}{50 \cdot 4} = \frac{500}{200} + \frac{60}{200} = \frac{560}{200} = \frac{2 \cdot 2 \cdot 2 \cdot 2 \cdot 5 \cdot 7}{2 \cdot 2 \cdot 2 \cdot 5 \cdot 5} = \frac{2 \cdot 7}{5} = \frac{14}{5}$$

*alternatively, from

$$\frac{100}{40} + \frac{15}{50} = \frac{2 \cdot 2 \cdot 5 \cdot 5}{2 \cdot 2 \cdot 2 \cdot 5} + \frac{3 \cdot 5}{2 \cdot 5 \cdot 5} = \frac{5}{2} + \frac{3}{10} = \frac{5 \cdot 5}{2 \cdot 5} + \frac{3}{10} = \frac{25}{10} + \frac{3}{10} = \frac{25+3}{10} = \frac{28}{10} = \frac{2 \cdot 2 \cdot 7}{2 \cdot 5} = \frac{2 \cdot 7}{5} = \frac{14}{5}$$

129. Mike pays $\frac{5}{8} \cdot 10 = \frac{5 \cdot 10}{8} = \frac{50}{8} = \frac{2 \cdot 5 \cdot 5}{2 \cdot 2 \cdot 2} = \frac{5 \cdot 5}{2 \cdot 2} = \frac{25}{4} = \$6.25$

Kathleen pays $10.00 − $6.25 = $3.75

131. Since each only pays for the amount they consume, and 6 slices were consumed in all,

Katy pays $\frac{1}{6} \cdot 15 = \frac{15}{6} = \$2.50$

Mike pays $\frac{2}{6} \cdot 15 = \frac{30}{6} = \$5.00$

Danny pays $\frac{3}{6} \cdot 15 = \frac{45}{6} = \$7.50$

3.2 Concepts and Vocabulary *(page 146)*

7. denominator **8.** False **9.** False **10.** $\frac{11}{2}$

3.2 Exercises *(page 146)*

11. $\frac{2x-4}{4x} = \frac{2(x-2)}{2 \cdot 2x} = \frac{x-2}{2x}$

13. $\frac{x^2-x}{x^2+x} = \frac{x(x-1)}{x(x+1)} = \frac{x-1}{x+1}$

15. $\frac{x^2+4x+4}{x^2+6x+8} = \frac{(x+2)(x+2)}{(x+2)(x+4)} = \frac{x+2}{x+4}$

17. $\frac{x^2-9}{x^2-6x+9} = \frac{(x-3)(x+3)}{(x-3)(x-3)} = \frac{x+3}{x-3}$

19. $\frac{x^2+x-2}{x^3+4x^2+4x} = \frac{(x-1)(x+2)}{x(x+2)(x+2)} = \frac{x-1}{x(x+2)}$

21. $\frac{5x+10}{x^2-4} = \frac{5(x+2)}{(x-2)(x+2)} = \frac{5}{x-2}$

23. $\frac{x^2-2x}{3x-6} = \frac{x(x-2)}{3(x-2)} = \frac{x}{3}$

25. $\frac{24x}{12x^2-6x} = \frac{24x}{6x(2x-1)} = \frac{6 \cdot 4x}{6x(2x-1)} = \frac{4}{2x-1}$

27. $\frac{y^2-25}{2y-10} = \frac{(y-5)(y+5)}{2(y-5)} = \frac{y+5}{2}$

29. $\frac{x^2+4x-5}{x-1} = \frac{(x+5)(x-1)}{(x-1)} = x+5$

31. $\frac{x^2-4}{x^2+5x+6} = \frac{(x-2)(x+2)}{(x+3)(x+2)} = \frac{x-2}{x+3}$

33. $\frac{x^2-x-2}{2x^2-3x-2} = \frac{(x-2)(x+1)}{(x-2)(2x+1)} = \frac{x+1}{2x+1}$

35. $\frac{x^3+x^2+x}{x^3-1} = \frac{x(x^2+x+1)}{(x-1)(x^2+x+1)} = \frac{x}{x-1}$

37. $\frac{3x^2+6x}{3x^2+5x-2} = \frac{3x(x+2)}{(x+2)(3x-1)} = \frac{3x}{3x-1}$

39. $\frac{x^4-1}{x^3-x} = \frac{(x^2-1)(x^2+1)}{x(x^2-1)} = \frac{x^2+1}{x}$

41. $\frac{(x^2-3x-10)(x^2+4x-21)}{(x^2+2x-35)(x^2+9x+14)} = \frac{(x-5)(x+2)(x+7)(x-3)}{(x-5)(x+7)(x+7)(x+2)} = \frac{x-3}{x+7}$

43. $\frac{x^2+5x-14}{2-x} = \frac{(x+7)(x-2)}{(-1)(x-2)} = -(x+7)$

45. $\frac{2x^3-x^2-10x}{x^3-2x^2-8x} = \frac{x(2x^2-x-10)}{x(x^2-2x-8)} = \frac{x(x+2)(2x-5)}{x(x-4)(x+2)} = \frac{2x-5}{x-4}$

47. $\frac{(x-4)^2-9}{(x+3)^2-16} = \frac{(x-4)^2-3^2}{(x+3)^2-4^2} = \frac{[(x-4)-3][(x-4)+3]}{[(x+3)-4][(x+3)+4]} = \frac{(x-7)(x-1)}{(x-1)(x+7)} = \frac{x-7}{x+7}$

49. $$\frac{6x(x-1)-12}{x^3-8-(x-2)^2} = \frac{6x^2-6x-12}{(x^3-2^3)-(x-2)^2} = \frac{6(x^2-x-2)}{(x-2)(x^2+2x+4)-(x-2)^2}$$
$$= \frac{6(x-2)(x+1)}{(x-2)[x^2+2x+4-(x-2)]} = \frac{6(x-2)(x+1)}{(x-2)[x^2+2x+4-x+2]}$$
$$= \frac{6(x-2)(x+1)}{(x-2)[x^2+x+6]} = \frac{6(x+1)}{x^2+x+6}$$

51. $\dfrac{x^2+1}{3x}$ at $x = 3$

$$\frac{x^2+1}{3x} = \frac{(3)^2+1}{3(3)} = \frac{9+1}{9} = \frac{10}{9}$$

$\uparrow$ $x = 3$

53. $\dfrac{x^2-4x+4}{x^2-25}$ at $x = -4$

$$\frac{x^2-4x+4}{x^2-25} = \frac{(-4)^2-4(-4)+4}{(-4)^2-25} = \frac{16-(-16)+4}{16-25} = \frac{16+16+4}{-9} = \frac{36}{-9} = \frac{2\cdot2\cdot3\cdot3}{-3\cdot3}$$
$$= \frac{2\cdot2}{-1} = -4$$

$\uparrow$ $x = -4$

55. $\dfrac{9x^2}{x^2+1}$ at $x = -1$

$$\frac{9x^2}{x^2+1} = \frac{9(-1)^2}{(-1)^2+1} = \frac{9\cdot1}{1+1} = \frac{9}{2}$$

$\uparrow$ $x = -1$

57. $\dfrac{x^2+x+1}{x^2-x+1}$ at $x = 1$

$$\frac{x^2+x+1}{x^2-x+1} = \frac{1^2+1+1}{1^2-1+1} = \frac{1+1+1}{1-1+1} = \frac{3}{1} = 3$$

$\uparrow$ $x = 1$

59. $\dfrac{1+x+x^2}{x^2}$ at $x = 3.21$

$$\frac{1+x+x^2}{x^2} = \frac{1+3.21+3.21^2}{3.21^2} = \frac{14.5141}{10.3041} \approx 1.4086$$

$\uparrow$ $x = 3.21$

The only reason for excluding a number in the domain for the following equations is if the denominator becomes zero. Therefore, check each problem's denominator at the value prescribed and if it is zero, the number must be excluded from the domain.

61. $\dfrac{x^2-1}{x}$

(a) At $x = 3$, the denominator is 3.
(b) At $x = 1$, the denominator is 1.
(c) At $x = 0$, the denominator is 0; this must be excluded from the domain.
(d) At $x = -1$, the denominator is -1.

63. $\dfrac{x}{x^2-9}$

(a) At $x = 3$, the denominator is $3^2 - 9 = 9 - 9 = 0$; this must be excluded from the domain.
(b) At $x = 1$, the denominator is $1^2 - 9 = 1 - 9 = -8$.
(c) At $x = 0$, the denominator is $0^2 - 9 = 0 - 9 = -9$.
(d) At $x = -1$, the denominator is $(-1)^2 - 9 = 1 - 9 = -8$.

65. $\dfrac{x^2}{x^2+1}$

(a) At $x = 3$, the denominator is $3^2 + 1 = 9 + 1 = 10$.
(b) At $x = 1$, the denominator is $1^2 + 1 = 1 + 1 = 2$.
(c) At $x = 0$, the denominator is $0^2 + 1 = 0 + 1 = 1$.
(d) At $x = -1$, the denominator is $(-1)^2 + 1 = 1 + 1 = 2$.

67. $\dfrac{x^2+5x-10}{x^3-x}$

(a) At $x = 3$, the denominator is $3^3 - 3 = 27 - 3 = 24$.
(b) At $x = 1$, the denominator is $1^3 - 1 = 1 - 1 = 0$; this must be excluded from the domain.
(c) At $x = 0$, the denominator is $0^3 - 0 = 0$; this must be excluded from the domain.
(d) At $x = -1$, the denominator is $(-1)^3 - (-1) = -1 + 1 = 0$; this must be excluded from the domain.

69. $\dfrac{x^2+x+1}{x^2-x+1}$

(a) At $x = 3$, the denominator is $3^2 - 3 + 1 = 9 - 3 + 1 = 7$.
(b) At $x = 1$, the denominator is $1^2 - 1 + 1 = 1 - 1 + 1 = 1$.
(c) At $x = 0$, the denominator is $0^2 - 0 + 1 = 0 - 0 + 1 = 1$.
(d) At $x = -1$, the denominator is $(-1)^2 - (-1) + 1 = 1 + 1 + 1 = 3$.

71. $\dfrac{5}{4-x} = \dfrac{5}{(-1)(x-4)} = \dfrac{-5}{x-4}$

73. $\dfrac{2x+3}{4-x} = \dfrac{2x+3}{(-1)(x-4)} = -\dfrac{2x+3}{x-4}$

75. $\dfrac{x}{x^2-1}$

(a) At $x = 1.1$, $\dfrac{x}{x^2-1} = \dfrac{1.1}{(1.1)^2-1} = \dfrac{1.1}{1.21-1} = \dfrac{1.1}{0.21} \approx 5.238$
$\uparrow$
$x = 1.1$

(b) At $x = 1.01$, $\dfrac{x}{x^2-1} = \dfrac{1.01}{(1.01)^2-1} = \dfrac{1.01}{1.0201-1} = \dfrac{1.01}{0.0201} \approx 50.249$
$\uparrow$
$x = 1.01$

(c) At $x = 1.001$, $\dfrac{x}{x^2-1} = \dfrac{1.001}{(1.001)^2-1} = \dfrac{1.001}{1.002001-1} = \dfrac{1.01}{0.002001} \approx 500.250$
$\uparrow$
$x = 1.001$

(d) At $x = 1.0001$,
$\dfrac{x}{x^2-1} = \dfrac{1.0001}{(1.0001)^2-1} = \dfrac{1.0001}{1.00020001-1} = \dfrac{1.0001}{0.00020001} \approx 5000.250$
$\uparrow$
$x = 1.0001$

(e) Let $x = 1$, $\dfrac{x}{x^2-1} = \dfrac{1}{1^2-1} = \dfrac{1}{1-1} = \dfrac{1}{0}$
$\uparrow$
$x = 1$

You cannot have a zero in the denominator. Therefore, x cannot equal one.

3.3 Concepts and Vocabulary *(page 152)*

7. numerators, denominators **8.** $\dfrac{x+2}{x^2-1}$

3.3 Exercises *(page 152)*

9. $\dfrac{3x}{4}\cdot\dfrac{12}{x} = \dfrac{3x\cdot 12}{4x} = \dfrac{3\cdot x\cdot 3\cdot 4}{4\cdot x} = \dfrac{3\cdot 3}{1} = 9$

11. $\dfrac{8x^2}{x+1}\cdot\dfrac{x^2-1}{2x} = \dfrac{8x^2\cdot(x^2-1)}{(x+1)\cdot 2x} = \dfrac{4\cdot 2\cdot x\cdot x(x-1)(x+1)}{(x+1)\cdot 2\cdot x} = \dfrac{4\cdot x\cdot(x-1)}{1} = 4x(x-1)$

13. $\dfrac{x+5}{6x}\cdot\dfrac{x^2}{2x+10} = \dfrac{(x+5)\cdot x^2}{6x\cdot(2x+10)} = \dfrac{(x+5)\cdot x\cdot x}{6\cdot x\cdot 2(x+5)} = \dfrac{x}{6\cdot 2} = \dfrac{x}{12}$

15. $\dfrac{4x-8}{3x+6}\cdot\dfrac{3}{2x-4} = \dfrac{(4x-8)\cdot 3}{(3x+6)(2x-4)} = \dfrac{4(x-2)\cdot 3}{3(x+2)\cdot 2(x-2)} = \dfrac{2\cdot 2\cdot 3(x-2)}{3\cdot 2\cdot(x+2)(x-2)} = \dfrac{2}{x+2}$

17. $\frac{3x+9}{2x-4}\cdot\frac{x^2-4}{6x}=\frac{(3x+9)\cdot(x^2-4)}{(2x-4)\cdot 6x}=\frac{3(x+3)(x-2)(x+2)}{2(x-2)\cdot 2\cdot 3\cdot x}$

$=\frac{(x+3)(x+2)}{2\cdot 2\cdot x}=\frac{(x+3)(x+2)}{4x}$

19. $\frac{3x-6}{5x}\cdot\frac{x^2-x-6}{x^2-4}=\frac{(3x-6)(x^2-x-6)}{5x\cdot(x^2-4)}=\frac{3(x-2)(x-3)(x+2)}{5x\cdot(x-2)(x+2)}=\frac{3(x-3)}{5x}$

21. $\frac{4x^2-1}{x^2-16}\cdot\frac{x^2-4x}{2x+1}=\frac{(4x^2-1)(x^2-4x)}{(x^2-16)(2x+1)}=\frac{(2x-1)(2x+1)\cdot x(x-4)}{(x-4)(x+4)(2x+1)}=\frac{(2x-1)\cdot x}{x+4}$

23. $\frac{4x-8}{-3x}\cdot\frac{12}{12-6x}=\frac{(4x-8)\cdot 12}{-3x\cdot(12-6x)}=\frac{(4x-8)\cdot 12}{-3x\cdot(-1)(6x-12)}=\frac{4(x-2)\cdot 3\cdot 2\cdot 2}{-3x\cdot(-1)\cdot 6(x-2)}$

$=\frac{4\cdot 2\cdot 2}{6x}=\frac{4\cdot 2\cdot 2}{3\cdot 2\cdot x}=\frac{4\cdot 2}{3\cdot x}=\frac{8}{3x}$

25. $\frac{x^2-3x-10}{x^2+2x-35}\cdot\frac{x^2+4x-21}{x^2+9x+14}=\frac{(x^2-3x-10)(x^2+4x-21)}{(x^2+2x-35)(x^2+9x+14)}$

$=\frac{(x-5)(x+2)(x+7)(x-3)}{(x+7)(x-5)(x+7)(x+2)}=\frac{x-3}{x+7}$

27. $\frac{x^2+x-12}{x^2-x-12}\cdot\frac{x^2+7x+12}{x^2-7x+12}=\frac{(x^2+x-12)(x^2+7x+12)}{(x^2-x-12)(x^2-7x+12)}$

$=\frac{(x+4)(x-3)(x+4)(x+3)}{(x-4)(x+3)(x-4)(x-3)}=\frac{(x+4)(x+4)}{(x-4)(x-4)}$

$=\frac{(x+4)^2}{(x-4)^2}$

29. $\frac{1-x^2}{1+x^2}\cdot\frac{x^3+x}{x^3-x}=\frac{(1-x^2)(x^3+x)}{(1+x^2)(x^3-x)}=\frac{(-1)(x^2-1)(x^3+x)}{(x^2+1)(x^3-x)}$

$=\frac{(-1)(x^2-1)\cdot x(x^2+1)}{(x^2+1)\cdot x(x^2-1)}=\frac{(-1)}{1}=-1$

31. $\frac{4x^2+4x+1}{x^2+4x+4}\cdot\frac{x^2-4}{4x^2-1}=\frac{(4x^2+4x+1)(x^2-4)}{(x^2+4x+4)(4x^2-1)}$

$=\frac{(2x+1)(2x+1)(x-2)(x+2)}{(x+2)(x+2)(2x-1)(2x+1)}$

$=\frac{(2x+1)(x-2)}{(x+2)(2x-1)}$

33. $\frac{2x^2+x-3}{2x^2-x-3}\cdot\frac{4x^2-9}{x^2-1}=\frac{(2x^2+x-3)(4x^2-9)}{(2x^2-x-3)(x^2-1)}$

$=\frac{(x-1)(2x+3)(2x-3)(2x+3)}{(x+1)(2x-3)(x-1)(x+1)}$

$=\frac{(2x+3)(2x+3)}{(x+1)(x+1)}=\frac{(2x+3)^2}{(x+1)^2}$

35. $$\frac{x^3-8}{25-4x^2}\cdot\frac{10+4x}{2x^2-9x+10}=\frac{(x^3-8)(10+4x)}{(25-4x^2)(2x^2-9x+10)}$$
$$=\frac{(x-2)(x^2+2x+4)\cdot 2(5+2x)}{(5-2x)(5+2x)(x-2)(2x-5)}$$
$$=\frac{2\cdot(x-2)(x^2+2x+4)(5+2x)}{(-1)(2x-5)(5+2x)(x-2)(2x-5)}$$
$$=\frac{2\cdot(x^2+2x+4)}{(-1)(2x-5)(2x-5)}$$
$$=\frac{-2(x^2+2x+4)}{(2x-5)^2}$$

37. $$\frac{8x^3+27}{1+x-6x^2}\cdot\frac{4-10x-6x^2}{4x^2+12x+9}=\frac{(8x^3+27)(4-10x-6x^2)}{(1+x-6x^2)(4x^2+12x+9)}$$
$$=\frac{(8x^3+27)\cdot(-2)(3x^2+5x-1)}{(-1)\cdot(6x^2-x-1)(4x^2+12x+9)}$$
$$=\frac{(2x+3)(4x^2-6x+9)\cdot(-1)(2)(x+2)(3x-1)}{(-1)\cdot(2x-1)(3x+1)(2x+3)(2x+3)}$$
$$=\frac{(-1)(2)\cdot(2x+3)(4x^2-6x+9)(x+2)(3x-1)}{(-1)\cdot(2x-1)(3x+1)(2x+3)(2x+3)}$$
$$=\frac{2\cdot(4x^2-6x+9)(x+2)(3x-1)}{(2x-1)(3x+1)(2x+3)}$$

39. $$\frac{\frac{2x}{9}}{\frac{x}{3}}=\frac{2x}{9}\div\frac{x}{3}=\frac{2x}{9}\cdot\frac{3}{x}=\frac{2x\cdot 3}{9\cdot x}=\frac{2\cdot x\cdot 3}{3\cdot 3\cdot x}=\frac{2}{3}$$

41. $$\frac{\frac{3x^2}{x+1}}{\frac{6x}{x^2-1}}=\frac{3x^2}{x+1}\div\frac{6x}{x^2-1}=\frac{3x^2}{x+1}\cdot\frac{x^2-1}{6x}=\frac{3x^2\cdot(x^2-1)}{(x+1)\cdot 6x}=\frac{3\cdot x\cdot x\cdot(x-1)(x+1)}{(x+1)\cdot 3\cdot 2\cdot x}=\frac{x(x-1)}{2}$$

43. $$\frac{\frac{12x}{x-5}}{\frac{2x^2}{3x-15}}=\frac{12x}{x-5}\div\frac{2x^2}{3x-15}=\frac{12x}{x-5}\cdot\frac{3x-15}{2x^2}=\frac{12x\cdot(3x-15)}{(x-5)\cdot 2x^2}$$
$$=\frac{6\cdot 2\cdot x\cdot 3(x-5)}{(x-5)\cdot 2\cdot x\cdot x}=\frac{6\cdot 3}{x}=\frac{18}{x}$$

45. $$\frac{\frac{4x-8}{3x+6}}{\frac{8}{x^2-4}}=\frac{4x-8}{3x+6}\div\frac{8}{x^2-4}=\frac{4x-8}{3x+6}\cdot\frac{x^2-4}{8}=\frac{(4x-8)(x^2-4)}{(3x+6)\cdot 8}$$
$$=\frac{4(x-2)(x-2)(x+2)}{3(x+2)\cdot 2\cdot 2\cdot 2}=\frac{2\cdot 2(x-2)(x-2)}{3\cdot 2\cdot 2\cdot 2}=\frac{(x-2)(x-2)}{3\cdot 2}=\frac{(x-2)^2}{6}$$

47. $$\frac{\frac{6x}{x^2-4}}{\frac{3x-9}{2x+4}} = \frac{6x}{x^2-4} \div \frac{3x-9}{2x+4} = \frac{6x}{x^2-4} \cdot \frac{2x+4}{3x-9} = \frac{6x \cdot (2x+4)}{(x^2-4)(3x-9)}$$

$$= \frac{3 \cdot 2 \cdot x \cdot 2(x+2)}{(x-2)(x+2) \cdot 3(x-3)} = \frac{2 \cdot x \cdot 2}{(x-2)(x-3)} = \frac{4x}{(x-2)(x-3)}$$

49. $$\frac{\frac{8x}{x^2-1}}{\frac{10x}{x+1}} = \frac{8x}{x^2-1} \div \frac{10x}{x+1} = \frac{8x}{x^2-1} \cdot \frac{x+1}{10x} = \frac{8x \cdot (x+1)}{(x^2-1) \cdot 10x}$$

$$= \frac{4 \cdot 2 \cdot x \cdot (x+1)}{(x-1)(x+1) \cdot 2 \cdot 5 \cdot x} = \frac{4}{(x-1) \cdot 5} = \frac{4}{5(x-1)}$$

51. $$\frac{\frac{4-x}{4+x}}{\frac{4x}{x^2-16}} = \frac{4-x}{4+x} \div \frac{4x}{x^2-16} = \frac{4-x}{4+x} \cdot \frac{x^2-16}{4x} = \frac{(4-x)(x^2-16)}{(4+x) \cdot 4x}$$

$$= \frac{(-1)(x-4)(x^2-16)}{(x+4) \cdot 4 \cdot x} = \frac{(-1)(x-4)(x-4)(x+4)}{(x+4) \cdot 4 \cdot x} = \frac{(-1)(x-4)^2}{4x}$$

53. $$\frac{\frac{x^2+7x+12}{x^2-7x+12}}{\frac{x^2+x-12}{x^2-x-12}} = \frac{x^2+7x+12}{x^2-7x+12} \div \frac{x^2+x-12}{x^2-x-12} = \frac{x^2+7x+12}{x^2-7x+12} \cdot \frac{x^2-x-12}{x^2+x-12}$$

$$= \frac{(x^2+7x+12)(x^2-x-12)}{(x^2-7x+12)(x^2+x-12)}$$

$$= \frac{(x+4)(x+3)(x-4)(x+3)}{(x-4)(x-3)(x+4)(x-3)}$$

$$= \frac{(x+3)(x+3)}{(x-3)(x-3)} = \frac{(x+3)^2}{(x-3)^2}$$

55. $$\frac{\frac{1-x^2}{1+x^2}}{\frac{x-x^3}{x+x^3}} = \frac{1-x^2}{1+x^2} \div \frac{x-x^3}{x+x^3} = \frac{1-x^2}{1+x^2} \cdot \frac{x+x^3}{x-x^3} = \frac{(1-x^2)(x+x^3)}{(1+x^2)(x-x^3)}$$

$$= \frac{(-1)(x^2-1)(x^3+x)}{(x^2+1) \cdot (-1)(x^3-x)} = \frac{(-1)(x^2-1) \cdot x(x^2+1)}{(x^2+1) \cdot (-1) \cdot x(x^2-1)} = 1$$

57. $$\frac{\frac{2x^2-x-28}{3x^2-x-2}}{\frac{4x^2+16x+7}{3x^2+11x+6}} = \frac{2x^2-x-28}{3x^2-x-2} \div \frac{4x^2+16x+7}{3x^2+11x+6} = \frac{2x^2-x-28}{3x^2-x-2} \cdot \frac{3x^2+11x+6}{4x^2+16x+7}$$

$$= \frac{(2x^2-x-28)(3x^2+11x+6)}{(3x^2-x-2)(4x^2+16x+7)} = \frac{(x-4)(2x+7)(x+3)(3x+2)}{(x-1)(3x+2)(2x+1)(2x+7)}$$

$$= \frac{(x-4)(x+3)}{(x-1)(2x+1)}$$

59. $$\frac{\dfrac{8x^2-6x+1}{4x^2-1}}{\dfrac{12x^2+5x-2}{6x^2-x-2}} = \frac{8x^2-6x+1}{4x^2-1} \div \frac{12x^2+5x-2}{6x^2-x-2} = \frac{8x^2-6x+1}{4x^2-1}\cdot\frac{6x^2-x-2}{12x^2+5x-2}$$
$$= \frac{(8x^2-6x+1)(6x^2-x-2)}{(4x^2-1)(12x^2+5x-2)} = \frac{(2x-1)(4x-1)(2x+1)(3x-2)}{(2x-1)(2x+1)(4x-1)(3x+2)} = \frac{3x-2}{3x+2}$$

61. $$\frac{\dfrac{9x^2+6x+1}{x^2+6x+9}}{\dfrac{9x^2-1}{x^2-9}} = \frac{9x^2+6x+1}{x^2+6x+9} \div \frac{9x^2-1}{x^2-9} = \frac{9x^2+6x+1}{x^2+6x+9}\cdot\frac{x^2-9}{9x^2-1}$$
$$= \frac{(9x^2+6x+1)(x^2-9)}{(x^2+6x+9)(9x^2-1)} = \frac{(3x+1)(3x+1)(x-3)(x+3)}{(x+3)(x+3)(3x-1)(3x+1)}$$
$$= \frac{(3x+1)(x-3)}{(x+3)(3x-1)}$$

63. $$\frac{\dfrac{9-4x^2}{1-x^2}}{\dfrac{2x^2-x-3}{2x^2+x-3}} = \frac{9-4x^2}{1-x^2} \div \frac{2x^2-x-3}{2x^2+x-3} = \frac{9-4x^2}{1-x^2}\cdot\frac{2x^2+x-3}{2x^2-x-3}$$
$$= \frac{(9-4x^2)(2x^2+x-3)}{(1-x^2)(2x^2-x-3)} = \frac{(-1)(4x^2-9)(2x^2+x-3)}{(-1)(x^2-1)(2x^2-x-3)}$$
$$= \frac{(-1)(2x-3)(2x+3)(x-1)(2x+3)}{(-1)(x-1)(x+1)(x+1)(2x-3)} = \frac{(2x+3)(2x+3)}{(x+1)(x+1)} = \frac{(2x+3)^2}{(x+1)^2}$$

65. $$\frac{\dfrac{4x+10}{2x^2-9x+10}}{\dfrac{25-4x^2}{8-x^3}} = \frac{4x+10}{2x^2-9x+10} \div \frac{25-4x^2}{8-x^3} = \frac{4x+10}{2x^2-9x+10}\cdot\frac{8-x^3}{25-4x^2}$$
$$= \frac{(4x+10)(8-x^3)}{(2x^2-9x+10)(25-4x^2)} = \frac{(4x+10)(-1)(x^3-8)}{(2x^2-9x+10)(-1)(4x^2-25)}$$
$$= \frac{2(2x+5)(-1)(x-2)(x^2+2x+4)}{(x-2)(2x-5)(-1)(2x-5)(2x+5)} = \frac{2(x^2+2x+4)}{(2x-5)(2x-5)} = \frac{2(x^2+2x+4)}{(2x-5)^2}$$

67. $$\frac{\dfrac{9x^2-12x+4}{3-8x-3x^2}}{\dfrac{27x^3-8}{2-3x-9x^2}} = \frac{9x^2-12x+4}{3-8x-3x^2} \div \frac{27x^3-8}{2-3x-9x^2} = \frac{9x^2-12x+4}{3-8x-3x^2}\cdot\frac{2-3x-9x^2}{27x^3-8}$$
$$= \frac{(9x^2-12x+4)(2-3x-9x^2)}{(3-8x-3x^2)(27x^3-8)} = \frac{(9x^2-12x+4)(-1)(9x^2+3x-2)}{(-1)(3x^2+8x-3)(27x^3-8)}$$
$$= \frac{(3x-2)(3x-2)(-1)(3x+2)(3x-1)}{(-1)(x+3)(3x-1)(3x-2)(9x^2+6x+4)} = \frac{(3x-2)(3x+2)}{(x+3)(9x^2+6x+4)}$$

69. $$\frac{3x}{x+2}\cdot\frac{x^2-4}{12x^3}\cdot\frac{18x}{x-2} = \frac{3x\cdot(x^2-4)\cdot 18x}{(x+2)\cdot 12x^3\cdot(x-2)} = \frac{3\cdot x\cdot(x-2)(x+2)\cdot 2\cdot 3\cdot 3\cdot x}{(x+2)\cdot 2\cdot 2\cdot 3\cdot x\cdot x\cdot x\cdot(x-2)} = \frac{3\cdot 3}{2\cdot x} = \frac{9}{2x}$$

71. $$\frac{5x^2-x}{3x+2}\cdot\frac{2x^2-x}{2x^2-x-1}\cdot\frac{10x^2+3x-1}{6x^2+x-2} = \frac{(5x^2-x)(2x^2-x)(10x^2+3x-1)}{(3x+2)(2x^2-x-1)(6x^2+x-2)}$$
$$= \frac{x(5x-1)\cdot x(2x-1)(2x+1)(5x-1)}{(3x+2)(x-1)(2x+1)(3x+2)(2x-1)}$$
$$= \frac{x\cdot x(5x-1)(5x-1)}{(3x+2)(3x+2)(x-1)} = \frac{x^2\cdot(5x-1)^2}{(3x+2)^2(x-1)}$$

73. $$\frac{\dfrac{x+1}{x+2}\cdot\dfrac{x+3}{x-1}}{\dfrac{x^2-1}{x^2+2x}} = \left(\frac{x+1}{x+2}\cdot\frac{x+3}{x-1}\right) \div \frac{x^2-1}{x^2+2x} = \frac{x+1}{x+2}\cdot\frac{x+3}{x-1}\cdot\frac{x^2+2x}{x^2-1}$$

$$= \frac{(x+1)(x+3)(x^2+2x)}{(x+2)(x-1)(x^2-1)} = \frac{(x+1)(x+3)\cdot x(x+2)}{(x+2)(x-1)(x-1)(x+1)}$$

$$= \frac{(x+3)\cdot x}{(x-1)(x-1)} = \frac{x(x+3)}{(x-1)^2}$$

75. $$\frac{\dfrac{x^2+3x+2}{x^2-9}\cdot\dfrac{x^2+9}{x^2+6x+9}}{\dfrac{x^2+4}{x^2}\cdot\dfrac{3x^2+27}{2x^2-18}} = \left(\frac{x^2+3x+2}{x^2-9}\cdot\frac{x^2+9}{x^2+6x+9}\right) \div \left(\frac{x^2+4}{x^2}\cdot\frac{3x^2+27}{2x^2-18}\right)$$

$$= \left(\frac{(x^2+3x+2)(x^2+9)}{(x^2-9)(x^2+6x+9)}\right) \div \left(\frac{(x^2+4)(3x^2+27)}{x^2\cdot(2x^2-18)}\right)$$

$$= \left(\frac{(x^2+3x+2)(x^2+9)}{(x^2-9)(x^2+6x+9)}\right)\cdot\left(\frac{x^2\cdot(2x^2-18)}{(x^2+4)(3x^2+27)}\right)$$

$$= \frac{(x^2+3x+2)(x^2+9)x^2\cdot(2x^2-18)}{(x^2-9)(x^2+6x+9)(x^2+4)(3x^2+27)}$$

$$= \frac{(x+2)(x+1)(x^2+9)\cdot x\cdot x\cdot 2(x^2-9)}{(x^2-9)(x+3)(x+3)(x^2+4)\cdot 3(x^2+9)}$$

$$= \frac{(x+2)(x+1)\cdot x\cdot x\cdot 2}{(x+3)(x+3)(x^2+4)\cdot 3} = \frac{2\cdot x^2\cdot(x+2)(x+1)}{3\cdot(x+3)^2(x^2+4)}$$

77. $$\frac{\dfrac{x^4-x^8}{x^2+1}\cdot\dfrac{3x^2}{(x-2)^2}}{\dfrac{x^3+x^6}{x^2-1}\cdot\dfrac{12x}{x^4-1}} = \left(\frac{x^4-x^8}{x^2+1}\cdot\frac{3x^2}{(x-2)^2}\right) \div \left(\frac{x^3+x^6}{x^2-1}\cdot\frac{12x}{x^4-1}\right)$$

$$= \left(\frac{(x^4-x^8)\cdot 3x^2}{(x^2+1)(x-2)^2}\right) \div \left(\frac{(x^3+x^6)\cdot 12x}{(x^2-1)(x^4-1)}\right)$$

$$= \left(\frac{(x^4-x^8)\cdot 3x^2}{(x^2+1)(x-2)^2}\right)\cdot\left(\frac{(x^2-1)(x^4-1)}{(x^3+x^6)\cdot 12x}\right)$$

$$= \frac{(x^4-x^8)\cdot 3x^2\cdot(x^2-1)(x^4-1)}{(x^2+1)(x-2)^2(x^3+x^6)\cdot 12x}$$

$$= \frac{(-1)(x^8-x^4)\cdot 3x^2\cdot(x^2-1)(x^4-1)}{(x^2+1)(x-2)^2(x^6+x^3)\cdot 12x}$$

$$= \frac{(-1)x^4(x^4-1)\cdot 3x^2\cdot(x^2-1)(x^4-1)}{(x^2+1)(x-2)^2\cdot x^3(x^3+1)\cdot 12x}$$

$$= \frac{(-1)x^4(x^2-1)(x^2+1)\cdot 3x^2\cdot(x^2-1)(x^2-1)(x^2+1)}{(x^2+1)(x-2)^2\cdot x^3\cdot(x+1)(x^2-x+1)\cdot 12x}$$

$$= \frac{(-1)x\cdot x\cdot x\cdot x\cdot(x-1)(x+1)(x^2+1)\cdot 3\cdot x\cdot x\cdot(x-1)(x+1)(x-1)(x+1)(x^2+1)}{(x^2+1)(x-2)(x-2)\cdot x\cdot x\cdot x\cdot(x+1)(x^2-x+1)\cdot 2\cdot 2\cdot 3\cdot x}$$

$$= \frac{(-1)\cdot(x-1)\cdot x\cdot x\cdot(x-1)(x+1)(x-1)(x+1)(x^2+1)}{(x-2)(x-2)\cdot(x^2-x+1)\cdot 2\cdot 2}$$

$$= \frac{-x^2\cdot(x-1)^3(x+1)^2(x^2+1)}{4\cdot(x-2)^2(x^2-x+1)}$$

3.4 Concepts and Vocabulary *(page 162)*

7. least common multiple **8.** $36x(x-1)(x+1)$ **9.** True **10.** $(2x+1)(x+5)$

3.4 Exercises *(page 162)*

11. $\frac{x^2}{2}+\frac{1}{2}=\frac{x^2+1}{2}$

13. $\frac{3}{x}+\frac{5}{x}=\frac{3+5}{x}=\frac{8}{x}$

15. $\frac{x}{x^2-4}+\frac{6}{x^2-4}=\frac{x+6}{x^2-4}=\frac{x+6}{(x-2)(x+2)}$

17. $\frac{x^2}{x^2+1}-\frac{1}{x^2+1}=\frac{x^2-1}{x^2+1}=\frac{(x-1)(x+1)}{x^2+1}$

19. $\frac{3}{x}-\frac{5}{x}=\frac{3-5}{x}=\frac{-2}{x}$

21. $\frac{x}{x+2}-\frac{x+1}{x+2}=\frac{x-(x+1)}{x+2}=\frac{x-x-1}{x+2}=\frac{-1}{x+2}$

23. $\frac{x^2}{x^2+4}-\frac{x^2+1}{x^2+4}=\frac{x^2-(x^2+1)}{x^2+4}=\frac{x^2-x^2-1}{x^2+4}=\frac{-1}{x^2+4}$

25. $\frac{3x+1}{x-2}+\frac{2x-3}{2-x}=\frac{3x+1}{x-2}+\frac{2x-3}{-(x-2)}=\frac{3x+1}{x-2}-\frac{2x-3}{x-2}=\frac{(3x+1)-(2x-3)}{x-2}$

$=\frac{3x+1-2x+3}{x-2}=\frac{x+4}{x-2}$

27. $\frac{2x-5}{x-7}-\frac{3x+4}{7-x}=\frac{2x-5}{x-7}-\frac{3x+4}{-(x-7)}=\frac{2x-5}{x-7}+\frac{3x+4}{x-7}=\frac{2x-5+3x+4}{x-7}=\frac{5x-1}{x-7}$

29. $\frac{x-4}{x-1}-\frac{x+2}{1-x}=\frac{x-4}{x-1}-\frac{x+2}{-(x-1)}=\frac{x-4}{x-1}+\frac{x+2}{x-1}=\frac{x-4+x+2}{x-1}=\frac{2x-2}{x-1}=\frac{2(x-1)}{(x-1)}=2$

31. $\frac{x-7}{x-5}+\frac{2x-9}{5-x}=\frac{x-7}{x-5}+\frac{2x-9}{-(x-5)}=\frac{x-7}{x-5}-\frac{2x-9}{x-5}=\frac{(x-7)-(2x-9)}{x-5}$

$=\frac{x-7-2x+9}{x-5}=\frac{-x+2}{x-5}=\frac{-(x-2)}{x-5}$

33. $\frac{3x+7}{x^2-4}-\frac{2x+1}{x^2-4}+\frac{x+3}{x^2-4}=\frac{(3x+7)-(2x+1)+(x+3)}{x^2-4}=\frac{3x+7-2x-1+x+3}{x^2-4}$

$=\frac{2x+9}{x^2-4}=\frac{2x+9}{(x-2)(x+2)}$

35. $\frac{x^2}{x-3}-\frac{x^2+2x}{3-x}+\frac{x^2-2x+4}{3-x}=\frac{x^2}{x-3}-\frac{x^2+2x}{-(x-3)}+\frac{x^2-2x+4}{-(x-3)}$

$=\frac{x^2}{x-3}+\frac{x^2+2x}{x-3}-\frac{x^2-2x+4}{x-3}$

$=\frac{x^2+(x^2+2x)-(x^2-2x+4)}{x-3}$

$=\frac{x^2+x^2+2x-x^2+2x-4}{x-3}$

$=\frac{x^2+4x-4}{x-3}$

37. $\frac{x}{2}+\frac{3}{x}=\frac{x\cdot x}{2\cdot x}+\frac{2\cdot 3}{2\cdot x}=\frac{x^2}{2x}+\frac{6}{2x}=\frac{x^2+6}{2x}$

39. $\frac{x-3}{4}+\frac{4}{x}=\frac{(x-3)\cdot x}{4\cdot x}+\frac{4\cdot 4}{4\cdot x}=\frac{x^2-3x}{4x}+\frac{16}{4x}=\frac{x^2-3x+16}{4x}$

41. $\frac{x-3}{x}-\frac{4}{3}=\frac{(x-3)\cdot 3}{x\cdot 3}-\frac{x\cdot 4}{x\cdot 3}=\frac{3x-9}{3x}-\frac{4x}{3x}=\frac{3x-9-4x}{3x}=\frac{-x-9}{3x}=\frac{-(x+9)}{3x}$

43. $\frac{x}{3}-\frac{x+1}{x}=\frac{x\cdot x}{3\cdot x}-\frac{3\cdot(x+1)}{3\cdot x}=\frac{x^2}{3x}-\frac{3x+3}{3x}=\frac{x^2-(3x+3)}{3x}=\frac{x^2-3x-3}{3x}$

45. $\dfrac{3}{x+1}+\dfrac{4}{x-2}=\dfrac{3\cdot(x-2)}{(x+1)\cdot(x-2)}+\dfrac{(x+1)\cdot 4}{(x+1)\cdot(x-2)}=\dfrac{3x-6}{(x+1)(x-2)}+\dfrac{4x+4}{(x+1)(x-2)}$

$=\dfrac{3x-6+4x+4}{(x+1)(x-2)}=\dfrac{7x-2}{(x+1)(x-2)}$

47. $\dfrac{4}{x-1}-\dfrac{1}{x+2}=\dfrac{4\cdot(x+2)}{(x-1)\cdot(x+2)}-\dfrac{(x-1)\cdot 1}{(x-1)\cdot(x+2)}=\dfrac{4x+8}{(x-1)\cdot(x+2)}-\dfrac{x-1}{(x-1)\cdot(x+2)}$

$=\dfrac{(4x+8)-(x-1)}{(x-1)(x+2)}=\dfrac{4x+8-x+1}{(x-1)(x+2)}=\dfrac{3x+9}{(x-1)(x+2)}=\dfrac{3(x+3)}{(x-1)(x+2)}$

49. $\dfrac{x}{x+1}+\dfrac{2x-3}{x-1}=\dfrac{x(x-1)}{(x+1)(x-1)}+\dfrac{(x+1)(2x-3)}{(x+1)(x-1)}$

$=\dfrac{x^2-x}{(x+1)(x-1)}+\dfrac{2x^2-x-3}{(x+1)(x-1)}$

$=\dfrac{x^2-x+2x^2-x-3}{(x+1)(x-1)}=\dfrac{3x^2-2x-3}{(x+1)(x-1)}$

51. $\dfrac{x-2}{x+2}-\dfrac{x+2}{x-2}=\dfrac{(x-2)(x-2)}{(x+2)(x-2)}-\dfrac{(x+2)(x+2)}{(x+2)(x-2)}$

$=\dfrac{x^2-4x+4}{(x+2)(x-2)}-\dfrac{x^2+4x+4}{(x+2)(x-2)}=\dfrac{(x^2-4x+4)-(x^2+4x+4)}{(x+2)(x-2)}$

$=\dfrac{x^2-4x+4-x^2-4x-4}{(x+2)(x-2)}=\dfrac{-8x}{(x+2)(x-2)}$

53. $\dfrac{x}{x^2-4}+\dfrac{1}{x}=\dfrac{x\cdot x}{(x^2-4)\cdot x}+\dfrac{(x^2-4)\cdot 1}{(x^2-4)\cdot x}=\dfrac{x^2}{(x^2-4)\cdot x}+\dfrac{x^2-4}{(x^2-4)\cdot x}$

$=\dfrac{x^2+x^2-4}{(x^2-4)\cdot x}=\dfrac{2x^2-4}{(x^2-4)\cdot x}=\dfrac{2(x^2-2)}{x(x-2)(x+2)}$

55. $\dfrac{x^3}{(x-1)^2}-\dfrac{x^2+1}{x}=\dfrac{x^3\cdot x}{(x-1)^2\cdot x}-\dfrac{(x-1)^2\cdot(x^2+1)}{(x-1)^2\cdot x}=\dfrac{x^4}{(x-1)^2\cdot x}-\dfrac{(x^2-2x+1)(x^2+1)}{(x-1)^2\cdot x}$

$=\dfrac{x^4}{(x-1)^2\cdot x}-\dfrac{(x^4-2x^3+2x^2-2x+1)}{(x-1)^2\cdot x}$

$=\dfrac{x^4-(x^4-2x^3+2x^2-2x+1)}{(x-1)^2\cdot x}=\dfrac{x^4-x^4+2x^3-2x^2+2x-1}{(x-1)^2\cdot x}$

$=\dfrac{2x^3-2x^2+2x-1}{(x-1)^2\cdot x}$

57. $\dfrac{x}{x+1}+\dfrac{x-2}{x-1}-\dfrac{x+1}{x-2}=\left(\dfrac{x}{x+1}+\dfrac{x-2}{x-1}\right)-\dfrac{x+1}{x-2}$

$=\left(\dfrac{x\cdot(x-1)}{(x+1)\cdot(x-1)}+\dfrac{(x+1)(x-2)}{(x+1)\cdot(x-1)}\right)-\dfrac{x+1}{x-2}$

$=\left(\dfrac{x^2-x}{(x+1)\cdot(x-1)}+\dfrac{x^2-x-2}{(x+1)\cdot(x-1)}\right)-\dfrac{x+1}{x-2}$

$=\dfrac{x^2-x+x^2-x-2}{(x+1)\cdot(x-1)}-\dfrac{x+1}{x-2}=\dfrac{2x^2-2x-2}{(x+1)\cdot(x-1)}-\dfrac{x+1}{x-2}$

$=\dfrac{(2x^2-2x-2)\cdot(x-2)}{(x+1)\cdot(x-1)\cdot(x-2)}-\dfrac{(x+1)\cdot(x-1)\cdot(x+1)}{(x+1)\cdot(x-1)\cdot(x-2)}$

$=\dfrac{2x^3-6x^2+2x+4}{(x+1)\cdot(x-1)\cdot(x-2)}-\dfrac{x^3+x^2-x-1}{(x+1)\cdot(x-1)\cdot(x-2)}$

$=\dfrac{(2x^3-6x^2+2x+4)-(x^3+x^2-x-1)}{(x+1)\cdot(x-1)\cdot(x-2)}$

$=\dfrac{2x^3-6x^2+2x+4-x^3-x^2+x+1}{(x+1)\cdot(x-1)\cdot(x-2)}=\dfrac{x^3-7x^2+3x+5}{(x+1)\cdot(x-1)\cdot(x-2)}$

59. $\dfrac{1}{x}+\dfrac{1}{x+1}-\dfrac{1}{x-1}=\dfrac{1\cdot(x+1)}{x\cdot(x+1)}+\dfrac{x\cdot 1}{x\cdot(x+1)}-\dfrac{1}{x-1}$

$$=\frac{x+1}{x\cdot(x+1)}+\frac{x}{x\cdot(x+1)}-\frac{1}{x-1}$$

$$=\frac{x+1+x}{x\cdot(x+1)}-\frac{1}{x-1}=\frac{2x+1}{x\cdot(x+1)}-\frac{1}{x-1}$$

$$=\frac{(2x+1)(x-1)}{x\cdot(x+1)(x-1)}-\frac{x\cdot(x+1)\cdot 1}{x\cdot(x+1)(x-1)}$$

$$=\frac{2x^2-x-1}{x\cdot(x+1)(x-1)}-\frac{x^2+x}{x\cdot(x+1)(x-1)}$$

$$=\frac{(2x^2-x-1)-(x^2+x)}{x\cdot(x+1)(x-1)}=\frac{2x^2-x-1-x^2-x}{x\cdot(x+1)(x-1)}$$

$$=\frac{x^2-2x-1}{x\cdot(x+1)(x-1)}$$

61. $x^2-4=(x-2)(x+2)$

$x^2-x-2=(x-2)(x+1)$

$\text{LCM}=(x-2)(x+2)(x+1)$

63. $x^3-x=x(x-1)(x+1)$

$x^2-x=x(x-1)$

$\text{LCM}=x(x-1)(x+1)$

65. $4x^3-4x^2+x=x(4x^2-4x+1)$

$=x(2x-1)^2$

$2x^3-x^2=x^2(2x-1)$

$x^3=x^3$

$\text{LCM}=x^3(2x-1)^2$

67. $x^3-x=x(x-1)(x+1)$

$x^3-2x^2+x=x(x^2-2x+1)$

$=x(x-1)^2$

$x^3-1=(x-1)(x^2+x+1)$

$\text{LCM}=x(x-1)^2(x+1)(x^2+x+1)$

69. $\dfrac{3}{x(x+1)}+\dfrac{4}{(x+1)(x+2)}$

$\text{LCM}=x(x+1)(x+2)$

$$\frac{3\cdot(x+2)}{x(x+1)(x+2)}+\frac{x\cdot 4}{x(x+1)(x+2)}=\frac{3x+6}{x(x+1)(x+2)}+\frac{4x}{x(x+1)(x+2)}$$

$$=\frac{3x+6+4x}{x(x+1)(x+2)}=\frac{7x+6}{x(x+1)(x+2)}$$

71. $\dfrac{x+3}{(x+1)(x-2)}+\dfrac{2x-6}{(x+1)(x+2)}$

$\text{LCM}=(x+1)(x-2)(x+2)$

$$\frac{(x+3)(x+2)}{(x+1)(x-2)(x+2)}+\frac{(x-2)(2x-6)}{(x-2)(x+1)(x+2)}$$

$$=\frac{x^2+5x+6}{(x+1)(x-2)(x+2)}+\frac{2x^2-10x+12}{(x+1)(x-2)(x+2)}$$

$$=\frac{x^2+5x+6+2x^2-10x+12}{(x+1)(x-2)(x+2)}=\frac{3x^2-5x+18}{(x+1)(x-2)(x+2)}$$

73. $\dfrac{2x}{(3x+1)(x-2)}-\dfrac{4x}{(3x+1)(x+2)}$

$\text{LCM}=(3x+1)(x-2)(x+2)$

$$\frac{2x\cdot(x+2)}{(3x+1)(x-2)(x+2)}-\frac{(x-2)\cdot 4x}{(x-2)(3x+1)(x+2)}$$

$$=\frac{2x^2+4x}{(3x+1)(x-2)(x+2)}-\frac{4x^2-8x}{(x-2)(3x+1)(x+2)}$$

$$=\frac{(2x^2+4x)-(4x^2-8x)}{(3x+1)(x-2)(x+2)}=\frac{2x^2+4x-4x^2+8x}{(3x+1)(x-2)(x+2)}$$

$$=\frac{-2x^2+12x}{(3x+1)(x-2)(x+2)}=\frac{-2x(x-6)}{(3x+1)(x-2)(x+2)}$$

75. $\dfrac{4x+1}{x^2+7x+12}+\dfrac{2x+3}{x^2+5x+4}$

$x^2+7x+12=(x+3)(x+4)$

$x^2+5x+4=(x+1)(x+4)$

$\text{LCM}=(x+3)(x+4)(x+1)$

$$\frac{(4x+1)(x+1)}{(x+3)(x+4)(x+1)}+\frac{(x+3)(2x+3)}{(x+3)(x+1)(x+4)}$$

$$=\frac{4x^2+5x+1}{(x+3)(x+4)(x+1)}+\frac{2x^2+9x+9}{(x+3)(x+1)(x+4)}$$

$$=\frac{4x^2+5x+1+2x^2+9x+9}{(x+3)(x+4)(x+1)}=\frac{6x^2+14x+10}{(x+3)(x+4)(x+1)}=\frac{2(3x^2+7x+5)}{(x+3)(x+4)(x+1)}$$

77. $\dfrac{x^2-1}{3x^2-5x-2}+\dfrac{x^2-x}{2x^2-3x-2}$

$3x^2-5x-2=(3x+1)(x-2)$

$2x^2-3x-2=(2x+1)(x-2)$

$\text{LCM}=(3x+1)(x-2)(2x+1)$

$$\frac{(x^2-1)(2x+1)}{(3x+1)(x-2)(2x+1)}+\frac{(x^2-x)(3x+1)}{(2x+1)(x-2)(3x+1)}$$

$$=\frac{2x^3+x^2-2x-1}{(3x+1)(x-2)(2x+1)}+\frac{3x^3-2x^2-x}{(2x+1)(x-2)(3x+1)}$$

$$=\frac{2x^3+x^2-2x-1+3x^3-2x^2-x}{(3x+1)(x-2)(2x+1)}=\frac{5x^3-x^2-3x-1}{(3x+1)(x-2)(2x+1)}$$

79. $\dfrac{x}{x^2-7x+6}-\dfrac{x}{x^2-2x-24}$

$x^2-7x+6=(x-6)(x-1)$

$x^2-2x-24=(x-6)(x+4)$

$\text{LCM}=(x-6)(x-1)(x+4)$

$$\frac{x(x+4)}{(x-6)(x-1)(x+4)}-\frac{x(x-1)}{(x-6)(x+4)(x-1)}$$

$$=\frac{x^2+4x}{(x-6)(x-1)(x+4)}-\frac{x^2-x}{(x-6)(x+4)(x-1)}$$

$$=\frac{(x^2+4x)-(x^2-x)}{(x-6)(x-1)(x+4)}=\frac{x^2+4x-x^2+x}{(x-6)(x-1)(x+4)}$$

$$=\frac{5x}{(x-6)(x-1)(x+4)}$$

81. $\dfrac{4}{x^2-4}-\dfrac{2}{x^2+x-6}$

$x^2-4=(x-2)(x+2)$

$x^2+x-6=(x+3)(x-2)$

$\text{LCM}=(x-2)(x+2)(x+3)$

$$\frac{4\cdot(x+3)}{(x-2)(x+2)(x+3)}-\frac{2\cdot(x+2)}{(x+3)(x-2)(x+2)}$$

$$=\frac{4x+12}{(x-2)(x+2)(x+3)}-\frac{2x+4}{(x+3)(x-2)(x+2)}$$

$$=\frac{(4x+12)-(2x+4)}{(x-2)(x+2)(x+3)}=\frac{4x+12-2x-4}{(x-2)(x+2)(x+3)}$$

$$=\frac{2x+8}{(x-2)(x+2)(x+3)}=\frac{2(x+4)}{(x-2)(x+2)(x+3)}$$

83. $\dfrac{3}{(x-1)^2(x+1)}+\dfrac{2}{(x-1)(x+1)^2}$

$\text{LCM}=(x-1)^2(x+1)^2$

$$\frac{3(x+1)}{(x-1)^2(x+1)(x+1)}+\frac{2(x-1)}{(x-1)(x-1)(x+1)^2}$$

$$=\frac{3x+3}{(x-1)^2(x+1)^2}+\frac{2x-2}{(x-1)^2(x+1)^2}$$

$$=\frac{3x+3+2x-2}{(x-1)^2(x+1)^2}=\frac{5x+1}{(x-1)^2(x+1)^2}$$

85. $\dfrac{x+4}{x^2-x-2}-\dfrac{2x+3}{x^2+2x-8}$

$x^2-x-2=(x-2)(x+1)$

$x^2+2x-8=(x+4)(x-2)$

$\text{LCM}=(x-2)(x+1)(x+4)$

$$\frac{(x+4)(x+4)}{(x-2)(x+1)(x+4)}-\frac{(2x+3)(x+1)}{(x+4)(x-2)(x+1)}$$
$$=\frac{x^2+8x+16}{(x-2)(x+1)(x+4)}-\frac{2x^2+5x+3}{(x+4)(x-2)(x+1)}$$
$$=\frac{x^2+8x+16-(2x^2+5x+3)}{(x-2)(x+1)(x+4)}=\frac{x^2+8x+16-2x^2-5x-3}{(x-2)(x+1)(x+4)}$$
$$=\frac{-x^2+3x+13}{(x-2)(x+1)(x+4)}=\frac{-(x^2-3x-13)}{(x-2)(x+1)(x+4)}$$

87. $\dfrac{1}{x}-\dfrac{2}{x^2+x}+\dfrac{3}{x^3-x^2}$

$x=x$

$x^2+x=x(x+1)$

$x^3-x^2=x^2(x-1)$

$\text{LCM}=x^2(x+1)(x-1)$

$$\frac{1\cdot x(x+1)(x-1)}{x\cdot x(x+1)(x-1)}-\frac{2\cdot x(x-1)}{x(x+1)\cdot x(x-1)}+\frac{3\cdot(x+1)}{x^2(x-1)\cdot(x+1)}$$
$$=\frac{x^3-x}{x^2(x+1)(x-1)}-\frac{2x^2-2x}{x^2(x+1)(x-1)}+\frac{3x+3}{x^2(x+1)(x-1)}$$
$$=\frac{x^3-x-(2x^2-2x)+3x+3}{x^2(x+1)(x-1)}=\frac{x^3-x-2x^2+2x+3x+3}{x^2(x+1)(x-1)}$$
$$=\frac{x^3-2x^2+4x+3}{x^2(x+1)(x-1)}$$

89. $\dfrac{1}{h}\left(\dfrac{1}{x+h}-\dfrac{1}{x}\right)$

$\text{LCM}=x(x+h)$

$$\frac{1}{h}\left(\frac{x\cdot 1}{x(x+h)}-\frac{1\cdot(x+h)}{x(x+h)}\right)$$
$$=\frac{1}{h}\left(\frac{x}{x(x+h)}-\frac{x+h}{x(x+h)}\right)=\frac{1}{h}\left(\frac{x-(x+h)}{x(x+h)}\right)=\frac{1}{h}\left(\frac{x-x-h}{x(x+h)}\right)$$
$$=\frac{1}{h}\left(\frac{-h}{x(x+h)}\right)=\frac{1\cdot(-h)}{h\cdot x(x+h)}=\frac{-1}{x(x+h)}$$

3.5 Concepts and Vocabulary *(page 170)*

3. False **4.** $4(x-1)$

3.5 Exercises *(page 170)*

5. $$\frac{\dfrac{x}{x+1}+\dfrac{4}{x+1}}{\dfrac{x+4}{2}}=\frac{\dfrac{x+4}{x+1}}{\dfrac{x+4}{2}}=\frac{x+4}{x+1}\div\frac{x+4}{2}=\frac{x+4}{x+1}\cdot\frac{2}{x+4}=\frac{(x+4)\cdot 2}{(x+1)(x+4)}=\frac{2}{x+1}$$

7. $$\frac{\frac{x-1}{3}}{\frac{x}{x+4}-\frac{1}{x+4}}=\frac{\frac{x-1}{3}}{\frac{x-1}{x+4}}=\frac{x-1}{3}\div\frac{x-1}{x+4}=\frac{x-1}{3}\cdot\frac{x+4}{x-1}=\frac{(x-1)(x+4)}{3\cdot(x-1)}=\frac{x+4}{3}$$

9. $$\frac{\frac{x}{x+1}+\frac{5}{x+1}}{\frac{x^2}{x-1}-\frac{25}{x-1}}=\frac{\frac{x+5}{x+1}}{\frac{x^2-25}{x-1}}=\frac{x+5}{x+1}\div\frac{x^2-25}{x-1}=\frac{x+5}{x+1}\cdot\frac{x-1}{x^2-25}=\frac{(x+5)(x-1)}{(x+1)\cdot(x^2-25)}$$
$$=\frac{(x+5)(x-1)}{(x+1)\cdot(x-5)(x+5)}=\frac{x-1}{(x+1)\cdot(x-5)}$$

11. $$\frac{\frac{1}{3}+\frac{2}{x}}{\frac{x+1}{4}}=\frac{\frac{1\cdot x}{3\cdot x}+\frac{3\cdot 2}{3\cdot x}}{\frac{x+1}{4}}=\frac{\frac{x}{3x}+\frac{6}{3x}}{\frac{x+1}{4}}=\frac{x+6}{3x}\div\frac{x+1}{4}=\frac{x+6}{3x}\cdot\frac{4}{x+1}=\frac{(x+6)\cdot 4}{3x(x+1)}$$

13. $$\frac{\frac{4}{x}+\frac{3}{x+1}}{\frac{4}{x}}=\frac{\frac{4\cdot(x+1)}{x\cdot(x+1)}+\frac{3\cdot x}{(x+1)\cdot x}}{\frac{4}{x}}=\frac{\frac{4x+4}{x(x+1)}+\frac{3x}{x(x+1)}}{\frac{4}{x}}=\frac{\frac{4x+4+3x}{x(x+1)}}{\frac{4}{x}}=\frac{\frac{7x+4}{x(x+1)}}{\frac{4}{x}}=\frac{7x+4}{x(x+1)}\div\frac{4}{x}=\frac{7x+4}{x(x+1)}\cdot\frac{x}{4}$$
$$=\frac{(7x+4)\cdot x}{x(x+1)\cdot 4}=\frac{(7x+4)\cdot x}{4x(x+1)}=\frac{7x+4}{4(x+1)}$$

15. $$\frac{\frac{x}{x+2}-\frac{3}{x+2}}{\frac{x}{x^2-4}-\frac{1}{x^2-4}}=\frac{\frac{x-3}{x+2}}{\frac{x-1}{x^2-4}}=\frac{x-3}{x+2}\div\frac{x-1}{x^2-4}=\frac{x-3}{x+2}\cdot\frac{x^2-4}{x-1}=\frac{(x-3)(x^2-4)}{(x+2)\cdot(x-1)}$$
$$=\frac{(x-3)(x-2)(x+2)}{(x+2)\cdot(x-1)}=\frac{(x-3)(x-2)}{x-1}$$

17. $$\frac{4+\frac{3}{x}}{1-\frac{2}{x}}=\frac{\frac{4\cdot x}{1\cdot x}+\frac{3}{x}}{\frac{1\cdot x}{1\cdot x}-\frac{2}{x}}=\frac{\frac{4\cdot x}{x}+\frac{3}{x}}{\frac{x}{x}-\frac{2}{x}}=\frac{\frac{4\cdot x+3}{x}}{\frac{x-2}{x}}=\frac{4x+3}{x}\div\frac{x-2}{x}=\frac{4x+3}{x}\cdot\frac{x}{x-2}$$
$$=\frac{(4x+3)\cdot x}{x\cdot(x-2)}=\frac{4x+3}{x-2}$$

19. $$\frac{x-\frac{1}{x}}{x+\frac{1}{x}}=\frac{\frac{x\cdot x}{1\cdot x}-\frac{1}{x}}{\frac{x\cdot x}{1\cdot x}+\frac{1}{x}}=\frac{\frac{x^2}{x}-\frac{1}{x}}{\frac{x^2}{x}+\frac{1}{x}}=\frac{\frac{x^2-1}{x}}{\frac{x^2+1}{x}}=\frac{x^2-1}{x}\div\frac{x^2+1}{x}=\frac{x^2-1}{x}\cdot\frac{x}{x^2+1}$$
$$=\frac{(x^2-1)\cdot x}{x\cdot(x^2+1)}=\frac{(x-1)(x+1)}{x^2+1}$$

21. $$\frac{3-\frac{x^2}{x+1}}{1+\frac{x}{x^2-1}}=\frac{\frac{3\cdot(x+1)}{1\cdot(x+1)}-\frac{x^2}{x+1}}{\frac{1\cdot(x^2-1)}{1\cdot(x^2-1)}+\frac{x}{x^2-1}}=\frac{\frac{3x+3}{x+1}-\frac{x^2}{x+1}}{\frac{x^2-1}{x^2-1}+\frac{x}{x^2-1}}=\frac{\frac{3x+3-x^2}{x+1}}{\frac{x^2-1+x}{x^2-1}}=\frac{3x+3-x^2}{x+1}\div\frac{x^2-1+x}{x^2-1}$$
$$=\frac{3x+3-x^2}{x+1}\cdot\frac{x^2-1}{x^2-1+x}=\frac{(3x+3-x^2)\cdot(x^2-1)}{(x+1)\cdot(x^2-1+x)}$$
$$=\frac{-(x^2-3x-3)(x-1)(x+1)}{(x+1)(x^2+x-1)}=\frac{-(x^2-3x-3)(x-1)}{x^2+x-1}$$

23. $$\frac{\dfrac{x+4}{x-2}-\dfrac{x-3}{x+1}}{x+1}=\frac{\dfrac{(x+4)\cdot(x+1)}{(x-2)\cdot(x+1)}-\dfrac{(x-2)\cdot(x-3)}{(x-2)\cdot(x+1)}}{x+1}=\frac{\dfrac{x^2+5x+4}{(x-2)\cdot(x+1)}-\dfrac{x^2-5x+6}{(x-2)\cdot(x+1)}}{x+1}=\frac{\dfrac{(x^2+5x+4)-(x^2-5x+6)}{(x-2)\cdot(x+1)}}{x+1}$$

$$=\frac{\dfrac{x^2+5x+4-x^2+5x-6}{(x-2)\cdot(x+1)}}{x+1}=\frac{\dfrac{10x-2}{(x-2)\cdot(x+1)}}{x+1}=\frac{\dfrac{2(5x-1)}{(x-2)\cdot(x+1)}}{x+1}=\frac{2(5x-1)}{(x-2)(x+1)}\div\frac{(x+1)}{1}$$

$$=\frac{2(5x-1)}{(x-2)(x+1)}\cdot\frac{1}{(x+1)}=\frac{2(5x-1)\cdot 1}{(x-2)(x+1)(x+1)}=\frac{2(5x-1)}{(x-2)(x+1)^2}$$

25. $$\frac{\dfrac{x-2}{x+2}+\dfrac{x-1}{x+1}}{\dfrac{x}{x+1}-\dfrac{2x-3}{x}}=\frac{\dfrac{(x-2)\cdot(x+1)}{(x+2)\cdot(x+1)}+\dfrac{(x+2)\cdot(x-1)}{(x+2)\cdot(x+1)}}{\dfrac{x\cdot x}{(x+1)\cdot x}-\dfrac{(x+1)\cdot(2x-3)}{(x+1)\cdot x}}=\frac{\dfrac{x^2-x-2}{(x+2)\cdot(x+1)}+\dfrac{x^2+x-2}{(x+2)\cdot(x+1)}}{\dfrac{x^2}{(x+1)\cdot x}-\dfrac{2x^2-x-3}{(x+1)\cdot x}}=\frac{\dfrac{x^2-x-2+x^2+x-2}{(x+2)\cdot(x+1)}}{\dfrac{x^2-(2x^2-x-3)}{(x+1)\cdot x}}$$

$$=\frac{\dfrac{2x^2-4}{(x+2)\cdot(x+1)}}{\dfrac{x^2-2x^2+x+3}{(x+1)\cdot x}}=\frac{\dfrac{2(x^2-2)}{(x+2)\cdot(x+1)}}{\dfrac{-x^2+x+3}{(x+1)\cdot x}}=\frac{2(x^2-2)}{(x+2)(x+1)}\div\frac{-x^2+x+3}{(x+1)x}$$

$$=\frac{2(x^2-2)}{(x+2)(x+1)}\cdot\frac{(x+1)x}{(-x^2+x+3)}=\frac{2(x^2-2)(x+1)x}{(x+2)(x+1)(-x^2+x+3)}$$

$$=\frac{2x(x^2-2)(x+1)}{(x+2)(x+1)(-1)(x^2-x-3)}=\frac{2x(x^2-2)}{-(x+2)(x^2-x-3)}$$

27. $$\frac{\dfrac{x}{x-5}-\dfrac{4}{5-x}}{\dfrac{3x}{x+1}+\dfrac{x}{x-1}-\dfrac{2}{x^2-1}}=\frac{\dfrac{x}{x-5}-\dfrac{4}{-(x-5)}}{\dfrac{3x}{x+1}+\dfrac{x}{x-1}-\dfrac{2}{(x-1)\cdot(x+1)}}=\frac{\dfrac{x}{x-5}+\dfrac{4}{x-5}}{\dfrac{3x\cdot(x-1)}{(x+1)\cdot(x-1)}+\dfrac{(x+1)\cdot x}{(x+1)\cdot(x-1)}-\dfrac{2}{(x-1)\cdot(x+1)}}$$

$$=\frac{\dfrac{x+4}{x-5}}{\dfrac{3x^2-3x}{(x+1)\cdot(x-1)}+\dfrac{x^2+x}{(x+1)\cdot(x-1)}-\dfrac{2}{(x-1)\cdot(x+1)}}=\frac{\dfrac{x+4}{x-5}}{\dfrac{3x^2-3x+x^2+x-2}{(x+1)\cdot(x-1)}}=\frac{\dfrac{x+4}{x-5}}{\dfrac{4x^2-2x-2}{(x+1)\cdot(x-1)}}$$

$$=\frac{\dfrac{x+4}{x-5}}{\dfrac{2(2x^2-x-1)}{(x+1)\cdot(x-1)}}=\frac{\dfrac{x+4}{x-5}}{\dfrac{2(2x+1)\cdot(x-1)}{(x+1)\cdot(x-1)}}=\frac{x+4}{x-5}\div\frac{2(2x+1)(x-1)}{(x+1)(x-1)}$$

$$=\frac{x+4}{x-5}\cdot\frac{(x+1)(x-1)}{2(2x+1)(x-1)}=\frac{(x+4)(x+1)(x-1)}{(x-5)2(2x+1)(x-1)}=\frac{(x+4)(x+1)}{2(x-5)(2x+1)}$$

29. $$\frac{1+\dfrac{1}{x}+\dfrac{1}{x^2}}{1-\dfrac{1}{x}+\dfrac{1}{x^2}}=\frac{\dfrac{1.x^2}{1.x^2}+\dfrac{1\cdot x}{x\cdot x}+\dfrac{1}{x^2}}{\dfrac{1.x^2}{1.x^2}-\dfrac{1\cdot x}{x\cdot x}+\dfrac{1}{x^2}}=\frac{\dfrac{x^2}{x^2}+\dfrac{x}{x^2}+\dfrac{1}{x^2}}{\dfrac{x^2}{x^2}-\dfrac{x}{x^2}+\dfrac{1}{x^2}}=\frac{\dfrac{x^2+x+1}{x^2}}{\dfrac{x^2-x+1}{x^2}}=\frac{x^2+x+1}{x^2}\div\frac{x^2-x+1}{x^2}$$

$$=\frac{x^2+x+1}{x^2}\cdot\frac{x^2}{x^2-x+1}=\frac{(x^2+x+1)\cdot x^2}{x^2\cdot(x^2-x+1)}=\frac{x^2+x+1}{x^2-x+1}$$

31. $$1-\frac{1}{1-\dfrac{1}{x}}=1-\frac{1}{\dfrac{1\cdot x}{1\cdot x}-\dfrac{1}{x}}=1-\frac{1}{\dfrac{x}{x}-\dfrac{1}{x}}=1-\frac{1}{\dfrac{x-1}{x}}=1-1\div\frac{x-1}{x}=1-1\cdot\frac{x}{x-1}$$

$$=1-\frac{x}{x-1}=\frac{1\cdot(x-1)}{x-1}-\frac{x}{x-1}=\frac{x-1-x}{x-1}=\frac{-1}{x-1}$$

33. $$\frac{\frac{x+h-2}{x+h+2}-\frac{x-2}{x+2}}{h}=\frac{\frac{(x+h-2)\cdot(x+2)}{(x+h+2)\cdot(x+2)}-\frac{(x+h+2)\cdot(x-2)}{(x+h+2)\cdot(x+2)}}{h}=\frac{\frac{x^2+hx+2h-4}{(x+h+2)\cdot(x+2)}-\frac{x^2+hx-2h-4}{(x+h+2)\cdot(x+2)}}{h}$$

$$=\frac{\frac{(x^2+hx+2h-4)-(x^2+hx-2h-4)}{(x+h+2)\cdot(x+2)}}{h}=\frac{\frac{(x^2+hx+2h-4-x^2-hx+2h+4)}{(x+h+2)\cdot(x+2)}}{h}=\frac{\frac{4h}{(x+h+2)\cdot(x+2)}}{h}$$

$$=\frac{4h}{(x+h+2)(x+2)}\div\frac{h}{1}=\frac{4h}{(x+h+2)(x+2)}\cdot\frac{1}{h}=\frac{4h\cdot 1}{(x+h+2)(x+2)\cdot h}$$

$$=\frac{4h}{(x+h+2)(x+2)\cdot h}=\frac{4}{(x+h+2)(x+2)}$$

35. $$\frac{x}{x+\frac{1}{x+\frac{1}{x+1}}}=\frac{x}{x+\frac{1}{\frac{x\cdot(x+1)}{1(x+1)}+\frac{1}{x+1}}}=\frac{x}{x+\frac{1}{\frac{x^2+x}{x+1}+\frac{1}{x+1}}}=\frac{x}{x+\frac{1}{\frac{x^2+x+1}{x+1}}}$$

$$=\frac{x}{x+1\div\frac{x^2+x+1}{x+1}}=\frac{x}{x+1\cdot\frac{x+1}{x^2+x+1}}=\frac{x}{x+\frac{x+1}{x^2+x+1}}=\frac{x}{\frac{x(x^2+x+1)}{1\cdot(x^2+x+1)}+\frac{x+1}{x^2+x+1}}$$

$$=\frac{x}{\frac{x^3+x^2+x}{x^2+x+1}+\frac{x+1}{x^2+x+1}}=\frac{x}{\frac{x^3+x^2+x+x+1}{x^2+x+1}}=\frac{x}{\frac{x^3+x^2+2x+1}{x^2+x+1}}=x\div\frac{x^3+x^2+2x+1}{x^2+x+1}$$

$$=x\cdot\frac{x^2+x+1}{x^3+x^2+2x+1}=\frac{x(x^2+x+1)}{x^3+x^2+2x+1}$$

37. $$R=\frac{1}{\frac{1}{R_1}+\frac{1}{R_2}}=\frac{1}{\frac{1\cdot R_2}{R_1\cdot R_2}+\frac{R_1\cdot 1}{R_1\cdot R_2}}=\frac{1}{\frac{R_2}{R_1\cdot R_2}+\frac{R_1}{R_1\cdot R_2}}=\frac{1}{\frac{R_2+R_1}{R_1\cdot R_2}}=1\div\frac{R_2+R_1}{R_1R_2}=1\cdot\frac{R_1R_2}{R_1+R_2}$$

$$R=\frac{R_1R_2}{R_1+R_2}$$

Let $R_1 = 6$ ohms, $R_2 = 10$ ohms

$$R=\frac{(6)(10)}{6+10}=\frac{60}{16}=\frac{15}{4}\text{ ohms}$$

39. $$1+\frac{1}{x}=\frac{1\cdot x}{1\cdot x}+\frac{1}{x}=\frac{x}{x}+\frac{1}{x}=\frac{x+1}{x}=\frac{ax+b}{bx+c}\quad\text{where } a=1, b=1, c=0$$

$$1+\frac{1}{1+\frac{1}{x}}=1+\frac{1}{\frac{x+1}{x}}=1+\frac{x}{x+1}=\frac{1(x+1)}{1(x+1)}+\frac{x}{x+1}=\frac{x+1}{x+1}+\frac{x}{x+1}=\frac{x+1+x}{x+1}$$

$$=\frac{2x+1}{x+1}=\frac{ax+b}{bx+c}\quad\text{where } a=2, b=1, c=1$$

$$1+\frac{1}{1+\frac{1}{1+\frac{1}{x}}}=1+\frac{1}{\frac{2x+1}{x+1}}=1+\frac{x+1}{2x+1}=\frac{1(2x+1)}{1(2x+1)}+\frac{x+1}{2x+1}=\frac{2x+1+x+1}{2x+1}$$

$$=\frac{3x+2}{2x+1}=\frac{ax+b}{bx+c}\quad\text{where } a=3, b=2, c=1$$

$$1+\frac{1}{1+\frac{1}{1+\frac{1}{1+\frac{1}{x}}}}=1+\frac{1}{\frac{3x+2}{2x+1}}=1+\frac{2x+1}{3x+2}=\frac{1(3x+2)}{1(3x+2)}+\frac{2x+1}{3x+2}=\frac{3x+2+2x+1}{3x+2}$$

$$=\frac{5x+3}{3x+2}=\frac{ax+b}{bx+c}\quad\text{where } a=5, b=3, c=2$$

Thus, the successive values of a,b and c are 1,1,0; 2,1,1; 3,2,1; 5,3,2; 8,5,3; 13,8,5; and so on. Can you discover the patterns that these values follow? Hint: Research Fibonacci numbers.

Chapter 3 Review Exercises *(page 172)*

1. $\dfrac{14}{5}$ and $\dfrac{84}{30}$

(a)
$$\begin{array}{r} 2.8 \\ 5\overline{)14.0} \\ \underline{10} \\ 40 \\ \underline{40} \\ 0 \end{array} \qquad \begin{array}{r} 2.8 \\ 30\overline{)84.0} \\ \underline{60} \\ 240 \\ \underline{240} \\ 0 \end{array}$$

Each fraction has the same decimal form, 2.8, so they are equivalent.

(b) $\dfrac{14}{5} = \dfrac{84}{30}$ since $(14)\cdot(30) = 420$ and $(5)\cdot(84) = 420$

(c) $\dfrac{84}{30} = \dfrac{14\cdot 6}{5\cdot 6} = \dfrac{14}{5}$ Cancel the 6's

3. $\dfrac{-5}{18}$ and $\dfrac{-30}{108}$

(a)
$$\begin{array}{r} -0.277\ldots \\ 18\overline{)-5.000} \\ \underline{36} \\ 140 \\ \underline{126} \\ 140 \\ \underline{126} \\ 14 \end{array} \qquad \begin{array}{r} -0.277\ldots \\ 108\overline{)-30.000} \\ \underline{216} \\ 840 \\ \underline{756} \\ 840 \\ \underline{756} \\ 84 \end{array}$$

Each fraction has the same decimal form, $-0.277\ldots$, so they are equivalent.

(b) $\dfrac{-5}{18} = \dfrac{-30}{108}$ since $(-5)\cdot(108) = -540$ and $(-30)\cdot(18) = -540$

(c) $\dfrac{-30}{108} = \dfrac{-5\cdot 6}{18\cdot 6} = \dfrac{-5}{18}$ Cancel the 6's

5. $\dfrac{-30}{-54}$ and $\dfrac{5}{9}$

(a)
$$\begin{array}{r} 0.555\ldots \\ -54\overline{)-30.000} \\ \underline{270} \\ 300 \\ \underline{270} \\ 300 \\ \underline{270} \\ 30 \end{array} \qquad \begin{array}{r} 0.555\ldots \\ 9\overline{)5.000} \\ \underline{45} \\ 50 \\ \underline{45} \\ 50 \\ \underline{45} \\ 5 \end{array}$$

Each fraction has the same decimal form, $0.555\ldots$, so they are equivalent.

(b) $\dfrac{-30}{-54} = \dfrac{5}{9}$ since $(-30)\cdot(9) = -270$ and $(-54)\cdot(5) = -270$

(c) $\dfrac{-30}{-54} = \dfrac{-6\cdot 5}{-6\cdot 9} = \dfrac{5}{9}$ Cancel the -6's

7. $\dfrac{36}{28} = \dfrac{2\cdot 2\cdot 3\cdot 3}{2\cdot 2\cdot 7} = \dfrac{3\cdot 3}{7} = \dfrac{9}{7}$

9. $\dfrac{-24}{-18} = \dfrac{(-1)\cdot 2\cdot 2\cdot 2\cdot 3}{(-1)\cdot 2\cdot 3\cdot 3} = \dfrac{2\cdot 2}{3} = \dfrac{4}{3}$

11. $\dfrac{3x^2}{9x} = \dfrac{3\cdot x\cdot x}{3\cdot 3\cdot x} = \dfrac{x}{3}$

13. $\dfrac{4x^2 + 4x + 1}{4x^2 - 1} = \dfrac{(2x+1)(2x+1)}{(2x-1)(2x+1)} = \dfrac{2x+1}{2x-1}$

15. $\dfrac{2x^2 + 11x + 14}{x^2 - 4} = \dfrac{(x+2)(2x+7)}{(x-2)(x+2)} = \dfrac{2x+7}{x-2}$

17. $\dfrac{x^3 + x^2}{x^3 + 1} = \dfrac{x^2(x+1)}{(x+1)(x^2 - x + 1)} = \dfrac{x^2}{x^2 - x + 1}$

19.
$$\begin{aligned} \frac{(4x - 12x^2)(x-2)^2}{(x^2-4)(1-9x^2)} &= \frac{(-1)(12x^2 - 4x)(x-2)^2}{(x^2-4)(-1)(9x^2-1)} \\ &= \frac{(-1)4x(3x-1)(x-2)(x-2)}{(x-2)(x+2)(-1)(3x-1)(3x+1)} \\ &= \frac{4x(x-2)}{(x+2)(3x+1)} \end{aligned}$$

21. $\dfrac{x^2 + x - 1}{(x-2)^2}$ at $x = 3$

$$\frac{x^2 + x - 1}{(x-2)^2} \underset{\substack{\uparrow \\ x=3}}{=} \frac{(3)^2 + 3 - 1}{(3-2)^2} = \frac{9+3-1}{(1)^2} = \frac{11}{1} = 11$$

23. $\dfrac{(x+1)^3}{x^2 + x + 1}$ at $x = 1$

$$\frac{(x+1)^3}{x^2 + x + 1} \underset{\substack{\uparrow \\ x=1}}{=} \frac{(1+1)^3}{(1)^2 + 1 + 1} = \frac{(2)^3}{1+1+1} = \frac{8}{3}$$

25. $\dfrac{x^3 - 3x^2 + 4}{(x+1)^2}$ at $x = -2$

$$\frac{x^3 - 3x^2 + 4}{(x+1)^2} \underset{\substack{\uparrow \\ x=-2}}{=} \frac{(-2)^3 - 3(-2)^2 + 4}{(-2+1)^2} = \frac{-8 - 3(4) + 4}{(-1)^2} = \frac{-8 - 12 + 4}{1} = \frac{-16}{1} = -16$$

27. $\dfrac{x^2 - 4}{x}$

(a) At $x = 2$, the denominator is 2
(b) At $x = 1$, the denominator is 1
(c) At $x = 0$, the denominator is 0; this must be excluded from the domain.
(d) At $x = -1$, the denominator is -1

29. $\dfrac{x(x-1)}{(x+2)(x+3)}$

(a) At $x = 2$, the denominator is $(2+2)(2+3) = (4)(5) = 20$
(b) At $x = 1$, the denominator is $(1+2)(1+3) = (3)(4) = 12$
(c) At $x = 0$, the denominator is $(0+2)(0+3) = (2)(3) = 6$
(d) At $x = -1$, the denominator is $(-1+2)(-1+3) = (1)(2) = 2$

31. $\dfrac{9}{8} \cdot \dfrac{2}{27} = \dfrac{9 \cdot 2}{8 \cdot 27} = \dfrac{3 \cdot 3 \cdot 2}{2 \cdot 2 \cdot 2 \cdot 3 \cdot 3 \cdot 3} = \dfrac{1}{2 \cdot 2 \cdot 3} = \dfrac{1}{12}$

33. $\dfrac{3}{4} + \dfrac{8}{4} = \dfrac{3+8}{4} = \dfrac{11}{4}$

35. $\dfrac{8}{3} - \dfrac{4}{3} = \dfrac{8-4}{3} = \dfrac{4}{3}$

37. $\dfrac{3}{4} - \dfrac{2}{3} = \dfrac{3 \cdot 3}{4 \cdot 3} - \dfrac{4 \cdot 2}{4 \cdot 3} = \dfrac{9}{12} - \dfrac{8}{12} = \dfrac{9-8}{12} = \dfrac{1}{12}$

39. $\dfrac{7}{24} + \dfrac{1}{9} - \dfrac{5}{6}$

$24 = 2 \cdot 2 \cdot 2 \cdot 3$
$9 = 3 \cdot 3$
$6 = 2 \cdot 3$
$\text{LCM} = 2 \cdot 2 \cdot 2 \cdot 3 \cdot 3 = 72$

$$\frac{7 \cdot 3}{24 \cdot 3} + \frac{1 \cdot 8}{9 \cdot 8} - \frac{5 \cdot 12}{6 \cdot 12} = \frac{21}{72} + \frac{8}{72} - \frac{60}{72} = \frac{21 + 8 - 60}{72} = \frac{-31}{72}$$

41. $$\frac{\frac{1}{2} + \frac{1}{3}}{\frac{3}{8} - \frac{1}{4}} = \frac{\frac{1 \cdot 3}{2 \cdot 3} + \frac{2 \cdot 1}{2 \cdot 3}}{\frac{3}{8} - \frac{1 \cdot 2}{4 \cdot 2}} = \frac{\frac{3}{6} + \frac{2}{6}}{\frac{3}{8} - \frac{2}{8}} = \frac{\frac{3+2}{6}}{\frac{3-2}{8}} = \frac{\frac{5}{6}}{\frac{1}{8}} = \frac{5}{6} \div \frac{1}{8} = \frac{5}{6} \cdot \frac{8}{1} = \frac{5 \cdot 8}{6 \cdot 1} = \frac{5 \cdot 2 \cdot 2 \cdot 2}{2 \cdot 3} = \frac{5 \cdot 2 \cdot 2}{3} = \frac{20}{3}$$

43. $$\frac{x+3}{8x} \cdot \frac{6x^2}{3x^2 + 9x} = \frac{(x+3) \cdot 6x^2}{8x \cdot (3x^2 + 9x)} = \frac{(x+3) \cdot 2 \cdot 3 \cdot x \cdot x}{2 \cdot 2 \cdot 2 \cdot x \cdot 3x(x+3)} = \frac{1}{2 \cdot 2} = \frac{1}{4}$$

45. $$\frac{x^2 + 7x + 6}{x^2 + 5x - 14} \cdot \frac{(x+2)^2}{1 - x^2} = \frac{(x+6)(x+1)}{(x+7)(x-2)} \cdot \frac{(x+2)(x+2)}{(-1)(x-1)(x+1)}$$
$$= \frac{(x+6)(x+1)(x+2)(x+2)}{(x+7)(x-2)(-1)(x-1)(x+1)}$$
$$= \frac{(x+6)(x+2)(x+2)}{-(x+7)(x-2)(x-1)}$$

47. $$\frac{6x^2+7x-3}{3x^2+11x-4}\cdot\frac{x^2-16}{9-4x^2}=\frac{(3x-1)(2x+3)}{(x+4)(3x-1)}\cdot\frac{(x+4)(x-4)}{(-1)(2x-3)(2x+3)}$$
$$=\frac{(3x-1)(2x+3)(x+4)(x-4)}{(x+4)(3x-1)(-1)(2x-3)(2x+3)}$$
$$=\frac{x-4}{-(2x-3)}$$

49. $$\frac{\frac{4x+20}{9x^2}}{\frac{x^2-25}{3x}}=\frac{\frac{4(x+5)}{3\cdot 3\cdot x\cdot x}}{\frac{(x-5)(x+5)}{3\cdot x}}=\frac{4(x+5)}{3\cdot 3\cdot x\cdot x}\div\frac{(x-5)(x+5)}{3\cdot x}=\frac{4(x+5)}{3\cdot 3\cdot x\cdot x}\cdot\frac{3\cdot x}{(x-5)(x+5)}$$
$$=\frac{4(x+5)\cdot 3\cdot x}{3\cdot 3\cdot x\cdot x\cdot(x-5)(x+5)}=\frac{4}{3\cdot x\cdot(x-5)}=\frac{4}{3x(x-5)}$$

51. $$\frac{\frac{4x^2+4x+1}{4-x^2}}{\frac{4x^2-1}{x^2+4x+4}}=\frac{\frac{(2x+1)\cdot(2x+1)}{(-1)\cdot(x-2)\cdot(x+2)}}{\frac{(2x-1)\cdot(2x+1)}{(x+2)\cdot(x+2)}}=\frac{(2x+1)(2x+1)}{(-1)(x-2)(x+2)}\div\frac{(2x-1)(2x+1)}{(x+2)(x+2)}$$
$$=\frac{(2x+1)(2x+1)}{(-1)(x-2)(x+2)}\cdot\frac{(x+2)(x+2)}{(2x-1)(2x+1)}=\frac{(2x+1)(2x+1)\cdot(x+2)(x+2)}{(-1)(x-2)(x+2)\cdot(2x-1)(2x+1)}$$
$$=\frac{(2x+1)(x+2)}{-(x-2)(2x-1)}$$

53. $$\frac{\frac{x^2-25}{x^2-x-6}\cdot\frac{9-x^2}{x^2+10x+25}}{\frac{5-x}{3+x}\cdot\frac{x^2+x}{x^2+4x+4}}=\frac{\frac{(x-5)\cdot(x+5)}{(x-3)\cdot(x+2)}\cdot\frac{(-1)\cdot(x-3)\cdot(x+3)}{(x+5)\cdot(x+5)}}{\frac{-(x-5)}{x+3}\cdot\frac{x(x+1)}{(x+2)\cdot(x+2)}}=\frac{\frac{(x-5)\cdot(x+5)\cdot(-1)\cdot(x-3)\cdot(x+3)}{(x-3)\cdot(x+2)\cdot(x+5)\cdot(x+5)}}{\frac{(-1)\cdot(x-5)\cdot x(x+1)}{(x+3)\cdot(x+2)\cdot(x+2)}}$$
$$=\frac{(-1)(x-5)(x+5)(x-3)(x+3)}{(x-3)(x+2)(x+5)(x+5)}\div\frac{(-1)(x-5)x(x+1)}{(x+3)(x+2)(x+2)}$$
$$=\frac{(-1)(x-5)(x+5)(x-3)(x+3)}{(x-3)(x+2)(x+5)(x+5)}\cdot\frac{(x+3)(x+2)(x+2)}{(-1)(x-5)x(x+1)}$$
$$=\frac{(-1)(x-5)(x+5)(x-3)(x+3)\cdot(x+3)(x+2)(x+2)}{(x-3)(x+2)(x+5)(x+5)\cdot(-1)(x-5)x(x+1)}$$
$$=\frac{(x+3)\cdot(x+3)(x+2)}{x(x+5)(x+1)}$$

55. $$\frac{3x+1}{x-2}+\frac{2x-1}{x-2}=\frac{3x+1+2x-1}{x-2}=\frac{5x}{x-2}$$

57. $$\frac{x^2-1}{x^2+4}-\frac{1-x+x^2}{x^2+4}=\frac{(x^2-1)-(1-x+x^2)}{x^2+4}=\frac{x^2-1-1+x-x^2}{x^2+4}=\frac{x-2}{x^2+4}$$

59. $$\frac{5x-3}{4-x}+\frac{1-2x}{x-4}=\frac{5x-3}{-(x-4)}+\frac{1-2x}{x-4}=\frac{-(5x-3)}{x-4}+\frac{1-2x}{x-4}$$
$$=\frac{-5x+3}{x-4}+\frac{1-2x}{x-4}=\frac{-5x+3+1-2x}{x-4}=\frac{-7x+4}{x-4}$$

61. $$\frac{x+1}{x}+\frac{x-1}{x+1}=\frac{(x+1)(x+1)}{x(x+1)}+\frac{x(x-1)}{x(x+1)}=\frac{x^2+2x+1}{x(x+1)}+\frac{x^2-x}{x(x+1)}$$
$$=\frac{x^2+2x+1+x^2-x}{x(x+1)}=\frac{2x^2+x+1}{x(x+1)}$$

63. $\dfrac{x}{(x-1)(x+1)} - \dfrac{4x}{(x-1)(x+2)}$

LCM $= (x-1)(x+1)(x+2)$

$$\frac{x(x+2)}{(x-1)(x+1)(x+2)} - \frac{4x(x+1)}{(x-1)(x+2)(x+1)} = \frac{x^2+2x}{(x-1)(x+1)(x+2)} - \frac{4x^2+4x}{(x-1)(x+2)(x+1)}$$

$$= \frac{(x^2+2x)-(4x^2+4x)}{(x-1)(x+1)(x+2)}$$

$$= \frac{x^2+2x-4x^2-4x}{(x-1)(x+1)(x+2)} = \frac{-3x^2-2x}{(x-1)(x+1)(x+2)}$$

$$= \frac{-x(3x+2)}{(x-1)(x+1)(x+2)}$$

65. $\dfrac{3x+5}{x^2+x-6} + \dfrac{2x-1}{x^2-9} = \dfrac{3x+5}{(x+3)(x-2)} + \dfrac{2x-1}{(x-3)(x+3)}$

LCM $= (x+3)(x-2)(x-3)$

$$\frac{(3x+5)(x-3)}{(x+3)(x-2)(x-3)} + \frac{(2x-1)(x-2)}{(x-3)(x+3)(x-2)}$$

$$= \frac{3x^2-4x-15}{(x+3)(x-2)(x-3)} + \frac{2x^2-5x+2}{(x-3)(x+3)(x-2)}$$

$$= \frac{3x^2-4x-15+2x^2-5x+2}{(x+3)(x-2)(x-3)} = \frac{5x^2-9x-13}{(x+3)(x-2)(x-3)}$$

67. $\dfrac{x^2+4}{3x^2-5x-2} - \dfrac{x^2+9}{2x^2-3x-2} = \dfrac{x^2+4}{(3x+1)(x-2)} - \dfrac{x^2+9}{(2x+1)(x-2)}$

LCM $= (3x+1)(x-2)(2x+1)$

$$\frac{(x^2+4)(2x+1)}{(3x+1)(x-2)(2x+1)} - \frac{(x^2+9)(3x+1)}{(2x+1)(x-2)(3x+1)}$$

$$= \frac{2x^3+x^2+8x+4}{(3x+1)(x-2)(2x+1)} - \frac{3x^3+x^2+27x+9}{(2x+1)(x-2)(3x+1)}$$

$$= \frac{(2x^3+x^2+8x+4)-(3x^3+x^2+27x+9)}{(3x+1)(x-2)(2x+1)} = \frac{2x^3+x^2+8x+4-3x^3-x^2-27x-9}{(3x+1)(x-2)(2x+1)}$$

$$= \frac{-x^3-19x-5}{(3x+1)(x-2)(2x+1)} = \frac{-(x^3+19x+5)}{(3x+1)(x-2)(2x+1)}$$

69. $\dfrac{\dfrac{x}{3}+\dfrac{4}{x}}{\dfrac{x+4}{x}} = \dfrac{\dfrac{x\cdot x}{3\cdot x}+\dfrac{3\cdot 4}{3\cdot x}}{\dfrac{x+4}{x}} = \dfrac{\dfrac{x^2}{3x}+\dfrac{12}{3x}}{\dfrac{x+4}{x}} = \dfrac{\dfrac{x^2+12}{3x}}{\dfrac{x+4}{x}} = \dfrac{x^2+12}{3x} \div \dfrac{x+4}{x} = \dfrac{x^2+12}{3x}\cdot\dfrac{x}{x+4}$

$$= \frac{(x^2+12)x}{3x(x+4)} = \frac{x^2+12}{3(x+4)}$$

71. $\dfrac{1-\dfrac{3}{x}}{1-\dfrac{2}{x}} = \dfrac{\dfrac{1\cdot x}{1\cdot x}-\dfrac{3}{x}}{\dfrac{1\cdot x}{1\cdot x}-\dfrac{2}{x}} = \dfrac{\dfrac{x}{x}-\dfrac{3}{x}}{\dfrac{x}{x}-\dfrac{2}{x}} = \dfrac{\dfrac{x-3}{x}}{\dfrac{x-2}{x}} = \dfrac{x-3}{x} \div \dfrac{x-2}{x} = \dfrac{x-3}{x}\cdot\dfrac{x}{x-2}$

$$= \frac{(x-3)x}{x(x-2)} = \frac{x-3}{x-2}$$

73. $$\frac{\frac{x^2}{3x-4}-\frac{1-x^2}{4-3x}}{\frac{x^2}{9x^2-16}}=\frac{\frac{x^2}{3x-4}-\frac{1-x^2}{-(3x-4)}}{\frac{x^2}{9x^2-16}}=\frac{\frac{x^2}{3x-4}+\frac{1-x^2}{3x-4}}{\frac{x^2}{9x^2-16}}=\frac{\frac{x^2+1-x^2}{3x-4}}{\frac{x^2}{9x^2-16}}=\frac{\frac{1}{3x-4}}{\frac{x^2}{(3x-4)(3x+4)}}$$
$$=\frac{1}{3x-4}\div\frac{x^2}{(3x-4)(3x+4)}=\frac{1}{3x-4}\cdot\frac{(3x-4)(3x+4)}{x^2}$$
$$=\frac{1\cdot(3x-4)(3x+4)}{(3x-4)\cdot x^2}=\frac{3x+4}{x^2}$$

75. $$\frac{\frac{x^2}{x-2}-\frac{x^2}{x+2}}{\frac{x}{x+2}+\frac{x}{x-2}}=\frac{\frac{x^2\cdot(x+2)}{(x-2)\cdot(x+2)}-\frac{(x-2)\cdot x^2}{(x-2)\cdot(x+2)}}{\frac{x\cdot(x-2)}{(x+2)\cdot(x-2)}+\frac{(x+2)\cdot x}{(x+2)\cdot(x-2)}}=\frac{\frac{x^3+2x^2}{(x-2)\cdot(x+2)}-\frac{x^3-2x^2}{(x-2)\cdot(x+2)}}{\frac{x^2-2x}{(x+2)\cdot(x-2)}+\frac{x^2+2x}{(x+2)\cdot(x-2)}}=\frac{\frac{(x^3+2x^2)-(x^3-2x^2)}{(x-2)\cdot(x+2)}}{\frac{x^2-2x+x^2+2x}{(x+2)\cdot(x-2)}}=\frac{\frac{x^3+2x^2-x^3+2x^2}{(x-2)\cdot(x+2)}}{\frac{2x^2}{(x+2)\cdot(x-2)}}$$
$$=\frac{\frac{4x^2}{(x-2)\cdot(x+2)}}{\frac{2x^2}{(x+2)\cdot(x-2)}}=\frac{4x^2}{(x-2)(x+2)}\div\frac{2x^2}{(x+2)(x-2)}=\frac{4x^2}{(x-2)(x+2)}\cdot\frac{(x+2)(x-2)}{2x^2}$$
$$=\frac{2\cdot2\cdot x\cdot x\cdot(x+2)(x-2)}{(x-2)(x+2)\cdot2\cdot x\cdot x}=\frac{2}{1}=2$$

77. $$\frac{3-\frac{x^2}{x^2+1}}{4-\frac{x^2}{x^2+1}}=\frac{\frac{3\cdot(x^2+1)}{1\cdot(x^2+1)}-\frac{x^2}{x^2+1}}{\frac{4\cdot(x^2+1)}{1\cdot(x^2+1)}-\frac{x^2}{x^2+1}}=\frac{\frac{3x^2+3}{x^2+1}-\frac{x^2}{x^2+1}}{\frac{4x^2+4}{x^2+1}-\frac{x^2}{x^2+1}}=\frac{\frac{3x^2+3-x^2}{x^2+1}}{\frac{4x^2+4-x^2}{x^2+1}}=\frac{\frac{2x^2+3}{x^2+1}}{\frac{3x^2+4}{x^2+1}}=\frac{2x^2+3}{x^2+1}\div\frac{3x^2+4}{x^2+1}$$
$$=\frac{2x^2+3}{x^2+1}\cdot\frac{x^2+1}{3x^2+4}=\frac{(2x^2+3)(x^2+1)}{(x^2+1)(3x^2+4)}=\frac{2x^2+3}{3x^2+4}$$

79. $$\frac{\frac{1}{x}-\frac{1}{2}}{x+1+\frac{1}{x}}=\frac{\frac{1\cdot2}{x\cdot2}-\frac{x\cdot1}{x\cdot2}}{\frac{x\cdot x}{1\cdot x}+\frac{1\cdot x}{1\cdot x}+\frac{1}{x}}=\frac{\frac{2}{2x}-\frac{x}{2x}}{\frac{x^2}{x}+\frac{x}{x}+\frac{1}{x}}=\frac{\frac{2-x}{2x}}{\frac{x^2+x+1}{x}}=\frac{2-x}{2x}\div\frac{x^2+x+1}{x}=\frac{2-x}{2x}\cdot\frac{x}{x^2+x+1}$$
$$=\frac{(2-x)\cdot x}{2\cdot x\cdot(x^2+x+1)}=\frac{2-x}{2\cdot(x^2+x+1)}$$

CHAPTER 4 Exponents, Radicals, Complex Numbers

4.1 Concepts and Vocabulary *(page 186)*

4. $\frac{1}{x^3}$ **5.** True **6.** False

4.1 Exercises *(page 186)*

7. $3^0 = 1$

9. $4^{-2}=\frac{1}{4^2}=\frac{1}{16}$

11. $\left(\frac{2}{3}\right)^2=\frac{2^2}{3^2}=\frac{4}{9}$

13. $3^0\cdot2^{-3}=\frac{3^0}{2^3}=\frac{1}{8}$

15. $2^{-3}+\left(\frac{1}{2}\right)^3=\frac{1}{2^3}+\left(\frac{1}{2}\right)^3$
$=\frac{1}{8}+\frac{1}{8}=\frac{2}{8}=\frac{1}{4}$

17. $3^{-6}\cdot3^4=3^{-6+4}=3^{-2}$
$=\frac{1}{3^2}=\frac{1}{9}$

19. $\frac{8^2}{2^3} = \frac{8^2}{8^1} = 8^{2-1} = 8^1 = 8$

21. $\left(\frac{2}{3}\right)^{-2} = \frac{2^{-2}}{3^{-2}} = \frac{3^2}{2^2} = \frac{9}{4}$

23. $\frac{2^3 \cdot 3^2}{2 \cdot 3^{-2}} = \frac{2^3}{2^1} \cdot \frac{3^2}{3^{-2}} = (2^{3-1}) \cdot (3^{2-(-2)}) = 2^2 \cdot 3^4 = 4 \cdot 81 = 324$

25. $\left(\frac{9}{2}\right)^{-2} = \frac{9^{-2}}{2^{-2}} = \frac{2^2}{9^2} = \frac{4}{81}$

27. $x^0y^2 = 1y^2 = y^2$

29. $x^{-2}y = \frac{y}{x^2}$

31. $(8x^3)^{-2} = 8^{-2}x^{3(-2)} = 8^{-2}x^{-6}$
$= \frac{1}{8^2x^6} = \frac{1}{64x^6}$

33. $-4x^{-1} = \frac{-4}{x^1} = \frac{-4}{x}$

35. $5x^0 = 5 \cdot 1 = 5$

37. $\frac{x^{-2}y^3}{xy^4} = \frac{y^3y^{-4}}{x^2x} = \frac{y^{3-4}}{x^{2+1}} = \frac{y^{-1}}{x^3} = \frac{1}{x^3y^1} = \frac{1}{x^3y}$

39. $x^{-1}y^{-1} = \frac{1}{x^1y^1} = \frac{1}{(xy)^1} = \frac{1}{xy}$

41. $\frac{x^{-1}}{y^{-1}} = \frac{y^1}{x^1} = \frac{y}{x}$

43. $\left(\frac{4x}{5y}\right)^{-2} = \left(\frac{5y}{4x}\right)^2 = \frac{5^2y^2}{4^2x^2} = \frac{25y^2}{16x^2}$

45. $x^{-1} + y^{-2} = \frac{1}{x} + \frac{1}{y^2} = \frac{y^2 + x}{xy^2}$

47. $\frac{x^{-1}y^{-2}z}{x^2yz^3} = \frac{z}{x^1y^2x^2yz^3} = \frac{z^{1-3}}{x^{1+2}y^{2+1}} = \frac{z^{-2}}{x^3y^3} = \frac{1}{x^3y^3z^2}$

49. $\frac{(-2)^3x^4(yz)^2}{3^2xy^3z^4} = \frac{-8x^4y^2z^2}{9xy^3z^4} = \frac{-8x^{4-1}y^{2-3}z^{2-4}}{9} = \frac{-8x^3y^{-1}z^{-2}}{9} = \frac{-8x^3}{9yz^2}$

51. $\frac{x^{-2}}{x^{-2} + y^{-2}} = \frac{\frac{1}{x^2}}{\frac{1}{x^2} + \frac{1}{y^2}} = \frac{\frac{1}{x^2}}{\frac{y^2 + x^2}{x^2y^2}} = \frac{1}{x^2} \cdot \frac{x^2y^2}{y^2 + x^2} = \frac{x^{2-2}y^2}{y^2 + x^2} = \frac{y^2}{y^2 + x^2}$

53. $\frac{\left(\frac{x}{y}\right)^{-2} \cdot \left(\frac{y}{x}\right)^4}{x^2y^3} = \frac{\left(\frac{y}{x}\right)^2 \cdot \left(\frac{y}{x}\right)^4}{x^2y^3} = \frac{\left(\frac{y}{x}\right)^{2+4}}{x^2y^3} = \left(\frac{y}{x}\right)^6 \cdot \frac{1}{x^2y^3} = \frac{y^6}{x^6} \cdot \frac{1}{x^2y^3} = \frac{y^{6-3}}{x^{6+2}} = \frac{y^3}{x^8}$

55. $\left(\frac{3x^{-1}}{4y^{-1}}\right)^{-2} = \left(\frac{4y^{-1}}{3x^{-1}}\right)^2 = \frac{(4y^{-1})^2}{(3x^{-1})^2} = \frac{16y^{-2}}{9x^{-2}} = \frac{16x^2}{9y^2}$

57. $\frac{(xy^{-1})^{-2}}{xy} = \frac{x^{-2}y^2}{xy} = \frac{y^2}{x^2xy} = \frac{y^{2-1}}{x^{2+1}} = \frac{y}{x^3}$

59. $\frac{\left(\frac{x^2}{y}\right)^3}{\left(\frac{x}{y^2}\right)^2} = \frac{\frac{x^6}{y^3}}{\frac{x^2}{y^4}} = \frac{x^6}{y^3} \cdot \frac{y^4}{x^2} = x^{6-2} \cdot y^{4-3} = x^4y^1 = x^4y$

61. $\left(\frac{x}{y^2}\right)^{-2} \cdot (y^2)^{-1} = \left(\frac{y^2}{x}\right)^2 \cdot y^{-2} = \frac{y^4}{x^2} \cdot y^{-2} = \frac{y^{4+(-2)}}{x^2} = \frac{y^2}{x^2}$

63. $(x^2y^3)^{-1}(xy)^5 = x^{-2}y^{-3} \cdot x^5y^5 = x^{-2+5}y^{-3+5} = x^3y^2$

Note: The keystrokes given for the following problems are those for a TI-83 calculator and may differ from those on the calculator you are using.

65. $(8.2)^5$

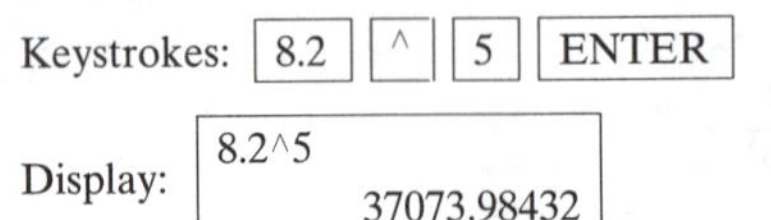

Rounded to three decimal places the answer is 37073.984

67. $(6.1)^{-3}$

Keystrokes: [6.1] [^] [(−)] [3] [ENTER]

Display: 6.1^−3 .0044056551

Rounded to three decimal places, the answer is 0.004

69. $(-2.8)^6$

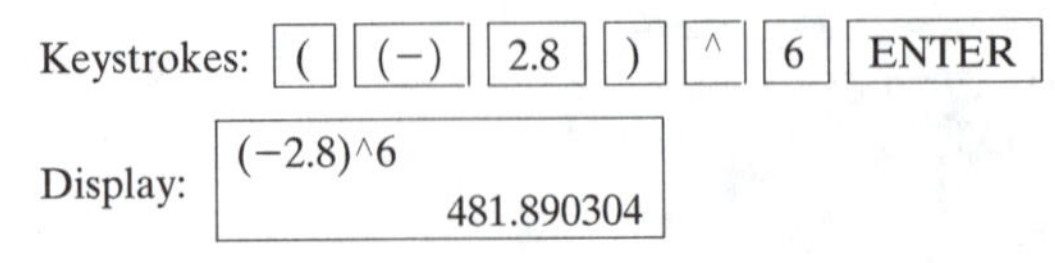

Rounded to three decimal places, the answer is 481.890

71. $-(9.25)^{-2}$

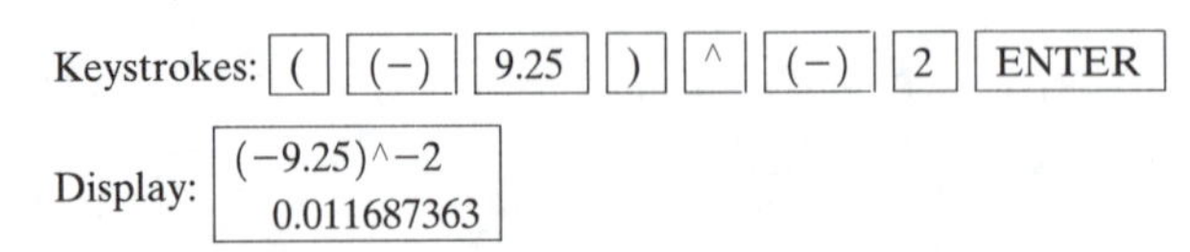

Rounded to three decimal places, the answer is 0.012

73. 454.2

The decimal point is between 4 and 2. We count

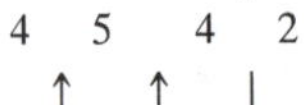

Stopping after 2 moves because 4.542 is a number between 1 and 10. Since 454.2 is a number between 100 and 1000, we write $454.2 = 4.542 \times 10^2$.

75. 0.013

The decimal point is between the two zeros. We count

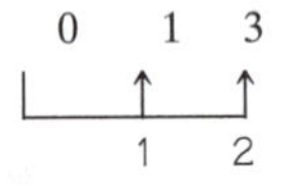

Stopping after two moves because 1.3 is a number between 1 and 10. Since .013 is a number between .001 and .01, we write $.013 = 1.3 \times 10^{-2}$.

77. 32, 155.0

The decimal is between the 5 and the 0. Thus, we count

3 2 1 5 5 0

4 3 2 1

Stopping after 4 moves because 3.2155 is a number between 1 and 10. Since 32,155 is between 10,000 and 100,000, we write $32{,}155 = 3.2155 \times 10^4$.

79. 0.000423

The decimal point is moved as follows:

0 0 0 0 4 2 3

1 2 3 4

$0.000423 = 4.23 \times 10^{-4}$.

81. $2.15 \times 10^4 = 2\ 1\ 5\ 0\ 0 \times 10^4 = 21{,}500$

1 2 3 4

83. $1.215 \times 10^{-3} = 0\ 0\ 1\ 215 \times 10^{-3} = 0.001215$

3 2 1

85. $1.1 \times 10^8 = 1\ 1\ 0\ 0\ 0\ 0\ 0\ 0\ 0 \times 10^8 = 110{,}000{,}000$

1 2 3 4 5 6 7 8

87. $8.1 \times 10^{-2} = 0\ 8\ 1 \times 10^{-2} = 0.081$

2 1

89. Speed of light = 186,000 miles/second

One light year = distance light will travel in one year.

$$\text{Distance} = \frac{186{,}000 \text{ miles}}{\text{second}} \cdot \frac{60 \text{ seconds}}{\text{minute}} \cdot \frac{60 \text{ minutes}}{\text{hour}} \cdot \frac{24 \text{ hours}}{\text{day}}$$
$$\cdot \frac{365 \text{ days}}{\text{year}} = 5.866 \times 10^{12} \text{ miles}$$

4.2 Concepts and Vocabulary *(page 198)*

5. False **6.** True **7.** $2 + \sqrt{3}$ **8.** False **9.** False **10.** False **11.** $3\sqrt{2} + 2\sqrt{3}$ **12.** 3

4.2 Exercises *(page 198)*

13. $\sqrt{4} = 2$ **15.** $\sqrt{\frac{1}{9}} = \frac{1}{3}$ **17.** $\sqrt{25} - \sqrt{9} = 5 - 3 = 2$

37. $\sqrt[4]{x^{12}y^8} = \sqrt[4]{(x^3)^4(y^2)^4} = x^3y^2$

39. $\sqrt[4]{\dfrac{x^9y^7}{xy^3}} = \sqrt[4]{x^9x^{-1}y^7y^{-3}}$
$= \sqrt[4]{x^{9-1}y^{7-3}} = \sqrt[4]{x^8y^4}$
$= \sqrt[4]{(x^2)^4(y)^4} = x^2y$

41. $\sqrt{36x} = \sqrt{6^2x} = 6\sqrt{x}$

43. $\sqrt{3x^2}\sqrt{12x} = \sqrt{36x^3}$
$= \sqrt{6^2x^2x^1} = 6x\sqrt{x}$

45. $\dfrac{\sqrt{3xy^3}\sqrt{2x^2y}}{\sqrt{6x^3y^4}} = \dfrac{\sqrt{6x^3y^4}}{\sqrt{6x^3y^4}} = 1$

47. $\sqrt{\dfrac{4}{9x^2y^4}} = \sqrt{\dfrac{2^2}{3^2x^2(y^2)^2}} = \dfrac{2}{3xy^2}$

49. $(\sqrt{5}\ \sqrt[3]{9})^2 = (\sqrt{5})^2(\sqrt[3]{(9)})^2$
$= (\sqrt{5})^2(\sqrt[3]{9^2})$
$= 5\sqrt[3]{81} = 5\sqrt[3]{27\cdot 3}$
$= 5\cdot 3\sqrt[3]{3} = 15\sqrt[3]{3}$

51. $(\sqrt[3]{4})^3 + (\sqrt[4]{5})^4 = \sqrt[3]{(4)^3} + \sqrt[4]{(5)^4}$
$= 4 + 5 = 9$

53. $3\sqrt[4]{2} + 2\sqrt[4]{2} - \sqrt[4]{2} = (3 + 2 - 1)\sqrt[4]{2}$
$= 4\sqrt[4]{2}$

55. $3\sqrt[3]{2} - \sqrt{18} + 2\sqrt{8} = 3\sqrt[3]{2} - \sqrt{9\cdot 2} + 2\sqrt{4\cdot 2} = 3\sqrt[3]{2} - 3\sqrt{2} + 4\sqrt{2}$
$= 3\sqrt[3]{2} + \sqrt{2}$

57. $\sqrt[3]{16} + 5\sqrt[3]{2} - 2\sqrt[3]{54} = \sqrt[3]{8\cdot 2} + 5\sqrt[3]{2} - 2\sqrt[3]{27\cdot 2}$
$= \sqrt[3]{(2)^3}\cdot\sqrt[3]{2} + 5\sqrt[3]{2} - 2\sqrt[3]{(3)^3}\ \sqrt[3]{2}$
$= 2\sqrt[3]{2} + 5\sqrt[3]{2} - 6\sqrt[3]{2} = \sqrt[3]{2}$

59. $\sqrt{8x^3} - 3\sqrt{50x} + \sqrt{2x^5} = \sqrt{4x^2}\ \sqrt{2x} - 3\sqrt{(25)(2x)} + \sqrt{2x^4x^1} = 2x\sqrt{2x} - 3(5)\sqrt{2x} + x^2\sqrt{2x}$
$= 2x\sqrt{2x} - 15\sqrt{2x} + x^2\sqrt{2x}$ or $(x^2 + 2x - 15)\sqrt{2x}$

61. $\sqrt[3]{16x^4y} - 3x\sqrt[3]{2xy} + 5\sqrt[3]{-2xy^4} = \sqrt[3]{8\cdot 2x^3xy} - 3x\sqrt[3]{2xy} + 5\sqrt[3]{(-1)2xy^3y}$
$= 2x\sqrt[3]{2xy} - 3x\sqrt[3]{2xy} - 5y\sqrt[3]{2xy}$
$= (2x - 3x - 5y)\sqrt[3]{2xy}$
$= (-x - 5y)\sqrt[3]{2xy}$

63. $(3\sqrt[3]{6})(2\sqrt[3]{9}) = 6\sqrt[3]{54} = 6\sqrt[3]{27\cdot 2} = 6\sqrt[3]{(3)^3\cdot 2} = 6\cdot 3\sqrt[3]{2} = 18\sqrt[3]{2}$

65. $(\sqrt[4]{9} + 3)(\sqrt[4]{9} - 3) = \sqrt[4]{81} - 3\sqrt[4]{9} + 3\sqrt[4]{9} - 9 = 3 - 9 = -6$

67. $(3\sqrt{7} + 3)(2\sqrt{7} + 2) = 6(\sqrt{7})^2 + 6\sqrt{7} + 6\sqrt{7} + 6 = 6(7) + 12\sqrt{7} + 6$
$= 42 + 6 + 12\sqrt{7} = 48 + 12\sqrt{7}$

69. $(\sqrt{x} - 1)^2 = (\sqrt{x} - 1)(\sqrt{x} - 1) = (\sqrt{x})^2 - \sqrt{x} - \sqrt{x} + 1 = x - 2\sqrt{x} + 1$

71. $(\sqrt[3]{x} - 1)^3 = (\sqrt[3]{x} - 1)(\sqrt[3]{x} - 1)(\sqrt[3]{x} - 1) = [(\sqrt[3]{x})^2 - \sqrt[3]{x} - \sqrt[3]{x} + 1](\sqrt[3]{x} - 1)$
$= (\sqrt[3]{x^2} - 2\sqrt[3]{x} + 1)(\sqrt[3]{x} - 1)$
$= \sqrt[3]{x^2}\sqrt[3]{x} - \sqrt[3]{x^2} - 2\sqrt[3]{x}\sqrt[3]{x} + 2\sqrt[3]{x} + \sqrt[3]{x} - 1$
$= \sqrt[3]{x^3} - \sqrt[3]{x^2} - 2\sqrt[3]{x^2} + 3\sqrt[3]{x} - 1$
$= x - 3\sqrt[3]{x^2} + 3\sqrt[3]{x} - 1$

73. $(2\sqrt{x} - 3\sqrt{y})(2\sqrt{x} + 5\sqrt{y})$
$= (2\sqrt{x})(2\sqrt{x}) + (2\sqrt{x})(5\sqrt{y}) - (3\sqrt{y})(2\sqrt{x}) - (3\sqrt{y})(5\sqrt{y})$
$= 4\sqrt{x^2} + 10\sqrt{xy} - 6\sqrt{xy} - 15\sqrt{y^2} = 4x + 4\sqrt{xy} - 15y$

75. 1.41 **77.** 1.59 **79.** 4.89 **81.** 2.15

4.4 Concepts and Vocabulary *(page 211)*

4. False **5.** True **6.** True

4.4 Exercises *(page 211)*

7. $\sqrt{3} = 3^{\frac{1}{2}}$

9. $\sqrt[3]{7} = 7^{\frac{1}{3}}$

11. $\sqrt[3]{5^2} = 5^{\frac{2}{3}}$

13. $\sqrt[7]{x^4} = x^{\frac{4}{7}}$

15. $\sqrt{x^5} = x^{\frac{5}{2}}$

17. $8^{\frac{2}{3}} = \left(\sqrt[3]{8}\right)^2 = 2^2 = 4$

19. $(-27)^{\frac{2}{3}} = \left(\sqrt[3]{-27}\right)^2 = (-3)^2 = 9$

21. $4^{-\frac{3}{2}} = \dfrac{1}{4^{\frac{3}{2}}} = \dfrac{1}{\left(\sqrt{4}\right)^3} = \dfrac{1}{2^3} = \dfrac{1}{8}$

23. $9^{-\frac{3}{2}} = \dfrac{1}{9^{\frac{3}{2}}} = \dfrac{1}{\left(\sqrt{9}\right)^3} = \dfrac{1}{3^3} = \dfrac{1}{27}$

25. $\left(\dfrac{9}{4}\right)^{\frac{3}{2}} = \left(\sqrt{\dfrac{9}{4}}\right)^3 = \left(\dfrac{\sqrt{9}}{\sqrt{4}}\right)^3 = \left(\dfrac{3}{2}\right)^3 = \dfrac{3^3}{2^3} = \dfrac{27}{8}$

27. $\left(\dfrac{4}{9}\right)^{-\frac{3}{2}} = \left(\dfrac{9}{4}\right)^{\frac{3}{2}} = \left(\sqrt{\dfrac{9}{4}}\right)^3 = \left(\dfrac{\sqrt{9}}{\sqrt{4}}\right)^3 = \left(\dfrac{3}{2}\right)^3 = \dfrac{3^3}{2^3} = \dfrac{27}{8}$

29. $4^{1.5} = 4^{\frac{3}{2}} = \left(\sqrt{4}\right)^3 = 2^3 = 8$

31. $\left(\dfrac{1}{4}\right)^{-1.5} = 4^{1.5} = 4^{\frac{3}{2}} = \left(\sqrt{4}\right)^3 = 2^3 = 8$

33. $\left(\sqrt{3}\right)^6 = 3^{\frac{6}{2}} = 3^3 = 27$

35. $\left(\sqrt{5}\right)^{-2} = 5^{\frac{-2}{2}} = 5^{-1} = \dfrac{1}{5}$

37. $3^{\frac{1}{2}} \cdot 3^{\frac{3}{2}} = 3^{\frac{1}{2}+\frac{3}{2}} = 3^2 = 9$

39. $\dfrac{7^{\frac{1}{3}}}{7^{\frac{4}{3}}} = 7^{\frac{1}{3}-\frac{4}{3}} = 7^{-1} = \dfrac{1}{7}$

41. $2^{\frac{1}{3}} \cdot 4^{\frac{1}{3}} = 2^{\frac{1}{3}} \cdot \left(2^2\right)^{\frac{1}{3}} = 2^{\frac{1}{3}} \cdot 2^{\frac{2}{3}} = 2^{\frac{1}{3}+\frac{2}{3}} = 2^1 = 2$

43. $\sqrt[4]{3} \cdot \sqrt[4]{27} = 3^{\frac{1}{4}} \cdot \left(3^3\right)^{\frac{1}{4}} = 3^{\frac{1}{4}} \cdot 3^{\frac{3}{4}} = 3^{\frac{1}{4}+\frac{3}{4}} = 3^1 = 3$

45. $\left(\sqrt[4]{2}\right)^{-4} = 2^{\frac{-4}{4}} = 2^{-1} = \dfrac{1}{2}$

47. $\left(\sqrt[3]{6}\right)^2 = 6^{\frac{2}{3}}$

49. $\sqrt{2} \cdot \sqrt[3]{2} = 2^{\frac{1}{2}} \cdot 2^{\frac{1}{3}} = 2^{\frac{1}{2}+\frac{1}{3}} = 2^{\frac{3}{6}+\frac{2}{6}} = 2^{\frac{5}{6}}$

51. $\sqrt[8]{x^4} = x^{\frac{4}{8}} = x^{\frac{1}{2}}$

53. $\sqrt{x^3} \cdot \sqrt[4]{x} = x^{\frac{3}{2}} \cdot x^{\frac{1}{4}} = x^{\frac{3}{2}+\frac{1}{4}} = x^{\frac{6}{4}+\frac{1}{4}} = x^{\frac{7}{4}}$

55. $x^{\frac{3}{2}} \cdot x^{-\frac{1}{2}} = x^{\frac{3}{2}+\frac{-1}{2}} = x^{\frac{2}{2}} = x^1 = x$

57. $\left(x^3y^6\right)^{\frac{2}{3}} = x^{3\left(\frac{2}{3}\right)} \cdot y^{6\left(\frac{2}{3}\right)} = x^2y^4$

59. $\left(x^2y\right)^{\frac{1}{3}}\left(xy^2\right)^{\frac{2}{3}} = x^{\frac{2}{3}}y^{\frac{1}{3}} \cdot x^{\frac{2}{3}}y^{\frac{4}{3}} = x^{\frac{4}{3}}y^{\frac{5}{3}}$

61. $\left(16x^2y^{-\frac{1}{3}}\right)^{\frac{3}{4}} = 16^{\frac{3}{4}} \cdot x^{2\left(\frac{3}{4}\right)} \cdot y^{-\frac{1}{3}\left(\frac{3}{4}\right)} = \left(\sqrt[4]{16}\right)^3 x^{\frac{3}{2}} y^{-\frac{1}{4}} = \dfrac{8x^{\frac{3}{2}}}{y^{\frac{1}{4}}}$

63. $\left(\dfrac{x^{\frac{2}{5}}y^{-\frac{1}{5}}}{x^{-\frac{1}{3}}}\right)^{15} = \dfrac{x^{\frac{2}{5}(15)}y^{-\frac{1}{5}(15)}}{x^{-\frac{1}{3}(15)}} = \dfrac{x^6y^{-3}}{x^{-5}} = \dfrac{x^6x^5}{y^3} = \dfrac{x^{11}}{y^3}$

4.5 Concepts and Vocabulary *(page 215)*

5. real part, imaginary part, imaginary unit **6.** $4i$ **7.** $3 - 5i$ **8.** False

4.5 Exercises *(page 215)*

9. In $8 + 3i$ the real part is 8 and the imaginary part is 3.

11. Since $6i = 0 + 6i$ the real part is 0 and the imaginary part is 6.

13. Since $42 = 42 + 0i$ the real part is 42 and the imaginary part is 0.

15. Since $5i - 12 = -12 + 5i$, the real part is -12 and the imaginary part is 5.

17. In $11 - i$ the real part is 11 and the imaginary part is -1.

19. $\sqrt{-9} = \sqrt{9(-1)} = \sqrt{9}\sqrt{-1} = 3i$

21. $\sqrt{-1} = i$

23. $\sqrt{-3} = \sqrt{3(-1)} = \sqrt{3}\sqrt{-1} = \sqrt{3}i$

25. $\sqrt{-20} = \sqrt{4(5)(-1)} = 2\sqrt{5}i$

27. $\sqrt{-75} = \sqrt{25(3\cdot)(-1)} = 5\sqrt{3}i$

29. $4 + \sqrt{-16} = 4 + 4i$

31. $\sqrt{-64} + 2 = 8i + 2 = 2 + 8i$

33. $6 + \sqrt{-15} = 6 + \sqrt{15}i$

35. $8 - \sqrt{-28} = 8 - \sqrt{4(7)(-1)} = 8 - 2\sqrt{7}i$

37. $\sqrt{-40} + 2 = 2 + \sqrt{4(10)(-1)} = 2 + 2\sqrt{10}i$

39. $\dfrac{2 + \sqrt{-12}}{6} = \dfrac{2}{6} + \dfrac{\sqrt{4(3)(-1)}}{6} = \dfrac{1}{3} + \dfrac{2\sqrt{3}}{6}i = \dfrac{1}{3} + \dfrac{\sqrt{3}}{3}i$

41. $\dfrac{4 - \sqrt{-16}}{4} = \dfrac{4}{4} - \dfrac{4i}{4} = 1 - i$

43. $\dfrac{5 - \sqrt{-8}}{6} = \dfrac{5 - \sqrt{4(2)(-1)}}{6} = \dfrac{5}{6} - \dfrac{2\sqrt{2}}{6}i = \dfrac{5}{6} - \dfrac{\sqrt{2}}{3}i$

Chapter 4 Review Exercises *(page 217)*

1. $(3)^{-1} + (2)^{-1} = \frac{1}{3} + \frac{1}{2} = \frac{2+3}{6} = \frac{5}{6}$

3. $\sqrt{25} - \sqrt[3]{-1} = \sqrt{(5)^2} = \sqrt[3]{(-1)^3}5 - (-1) = 6$

5. $9^{1/2}\cdot(-3)^2 = \sqrt[2]{9}\cdot(-3)^2 = \sqrt[2]{3^2}\cdot(-3)^2 = 3\cdot 9 = 27$

7. $5^2 - 3^3 = 5\cdot 5 - 3\cdot 3\cdot 3 = 25 - 27 = -2$

9. $\frac{2^{-3}(5^0)}{4^2} = \frac{5^0}{2^3(4^2)} = \frac{1}{8(16)} = \frac{1}{128}$

11. $\sqrt{12}\sqrt{75} = \sqrt{2^2\cdot 3}\sqrt{5^2\cdot 3} = 2\sqrt{3}\cdot 5\sqrt{3} = 2\cdot 5(\sqrt{3})^2 = 10\cdot 3 = 30$

13.
$$\begin{aligned}(2\sqrt{5}-2)(2\sqrt{5}+2) &= (2\sqrt{5})(2\sqrt{5}) + 2\cdot 2\sqrt{5} - 2\cdot 2\sqrt{5} - 2(2)\\ &= 4(\sqrt{5})^2 + 4\sqrt{5} - 4\sqrt{5} - 4 = 4(5) - 4\\ &= 20 - 4 = 16\end{aligned}$$

15. $3\sqrt{24} + 2\sqrt{6} = 3\sqrt{2^2\cdot 6} + 2\sqrt{6} = 3\cdot 2\sqrt{6} + 2\sqrt{6} = 6\sqrt{6} + 2\sqrt{6} = 8\sqrt{6}$

17. $\frac{\sqrt{54}}{\sqrt{18}} = \sqrt{\frac{54}{18}} = \sqrt{\frac{3}{1}} = \sqrt{3}$

19. $\sqrt{\frac{98}{8}} = \sqrt{\frac{49}{4}} = \frac{\sqrt{49}}{\sqrt{4}} = \frac{7}{2}$

21. $\left(\frac{8}{9}\right)^{-\frac{2}{3}} = \left(\frac{9}{8}\right)^{\frac{2}{3}} = \frac{9^{\frac{2}{3}}}{8^{\frac{2}{3}}} = \frac{\sqrt[3]{9^2}}{\sqrt[3]{8^2}} = \frac{\sqrt[3]{81}}{\sqrt[3]{64}} = \frac{3\sqrt[3]{3}}{4}$

23. $(\sqrt[3]{2})^{-3} = ((2)^{1/3})^{-3} = (2)^{-1} = \frac{1}{2}$

25. $|5 - 8^{1/3}| = |5 - (2^3)^{1/3}| = |5 - 2| = |3| = 3$

27. $\sqrt{|3^2 - 5^2|} = \sqrt{|9 - 25|} = \sqrt{|-16|} = \sqrt{16} = \sqrt{4^2} = 4$

29. $\frac{x^{-2}}{y^{-2}} = \frac{y^2}{x^2}$

31. $\frac{(x^2y)^{-4}}{(xy)^{-3}} = \frac{(xy)^3}{(x^2y)^4} = \frac{x^3y^3}{x^8y^4} = x^{3-8}y^{3-4} = x^{-5}y^{-1} = \frac{1}{x^5y}$

33.
$$\begin{aligned}(25x^{-4/3}y^{-2/3})^{3/2} &= 25^{3/2}(x^{-4/3})^{3/2}(y^{-2/3})^{3/2} = \left(\sqrt{5^2}\right)^3x^{-2}y^{-1}\\ &= \frac{5^3}{x^2y} = \frac{125}{x^2y}\end{aligned}$$

35. $\left(\frac{2x^{-\frac{1}{2}}}{y^{-\frac{3}{4}}}\right)^{-4} = \frac{2^{-4}(x^{-\frac{1}{2}})^{-4}}{(y^{-\frac{3}{4}})^{-4}} = \frac{x^2}{2^4y^3} = \frac{x^2}{16y^3}$

37. 327500

39. $\sqrt[4]{12} \approx 1.861$

41. $\frac{2}{\sqrt{3}} = \frac{2}{\sqrt{3}}\cdot\frac{\sqrt{3}}{\sqrt{3}} = \frac{2\sqrt{3}}{(\sqrt{3})^2} = \frac{2\sqrt{3}}{3}$

43.
$$\begin{aligned}\frac{2}{1-\sqrt{2}} &= \frac{2}{1-\sqrt{2}}\cdot\frac{1+\sqrt{2}}{1+\sqrt{2}} = \frac{2(1+\sqrt{2})}{1\cdot 1 + 1\cdot\sqrt{2} - 1\cdot\sqrt{2} - \sqrt{2}\cdot\sqrt{2}}\\ &= \frac{2(1+\sqrt{2})}{1-(\sqrt{2})^2} = \frac{2(1+\sqrt{2})}{1-2} = \frac{2(1+\sqrt{2})}{-1} = -2(1+\sqrt{2})\end{aligned}$$

45. $\dfrac{1+\sqrt{5}}{1-\sqrt{5}} = \dfrac{1+\sqrt{5}}{1-\sqrt{5}}\cdot\dfrac{1+\sqrt{5}}{1+\sqrt{5}} = \dfrac{1\cdot 1 + 1\cdot\sqrt{5} + 1\cdot\sqrt{5} + \sqrt{5}\cdot\sqrt{5}}{1\cdot 1 + 1\cdot\sqrt{5} - 1\cdot\sqrt{5} - \sqrt{5}\cdot\sqrt{5}}$

$$= \frac{1+2\sqrt{5}+\left(\sqrt{5}\right)^2}{1-\left(\sqrt{5}\right)^2} = \frac{1+2\sqrt{5}+5}{1-5} = \frac{6+2\sqrt{5}}{-4}$$

$$= \frac{2\left(3+\sqrt{5}\right)}{-4} = \frac{-\left(3+\sqrt{5}\right)}{2}$$

47. $\sqrt{-81} = 9i$

49. $3 + \sqrt{-4} = 3 + 2i$

51. $3 + \sqrt{-4} = 3 + \sqrt{4\cdot -1} = 3 + 2i$

53. $\sqrt{-8} + 10 = 10 + \sqrt{4(2)(-1)} = 10 + 2\sqrt{2}i$

55. $\dfrac{12+\sqrt{-20}}{4} = \dfrac{12}{4} + \dfrac{\sqrt{4(5)(-1)}}{4} = 3 + \dfrac{2\sqrt{5}}{4}i = 3 + \dfrac{\sqrt{5}}{2}i$